£45:=

Ian Silverman

March 1993

Rev for IJES

D1740564

Landscape Ecology of a
Stressed Environment

Landscape Ecology of a Stressed Environment

Edited by

Claire C. Vos

Dutch Association for Landscape Ecology and DLO-Institute for
Forestry and Nature Research, Department of Landscape Ecology

Paul Opdam

DLO-Institute for Forestry and Nature Research, Department of
Landscape Ecology

CHAPMAN & HALL
London · Glasgow · New York · Tokyo · Melbourne · Madras

Published by Chapman & Hall, 2–6 Boundary Row, London SE1 8HN, UK

Chapman & Hall, 2–6 Boundary Row, London SE1 8HN, UK

Blackie Academic & Professional, Wester Cleddens Road, Bishopbriggs, Glasgow G64 2NZ, UK

Chapman & Hall, 29 West 35th Street, New York NY10001, USA

Chapman & Hall Japan, Thomson Publishing Japan, Hirakawacho Nemoto Building, 6F, 1-7-11 Hirakawa-cho, Chiyoda-ku, Tokyo 102, Japan

Chapman & Hall Australia, Thomas Nelson Australia, 102 Dodds Street, South Melbourne, Victoria 3205, Australia

Chapman & Hall India, R. Seshadri, 32 Second Main Road, CIT East, Madras 600 035, India

First edition 1993

© 1993 Chapman & Hall

Typeset in 10/12pt Palatino by ROM-Data Corporation Ltd., Falmouth Cornwall

Printed in Great Britain at the University Press, Cambridge

ISBN 0 412 44820 3

A catalogue record for this book is available from the British Library

Library of Congress Cataloging-in-Publication data
Landscape ecology of a stressed environment/edited by Claire C. Vos and Paul Opdam.
      p.   cm.
   Includes bibliographical references (p.   ) and index.
   ISBN 0-412-44820-3
   1. Natural resources—Management. 2. Conservation of natural resources. 3. Land use—Planning. 4. Landscape ecology. I. Vos Claire C. II. Opdam, Paul.
HC59.L334 1992   92–34190
333. 7'2—dc20   CIP

Printed on permanent acid-free text paper, manufactured in accordance with the proposed ANSI/NISO Z 39.48-199X and ANSI Z 39.48-1984

Contents

Contributors

Rien Aerts
Department of Plant Ecology and Evolutionary Biology, University of Utrecht, PO Box 80084, 3508 TB Utrecht, The Netherlands

Theo W.M. Bakker
NV Duinwaterbedrijf Zuid-Holland, Postbus 710, 2501 CS The Hague, The Netherlands

Aat Barendregt
Department of Environmental Studies, University of Utrecht, PO Box 80115, 3508 TC Utrecht, The Netherlands

Frank Berendse
DLO-Centre for Agrobiological Research, PO Box 14, 6700 AA Wageningen, The Netherlands

Roland Bobbink
Department of Plant Ecology and Evolutionary Biology, University of Utrecht, PO Box 80084, 3508 TB Utrecht, The Netherlands

Jaques T. De Smidt
Department of Environmental Studies, University of Utrecht, PO Box 80115, 3508 TC Utrecht, The Netherlands

Nico P.J. De Vries
Ecological Bureau Everts and De Vries, Bieslookstraat 31, 9731 BM Groningen, The Netherlands

Harm Duel
TNO Policy Research, Institute for Spatial Organization, Ecology Division, PO Box 6041, 2600 JA Delft, The Netherlands

Roel During
TNO Policy Research, Institute for Spatial Organization, Ecology Division, PO Box 6041, 2600 JA Delft, The Netherlands

Henk F. Everts
Ecological Bureau Everts and De Vries, Bieslookstraat 31, 9731 BM Groningen, The Netherlands

Bert H. Harms
DLO-Staring Centre, PO Box 125, 6700 AC Wageningen, The Netherlands

Ab. P. Grootjans
Department of Plant Ecology, University of Groningen, PO Box 14, 9750 AA Haren, The Netherlands

Jan Kalkhoven
DLO-Institute for Forestry and Nature Research, Department of Landscape Ecology, PO Box 46, 3956 ZR Leersum, The Netherlands

Rolf M. Kemmers
DLO-Staring Centre, PO Box 125, 6700 AC Wageningen, The Netherlands

Klaas Kerkstra
Department of Physical Planning and Rural Development, Section of Landscape Architecture, Agricultural University, Gen. Foulkesweg 13, 6703 BJ Wageningen, The Netherlands

Jan P. Knaapen
DLO-Staring Centre, PO Box 125, 6700 AC Wageningen, The Netherlands

Willem Koerselman
KIWA NV, PO Box 1072, 3430 BB Nieuwegein, The Netherlands

Cees Kwakernaak
TNO Policy Research, Institute for Spatial Organization, Ecology Division, PO Box 6041, 2600 JA Delft, The Netherlands

Evert Meelis
Institute of Theoretical Biology, Leiden University, PO Box 9516, 2300 RA Leiden, The Netherlands

Dick (Th.) C.P. Melman
Government Service for Land Management, Ministry of Agriculture, Nature Management and Fisheries, PO Box 20022, 3500 LA Utrecht, The Netherlands

Johan A.J. Metz
Institute of Theoretical Biology, Leiden University, PO Box 9516, 2300 RA Leiden, The Netherlands

Paul Opdam
DLO-Institute for Forestry and Nature Research, Department of Landscape Ecology, PO Box 46, 3956 ZR Leersum, The Netherlands

Jos G. Rademakers
Staring Centre, PO Box 125, 6700 AC Wageningen, The Netherlands

Piet C. Schipper
State Forestry Commission, PO Box 1300, 3970 BH Driebergen, The Netherlands

Alex Schotman
DLO-Institute for Forestry and Nature Research, Department of Landscape Ecology, PO Box 46, 3956 ZR Leersum, The Netherlands

Jan Streefkerk
State Forestry Commission, PO Box 1300, 3970 BH Driebergen, The Netherlands

Pieter J.Stuyfzand
KIWA NV, PO Box 1072, 3430 BB Nieuwegein, The Netherlands

Sybrand P. Tjallingii
DLO-Institute for Forestry and Nature Research, PO Box 14, Wageningen, The Netherlands

Rob Van Apeldoorn
DLO-Institute for Forestry and Nature Research, Department of Landscape Ecology, PO Box 46, 3956 ZR Leersum, The Netherlands

Michaël Van Buuren
Department of Physical Planning and Rural Development, Section of Landscape Architecture, Agricultural University, Gen. Foulkesweg 13, 6703 BJ Wageningen, The Netherlands

Rudy Van Diggelen
Department of Plant Ecology, University of Groningen, PO Box 14, 9750 AA Haren, The Netherlands

Arco J. Van Strien
Department of Environmental Biology, Leiden University, PO Box 9516, 2300 RA Leiden, The Netherlands

Jana Verboom
DLO-Institute for Forestry and Nature Research, Department of Landscape Ecology, PO Box 46, 3956 ZR Leersum, The Netherlands

Jos T.A. Verhoeven
Department of Plant Ecology and Evolutionary Biology, University of Utrecht, PO Box 80084, 3508 TB Utrecht, The Netherlands

Claire C. Vos
Dutch Association for Landscape Ecology, PO Box 9201, 6800 HB Arnhem (Present address: DLO-Institute for Forestry and Nature Research, Dept. of Landscape Ecology, PO Box 46, 3956 ZR Leersum, The Netherlands)

Martin J. Wassen
Department of Environmental Studies, University of Utrecht, PO Box 80115, 3508 TC Utrecht, The Netherlands

Azing Wierda
Department of Plant Ecology, University of Groningen, PO Box 14, 9750 AA Haren, The Netherlands

Jan I.S. Zonneveld
Graaf Janlaan 24, 3708 GM Zeist, The Netherlands

Series foreword

This series presents studies that have used the paradigm of landscape ecology. Other approaches, both to landscape and landscape ecology are common, but in the last decade landscape ecology has become distinct from its predecessors and its contemporaries. Landscape ecology addresses the relationships among spatial patterns, temporal patterns and ecological processes. The effect of spatial configurations on ecological processes is fundamental. When human activity is an important variable affecting those relationships, landscape ecology includes it. Spatial and temporal scales are as large as needed for comprehension of system processes and the mosaic included may be very heterogeneous. Intellectual utility and applicability of results are valued equally. The International Association for Landscape Ecology sponsors this series of studies in order to introduce and disseminate some of the new knowledge that is being produced by this exciting new environmental science.

Gray Merriam
Ottawa, Canada

Preface

In Europe, during the seventies, landscape ecology emerged as a fusion of the spatial approach of geographers and the functional approach of ecologists. The latter focused on ecosystem functioning, regarding ecosystems as homogeneous, almost abstract units in space, with input and output of energy and matter to and from the undefined surroundings. Geographers considered the heterogeneous pattern of the surface of the earth, on various levels of spatial scale, and described the changes in that pattern due to human land use. A more thorough knowledge on the functioning of such heterogeneous systems, called landscapes (in a functional sense), became urgent with the growing significance of spatial planning.

In Western Europe, and in The Netherlands first of all, a growing awareness of the deterioration of landscapes and a loss of natural diversity coincided with a high population density and an intensifying land use. The Netherlands have been the experimental garden where landscape ecology developed very fast as an applied science, aiming at solving problems arising from spatial interactions between land units with contrasting functions, e.g. agricultural areas and nature reserves.

Conservation of natural values is usually a function that becomes disrupted first while increasing the stress of human land use. Landscape ecology then becomes an important basic science for nature conservation and management, and a basis for spatially integrating conservation aims in an urban and agricultural space. It is this problem-solving role of landscape ecology that is especially featured in this first volume of the IALE-series on landscape ecology. In a highly artificial landscape, like the Northwesteuropean agricultural landscape, where land use is largely independent of landscape characteristics, the role of landscape ecology as a basis for land-use evaluation is less significant compared to less man-dominated landscapes.

Landscape ecological research plays different roles in various stages of problem solving. In an early stage, problems are detected and brought to the attention of policy makers, spatial planners and nature conservationists. Next, measures to solve the problem have to be detected and indicated. To implement such measures in a landscape where land use is extremely intensive and space is in short supply, they have to be spatially explicit. This is even more so where the policy of nature conservation develops into a more offensive strategy of nature restoration. Various conflicting land-use types have to be combined, and often such functions have to be spatially separated. To assess whether a specific mosaic of land-use types is ecologically sustainable, and whether landscape restoration plans are effective and efficient, a thorough quantitative knowledge of spatial relations is vital.

So, in this volume, we want to present landscape ecology as an applied science. But there is a second aim. The urge for applicable knowledge and spatially explicit methods of application has, in our view, an essential benefit to the development of landscape ecology as the science seeking to understand how landscape functioning relates to landscape pattern, and vice versa. It is this beneficial role of questions by landscape planners that we want to highlight here especially.

Often, planning problems are represented in spatially explicit images of future landscape patterns. In scientific terms, landscape plans include spatially explicit hypotheses on how landscape functioning will change with a changing land-use pattern. The underlying assumptions are that the proposed heterogeneous ecological system functions in a sustainable way. In fact, landscape ecologists are challenged to test these hypotheses, to develop tools to make predictions on the expected functioning of the prospected landscapes, and to evaluate whether the planning aims will be fulfilled. In other words, landscape ecologists are challenged to translate the prospected pattern of land units into spatial relations and predict their effect on the functioning of these units. Also, they are urged to develop predictive tools to discriminate between alternative plans and assess them with respect to conservation aims, sustainability and effectiveness. This integration of pattern and process knowledge is essential, in our view, to the development of landscape ecology. Questions by spatial planners are not indispensable to this essential development, but surely are stimulating. Moreover, they prevent landscape researchers from getting lost in the detailed analysis of processes which cannot be related to landscape pattern. In other words, they keep researchers inside the landscape ecological domain and withhold them from entering ecology.

The book is organized in four parts and an introductory chapter which introduces the study area, characterizes its major landscape ecological patterns and spatial relations, and indicates type of human land use as well as its detrimental effects on natural and seminatural ecosystems. The three following parts are devoted to spatial relations and the way they influence land units. Spatial relations are distinguished by water flows, by air flows and by organisms. The final part highlights methods of landscape planning in which the integration of ecological knowledge plays an important role.

We like to end this preface with thanking all the authors, who took much trouble to meet our wishes for the final version of their chapter. Each chapter of this volume was peer-reviewed, and we would like to thank the following persons for their constructive comments: H. Andrén, T.W.M. Bakker, D. Boeye, D.L. DeAngelis, G. de Blust, W.H.O. Ernst, H. Farjon, C.H. Gimingham, I. Hanski, M.H. Jalink, F. Kleefman, J.A. Klijn, J.A. Lee, F. van der Meulen, G. Merriam, M. Numata, B. Pedroli, D. Prins, F.R. Steiner, N.G.J. Straathof, M.G. Turner, F. van der Vegte, D. Verkaar, B.D. Wheeler.

Paul Opdam
Claire Vos

Patterns and processes in a landscape under stress: the study area

1

Claire C. Vos and Jan I.S. Zonneveld

1.1 INTRODUCTION

The function of this chapter is twofold. First, it forms an introduction to the study area of this book, describing the structure and genesis of the main landscape types that can be found in The Netherlands nowadays. The second function of this chapter is to give a general overview of the effects of modern land use on the landscape. The overexploitation of the Dutch landscape is placed in an international context of other highly industrialized countries.

1.1.1 LANDSCAPE FORMING FACTORS

The territory of The Netherlands is situated in the deltaic area of the rivers Rhine, Meuse and Scheldt (Figure 1.1). It takes up the western-most part of the North European Plain, a lowland area that stretches out from the North Sea in the West as far as the East European Lowlands in Russia. Like all landscapes its character is determined by the inter-action of a set of constituting components of abiotic, biotic and also anthropogenic character. In the Netherlands the substratum mainly consists of quaternary sediments like sand, clay and gravel. This mat-

Landscape Ecology of a Stressed Environment
Edited by Claire C. Vos and Paul Opdam
Published in 1993 by Chapman and Hall, London. ISBN 0 412 44820 3

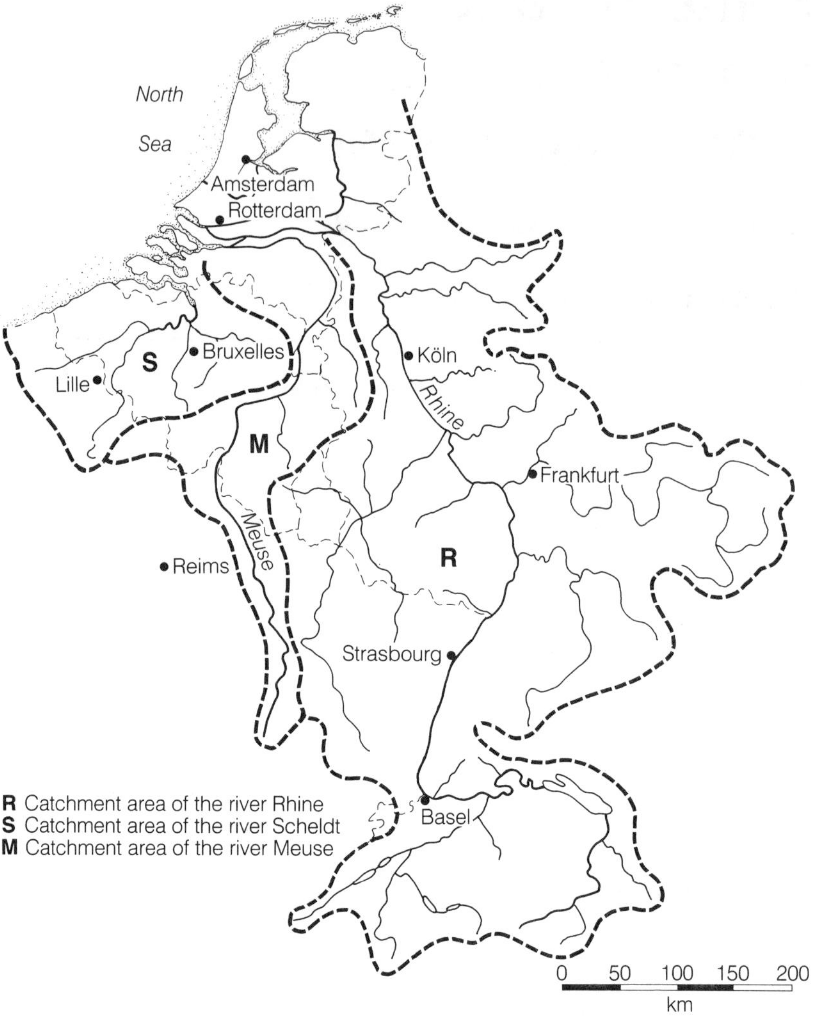

Figure 1.1 The Netherlands is the delta of some major West European rivers. The catchment areas of the rivers Rhine, Meuse and Scheldt are given.

erial was supplied and shaped into landforms by land ice masses during the pleistocene ice ages reaching down from Scandinavia, by rivers taking their rise in Western and Central European mountains, by currents and waves of the North sea and by predominating westerly winds. In areas where drainage was impeded large peat masses were formed. The differences in altitude of the landforms generally do not exceed a few metres. Only in the central and eastern parts of the country were low, glacially moulded hills (up to approx. 100 m above sea level) formed and in the extreme south the landscape is part of the transition zone between the North European Plain and the Central European mountainous regions. Here the relief, forming a hilly river-incised landscape, shows elevations up to 320 m. The territory of The Netherlands belongs to only one climatic zone: Köppen's Cfb climatic zone, characterized by an excess of precipitation over evaporation, by mild winters and relatively cool summers.

Apart from these independent conditional factors, the natural landscape includes a set of dependent and more mobile factors: hydrology, soil type, microclimate, flora and fauna. The hydrological situation in this part of Europe is basically determined by the above-mentioned precipitation excess. The depth of the groundwater table depends closely on the relief and the coarseness of the subsoil materials. Therefore deep groundwater levels are present in the sandy hills, whereas in the lowlands groundwater levels are near the surface. However, the hydrology of The Netherlands not only depends on the local climatological and geological conditions. Much water (surface as well as groundwater) is supplied by the rivers Rhine and Meuse draining an important part of West and Central Europe (Figure 1.1). The soil condition in its turn depends on the nature of the subsoil, the climate, the hydrology, the vegetation and also on human (agrarian) activities. In the sandy areas podzolic soils have developed. In some Holocene sediments however, complete soil formation could not yet take place. The natural vegetation and fauna belong to the Holarctic biogeographical realm, but species composition has been greatly influenced by human activities. Microclimatic conditions are determined by nearly all factors mentioned.

A very important factor in Dutch landscapes is human activity, particularly land use, by which the natural landscape has been reshaped into an almost completely 'cultural landscape'. Land use in fact is applying the various potentials or 'functions' offered by the environment. These functions are (Van der Maarel and Dauvellier, 1978):

1. *support functions*: offering a foundation for human constructions and transport;
2. *production functions*: providing mineral materials and giving the opportunity to produce agricultural products;
3. *regulation functions*: guaranteeing a sustainable environment based on the resilience of ecological systems;
4. *information functions*: offering information regarding the various potentials and linking up with aesthetic, scientific and historical interests of humankind.

For a long period of time human exploitation of the land was closely adapted to the natural conditions of subsoil, relief and the hydrological situation as well as to sedimentological (fluvial and coastal) processes. The support and production functions of the environment were exploited extensively, without a large impact on the information and regulation functions. The industrial and economic evolution of the last centuries, however, has led to an overexploitation of natural resources. Support and production functions have been used to such an extent that the regulation capacity of the natural systems was exceeded and information functions were destroyed. The overexploitation has given rise to serious environmental problems, eventually backfiring on the production functions themselves. Therefore, The Netherlands can, like other highly industrialized countries, be seen as an example of a landscape under stress.

1.2 LANDSCAPE TYPES

This section describes the genesis of the main landscape types in The Netherlands. The variety in ecological conditions resulted in many different landscape types on a relatively small scale. All these landscapes have to a smaller or larger extent been influenced by human activities past and present (Zonneveld, 1985).

Four main regions can be distinguished (Figure 1.2):

1. *The loess-area,* a region with a distinct plateau-and-valley topography, and with outcrops of prequaternary rocks;
2. *The sand region* in which the fine periglacial sands deposited during the last ice age play a predominant role;
3. *The fluvial district* characterized by holocene fluvial deposits and landforms;
4. *The marine-influenced part of The Netherlands* consisting of clay, sand and peat and mainly formed under holocene marine and perimarine influences.

1.2.1 THE LOESS-AREA

This area in the southern part of the country is a part of the transition zone between the North European Plain and the more elevated regions of the Ardennes (Figure 1.2). The presence of a relatively significant relief with plateaux and valleys and the occurrence of limestone have a great influence on the hydrological situation. Moreover, the area is part of the broad European loess belt, reaching from Northwest France to East Europe, where, during the penultimate and last glacial periods, substantial layers of aeolian dust were deposited. The valley slopes are

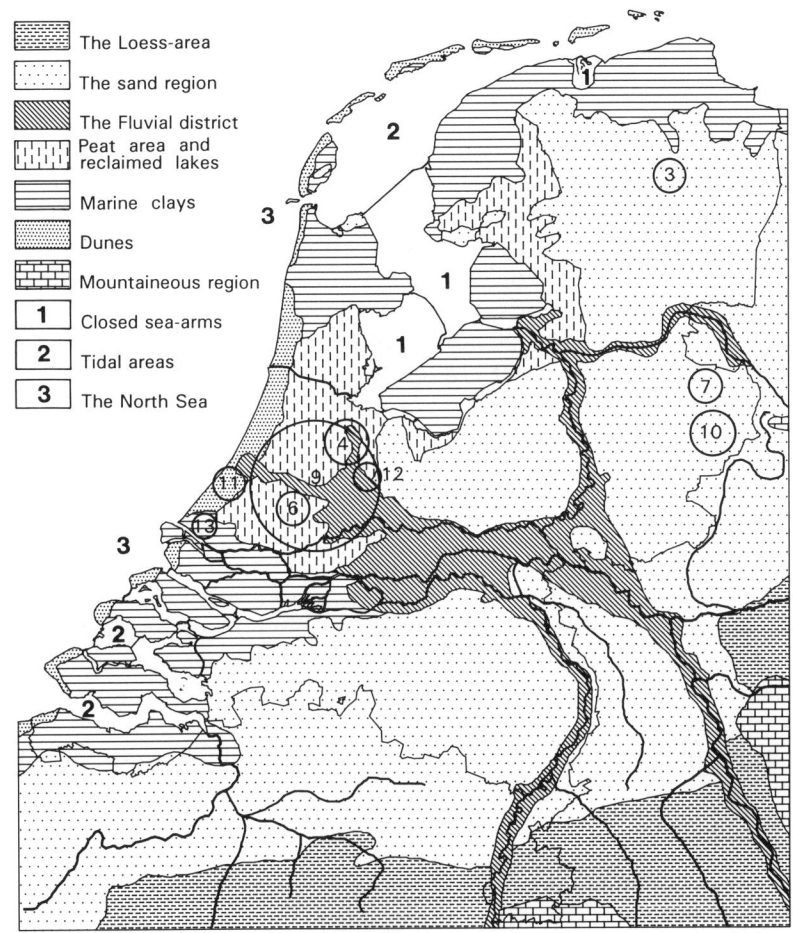

The Loess-area	
The sand region	
The Fluvial district	
Peat area and reclaimed lakes	
Marine clays	
Dunes	
Mountaineous region	
1 Closed sea-arms	
2 Tidal areas	
3 The North Sea	

Figure 1.2 Physicogeographical regions of The Netherlands and the surrounding countries: the loess area; the sand region, the fluvial district, the marine influenced district (peat area, reclaimed lakes, marine clays and dunes), closed sea-arms, tidal areas, the North Sea and the mountainous region. The numbers indicate the location of the study areas of the corresponding chapters of this volume.

steep enough to give rise to soil erosion: during the many ages of agricultural practice, the valley bottoms were raised by thick layers of colluviated loess. The grasslands on lime-rich soils (chalk meadows) and the forests on steep slopes are considered valuable (semi)natural ecosystems.

1.2.2 THE SANDY REGION

The sandy region includes almost half the surface of The Netherlands.

It links up with comparable regions situated north of the European loess belt in Belgium and Germany (Figure 1.2). Four districts can be distinguished, all characterized by the presence of more or less thick layers of fine sands (coversands), deposited during the last ice age under periglacial conditions.

Broad, but shallow valleys with holocene brook deposits have been carved out by the small local rivers. The relief of the cover sands is low, and was formed mainly by snow drifts during the last ice age. In the central and eastern parts of The Netherlands some low hills, consisting of pleistocene fluvial sands, pushed up by the ice masses of the penultimate glaciation, attain elevations of some tens of metres above sea level (the highest point is at 110 m).

The soils are poor. Agriculture originally began on the footslopes of hills and on the valley slopes. In these districts agriculture was, for centuries, based on the so-called 'potstal-system', using manure supplied by sheep grazing by day on the (heather-) moorland of the plateaux and put up each night in stables (the 'potstal'). Sods were taken from the heathland and from the meadows in the valley bottom. The arable fields in this way acquired a more or less thick layer of 'plaggen soil'. In the regions where the sheep were grazing and heather sods were taken away locally, the vegetation cover was sometimes damaged to such an extent that wind erosion and dune formation could take place.

The drainage in many coversand areas, especially in the valleys but also in water divides with a less permeable subsoil, was originally rather poor. Locally it gave rise to large peat moors. Due to peat digging over the last centuries only a very small part of the original moors is left, now functioning as nature reserves. Regionally the higher parts of the coversand areas may show dry conditions.

The complexity of abiotic conditions, in particular the transition zones, resulted in a great natural diversity in this region. The remaining heathlands, peat moors, fens, forests and the wet meadows in the catchment areas are considered valuable (semi-)natural ecosystems.

1.2.3 THE FLUVIAL REGION

The Netherlands are a depositional area built up during pleistocene times by some mid- and west European rivers. In the present landscape, especially the rivers Rhine and Meuse, have produced a region characterized by typical fluvial landscape types (Figure 1.2). In the eastern part of this region some remnants of the braided river patterns are visible that were formed during the last glacial period. More to the west these forms and sediments have been covered by holocene deposits, laid down by meandering rivers. Sandy levees and (clayey) backswamps are present in a great variety. Only locally the summits of some late-pleistocene river dunes are still protruding through these sediments. In the west, marine influences are conspicuous. Here the fluvial region passes into the 'marine influenced lowlands'.

Dikes were built parallel to the river courses and at some distance from the river bank itself, to give room to high discharges. Nevertheless, from time to time the dikes collapsed, especially during floods caused by dams of clogging ice during periods of ice drift in the rivers. Nearly all these disasters caused large potholes, known as 'wielen'. In the periods with high discharge of the rivers seepage may take place, especially in areas where the dikes were built upon sandy (levee) materials.

The former riverbeds, potholes, old dikes and river dunes are, among others, landscape elements with high natural values.

1.2.4 THE MARINE-INFLUENCED PART OF THE NETHERLANDS

The Southwestern, Western and Northern parts of The Netherlands belong to the coastal lowlands, bordering the North Sea ranging from the English Channel in the southwest to Denmark in the northeast. This area developed during the holocene, mainly under direct or indirect marine influences. During this period the sea showed transgressions and regressions. These periods with respectively more and less marine aggression were partly due to differences in sea level rise and, especially during the last 4000 years, due to differences in the frequency of storm floods. A set of 'barrier islands' was built up between a tidal-flat area and the open sea. The rivers formed their estuaries in this region and, during the regressions of the Subboreal period, extensive peat masses developed between the river branches. Finally, at the end of the tenth century, a dune landscape of different widths developed along the central coast and on the islands in the Southwestern, Western and Northern parts.

During this development the barrier islands and the regions along the estuaries were already inhabited by man. About 1000 AD, during a transgressive period and in answer to renewed aggression of the sea, the first dikes were built in Zeeland (in the Southwest) and in Friesland (in the North). Apparently the first dikes were no more than raised roads, connecting some settlements in a region suffering from increased flooding. Rather soon, however, they were also used to defend the enclosed areas against the rising water. Dike building and reclamation of silted-up areas in the estuaries and salt marshes became a much practised activity, highly influencing the topographical patterns as well as the sedimentological and hydrological processes and conditions (Zonneveld, 1991). In the 17th century not only was land-reclamation practised by building dams around areas that had been silted-up to an appropriate height, but it also appeared possible to drain lakes. Lakes were surrounded by dikes and subsequently pumped dry, thus changing them into 'Droogmakerijen'. Finally, in the 20th century this procedure was also applied for parts of the Zuiderzee.

Until about 1000–1200 AD most peat moors of the Southwest and North Netherlands had been removed by (marine) erosion and covered by sedimentation. The remaining peat moors in the west became the scene of intensive reclamation activities. From the 16th century onwards, however, a start was made on the exploitation of the peat as fuel for the heating of houses and on behalf of developing industries in the surrounding towns. Due to the growing shortage of firewood and the increasing need for fuel in the towns, peat was used increasingly. This activity went on until the first half of the 20th century.

The lower part of the marine-influenced region thus includes a great variety of landscapes, practically all of them highly influenced by human activities, some even completely 'man made'. Nevertheless, in this region many natural values are present. In this respect the peat moor area, a landscape almost completely restricted to The Netherlands, has an international significance for its plant- and wetland bird communities (Figure 1.2). The young dune area, which has been formed since the 10th century, can be regarded an important ecosystem, influenced by man, but nevertheless with high natural values (Figure 1.2). High ecological diversity is the result of a great variety in abiotic gradients, ranging from dynamic to stable, wet to dry conditions, lime-rich to acid soils and salt to fresh water conditions.

1.3 MODERN LAND USE

This section gives an overview of the developments in modern land use which have led to changes in patterns and processes in the landscape, affecting the functions of the landscape. Continued overexploitation not only affects the information functions by loss of species diversity. The over-burden and over-charge of the regulation functions have resulted in worldwide air and water pollution as well and these effects do not halt at the country's borders. When in a landscape under stress the production functions are eventually also affected – which can be illustrated by the collapse of fishery (Daan et al., 1990) or loss of agricultural yield (Eijsackers and Quispel, 1988) – mankind is hit in the core of its existence and environmental pollution becomes a major political issue. The situation in The Netherlands may be regarded as an illustration of the environmental problems in the West European industrialized world.

Table 1.1 Indicators for environmental pollution in 1987 per km^2 (a) and per capita (b). The Netherlands have the highest environmental pressure per km^2. In terms of environmental pollution per capita its position is not extreme (Maas, 1991)

(a)	Population density	Gross National Product	Energy Consumption	Water Consumption	Transport	LiveStock
	(persons/ km^2)	(1000 \$/ km^2)	(toe[a]/ km^2)	(1000 m^3/ km^2)	(1000 km[c]/ km^2)	(ae[b]/ km^2)
Belgium	322	4551	1096	296	1606	112
Denmark	119	2350	362	33	759	75
France	102	1609	252	61	693	37
UK	233	2736	597	116	1269	57
Germany[d]	246	4497	801	167	1593	63
The Netherlands	393	5714	1433	381	2219	174
USA	26	477	142	50	320	11
Japan	328	6374	706	290	1312	22
Poland	125	852	404	55	161	38
India	280	83	49	128	15	57

(b)	Land area	Gross National Product	Energy Consumption	Water Consumption	Transport	LiveStock
	(m^2 per capita)	(\$ per capita)	(toe[a] per capita)	(m^3 per capita)	(km[c] per capita)	(ae[b] per capita)
Belgium	3104	14124	3.4	918	4983	0.35
Denmark	8401	19739	3.0	273	6378	0.63
France	9833	15817	2.5	599	6813	0.37
UK	4300	11765	2.6	498	5458	0.24
Germany[d]	4062	18265	3.3	676	6472	0.25
The Netherlands	2544	14538	3.6	968	5646	0.44
USA	38426	18338	5.4	1915	12303	0.41
Japan	3049	19437	2.2	883	4001	0.07
Poland	7977	6799	3.2	440	1284	0.31
India	3567	295	0.2	456	55	0.20

[a] toe = tons oil equivalent.
[b] ae = animal unit based on nutritional needs.
[c] Motorized transport.
[d] Former German Federal Republic territory.

Table 1.2 Deposition figures (mol/ha/year) of acidifying compounds for the US, Europe and The Netherlands. For The Netherlands the deposition from natural sources is also listed. Total potential acid = $2SO_x + NO_y + NH_x$

(modified after Heij and Schneider, 1991)

	SO_x			NO_y			NH_x			Total potential acid
	Dry	Wet	Total	Dry	Wet	Total	Dry	Wet	Total	
US	140	130	270	160	165	325	-	-	-[a]	865[b]
Europe[c]	320	160	480	185	95	280	265	135	400	1640
The Netherlands[c]	450	220	670	850	310	1160	1550	640	2190	4700
Natural sources in The Netherlands[c]	24	84	108	13	36	49	48	27	75	340

[a] Not estimated.
[b] Minus NH_x
[c] dry/wet ratio 2:3.

1.3.1 URBAN AND INDUSTRIAL DEVELOPMENT

The triangle London–Paris–Ruhr area forms the industrial and commercial core of Northwest Europe. The population density in this triangle is among the highest of the world. During the last century the population of The Netherlands has increased considerably, from 5 million in 1900 to 15 million in 1990. A continued growth is expected until 2005, resulting in a population density up to 475 inhabitants per km^2 (Langeweg, 1989). Table 1.1 compares the intensity of human activities in some highly industrialized or densely populated countries. The high population density, the high energy consumption, the intense industrial activities and the increasing mobility all concentrated in a limited area, have led to a variety of environmental problems. The air pollution caused by these activities is illustrated by the deposition of SO_2 and NO_x in some industrialized countries (Table 1.2).

The drainage basins of the large rivers Rhine, Meuse and Scheldt include some of the most important industrial regions of Western Europe. The Netherlands being the delta where these rivers discharge into the sea receives – despite the purification that takes place – pollutants of various kinds. Table 1.3 shows the changes in water quality of these rivers over the last 20 years. Water and air pollution affect the environment in various ways resulting in eutrophication, acidification and the spreading of toxic compounds, which have negative impacts on all functions of the landscape.

Due to population growth, there has been an important expansion of

Table 1.3 Concentrations of nutrients and heavy metals in the rivers Rhine (R), Meuse (M) and Scheldt (S) in the years 1972–1974, 1981–1983, 1987–1989 and the basic surface water quality that is being used as a goal (limit) for environmental management (Ministerie van Verkeer en Waterstaat, 1989)

	1972–1974			1981–1983			1987–1989			Basic Quality
	R	M	S	R	M	S	R	M	S	
Oxygen (mg/l)	5.2	9.1	1.5	8.3	8.9	4.2	9.3	8.8	3.1	>5.0
Nitrogen total (mg/l)	4.3	2.3	7.8	4.9	4.0	-	5.0	4.1	-	<2.2
Phosphate total (mg/l)	0.88	0.91	1.43	0.58	0.49	0.94	0.33	0.53	0.89	<0.15
Chlorophyll-A (µg/l)	-	-	-	67	36	33	38	28	14	<100
Chloride (mg/l)	213	45	5161	159	46	3859	157	49	3291	<200
Mercury total (µg/l)	1.48	0.63	0.77	0.12	0.15	0.15	0.06	0.06	0.12	<0.03
Cadmium total (µg/l)	3.5	8.6	2.4	0.8	1.5	1.5	0.1	1.4	0.7	<0.2
Copper total (µg/l)	33.2	28.1	20.1	10.0	8.4	10.3	5.5	5.3	10.7	<3.0
Lead total (µg/l)	39.6	58.4	31.2	8.9	11.9	14.4	4.1	7.2	11.1	<25
Chromium total (µg/l)	55.8	18.6	49.0	11.6	8.8	18.1	7.4	7.2	19.1	<25
Zinc total (µg/l)	260	526	104	69	140	67	33	74	50	<30

Nitrogen, Phosphate and Chlorophyll-A summer average.

the built-up area from 1871 km^2 in 1977 to 2052 km^2 in 1983, which amounts to 6% of the total land area of The Netherlands. Urbanization not only takes space at the expense of natural areas, it also prevents precipitation water from penetrating into the subsoil. Large masses of water that otherwise would have replenished the subterranean aquifers are now discharged into the sewer drain systems and thus the regulation function is disturbed.

The development of industrial and general economic activities have resulted in an increase in freight transport. Moreover, higher standards of living have also resulted in an increase of mobility, due to commuter traffic and outdoor recreation. During the last 15 years the highway system has increased by 15%, and car density has doubled. The Neth-

(a)

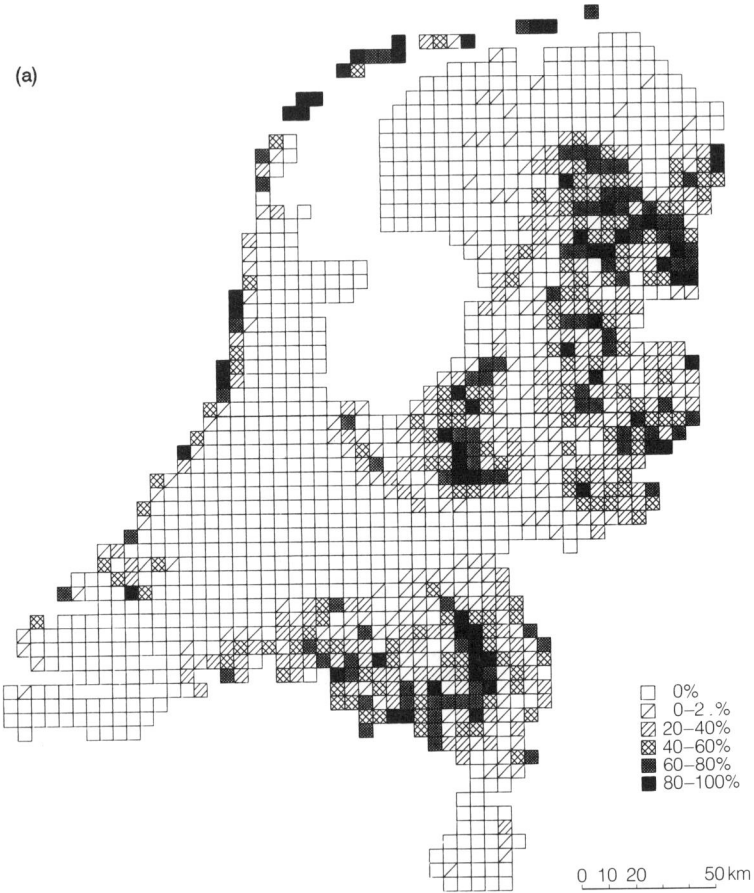

	0%
	0–2 .%
	20–40%
	40–60%
	60–80%
	80–100%

0 10 20   50 km

Figure 1.3 The percentage of unexploited grounds per 5×5 km^2 in (a) 1900 and (b) 1980 (Barends, 1989).

erlands have, with 128 cars per km^2 the highest car density of the world. It is expected that the number of cars will continue to increase from 5 million in 1985 to nearly 8 million in 2010 (Langeweg, 1989). Apart from energy consumption and air pollution, highways also cause disturbance for various kinds of organisms and form non-crossable barriers for many organisms.

1.3.2 AGRICULTURAL DEVELOPMENT

In former times when agricultural methods were determined by the natural conditions of the substratum, people already greatly influenced

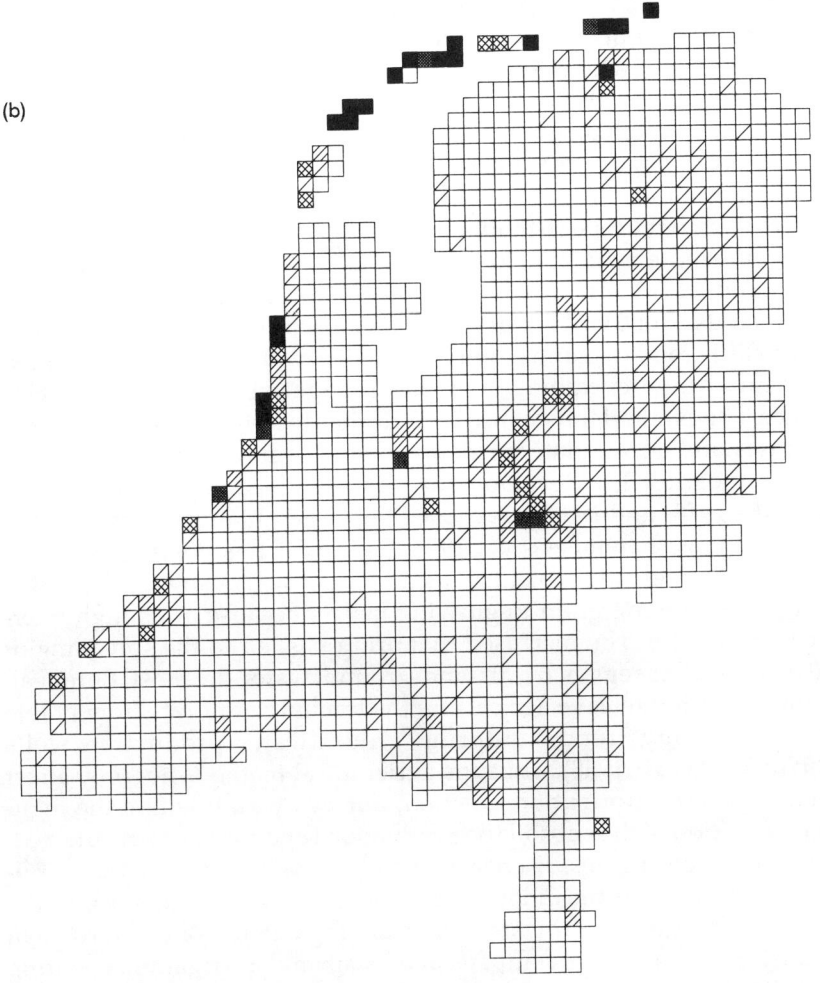

(b)

natural landscapes. Because these activities took place on a small scale and in close relation with the natural conditions of the substratum, there was an overall positive effect on the biotic diversity of landscapes. Without human impact the landscape in this region would have consisted in large part of deciduous woods, however, the agricultural activities gave room to species related to more open landscape types. Thus, land use widened the biotic diversity of the landscape, without influencing the abiotic conditions to a large extent. With regard to the landscape functions this means that the production function was integrated with the regulation function and enough space was left for information functions.

The introduction of artificial fertilizers in the second half of the 19th century, however, made man less dependent on field conditions. This was the starting point for the present highly intensified and industrialized agriculture, which has led to considerable scale expansion. Formerly unsuitable land could be exploited by the use of fertilizers, draining etc., which resulted in an almost total disappearance of (semi-) natural areas. Figure 1.3 illustrates these changes by comparing unexploited grounds in 1900 and 1980. Today The Netherlands are one of the greatest exporters of food and agricultural products. The Netherlands contribute nearly 5% to the world's food production of meat and dairy products on less than 0.05% of the world's grassland area. In the Netherlands 20 000 km^2 is being used for agricultural purposes. To meet Dutch cattle food demands, however, another 100 000 km^2 is exploited in (mainly) the developing countries (Langeweg, 1989). As a consequence The Netherlands contribute to the depletion of soils in other countries and accumulate nutrients in the form of manure within its own boundaries. The development of land-independent factory farming in the 1970s and 1980s has caused the largest manure surplus in the sandy regions (Berendse et al., 1993, Figure 5.1 this volume). In these regions, most sensitive to eutrophication, farmers changed on a large-scale from growing grain to growing maize, because of its high tolerance for manure. The increased manuring has led to the spreading of nutrients in the region by air and ground water. In most European countries with intensive agriculture the leaching of nitrate from arable soils is a rising trend, polluting groundwater production wells (Duynisveld et al., 1988). Thus the agricultural methods not only affect natural values (information function) but by overburdening the regulation function of the subsoil the production functions are also affected.

In the sandy regions discharge of surplus water in spring is stimulated by cutting off meanders and canalization of rivers. Also, in the polder landscape groundwater levels are regulated. Water shortage in fields during summer is compensated by sprinkler irrigation, resulting in additional lowering of the groundwater table. This increased regulation of (ground)water levels for agricultural purposes has caused a considerable lowering of groundwater tables. Often the resulting regional water shortage is compensated by the intake of highly polluted Rhine water, thus changing the region's specific water quality. The problems are increased by the increasing demands for drinking and industrial water. All these activities have far-reaching effects on the former natural conditions of the landscape (affecting information and regulation functions).

These changes in the original environmental quality affect the habitats of many species. Moreover, species are also affected more directly in various ways, for instance by the application of pesticides, herbicides and fungicides. Each year 21 million kg is used for agricultural purposes, of which 4–5 million kg spreads into the environment (Ministerie van Landbouw, Natuurbeheer en Visserij, 1991). The application of herbicides has caused a decline of 30–50% in species number of 'weed flora' (Hurle, 1988). Harvesting is another way in which species can be affected. For instance, the advanced methods of fishing, probably in combination with elevated nutrient loads, resulted in the breaking down of the North Sea population of herring (*Clupea harengus*) (Daan *et al.*, 1990). This illustrates that the detoriation of information and regulation functions eventually leads to the breaking down of production functions.

Reallotment projects were stimulated by the government from the 1950s onwards to meet mechanical land use demands. These have changed the formerly small-scale agricultural landscapes, especially in the sandy regions. Irregularly shaped plots have been adjusted and adjoining parcels have been put together. In addition wooded banks were neglected or cleared away because of their loss of economical function. As a result the amounts of hedgerows and wooded embankments have diminished drastically. Figure 1.4 illustrates the disappearance of natural elements along plot edges between 1900 and 1980, and the levelling out of former regional differences. The increased scale of farming thus has a direct negative effect on the information functions.

1.4 SPATIAL RELATIONS

Landscape ecological (inter)relations are based on fluxes of energy, matter and/or information between the composing attributes of an ecosystem. A change in land use may result in a completely different type of (man induced) ecosystem and, therefore, in a different type of landscape. Besides these 'vertical' or 'topological' processes, also 'horizontal', 'chorological' or 'spatial' landscape ecological relations play an important role. Processes operating in one land

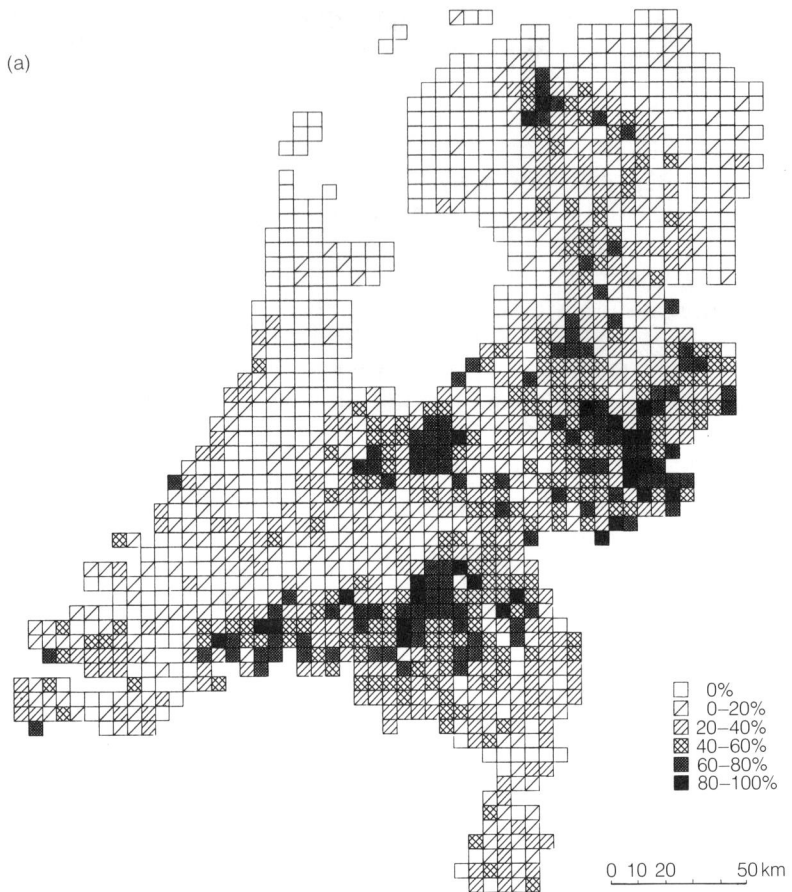

(a)

0%
0–20%
20–40%
40–60%
60–80%
80–100%

0 10 20   50 km

Figure 1.4 The percentage of natural elements along plot edges in 1900 (a) and 1980 (b). The classes give the percentage of plots per 5×5 km^2 in which more than 50% of the plot edge consists of natural elements (Barends, 1989).

unit may influence the situation in other areas. These spatial relations in the landscape generally appear in the form of (matter) flows of air and water and movements of organisms. And again these flows can be regulated by human activities. The human impact on the landscape may therefore reach over long distances in space and time and is not restricted to the region where the activity takes place.

1.4.1 SPATIAL RELATIONS BY AIR

Among the different spatial relations the transport of matter by air currents is paramount. Transport of dust and polluting substances over

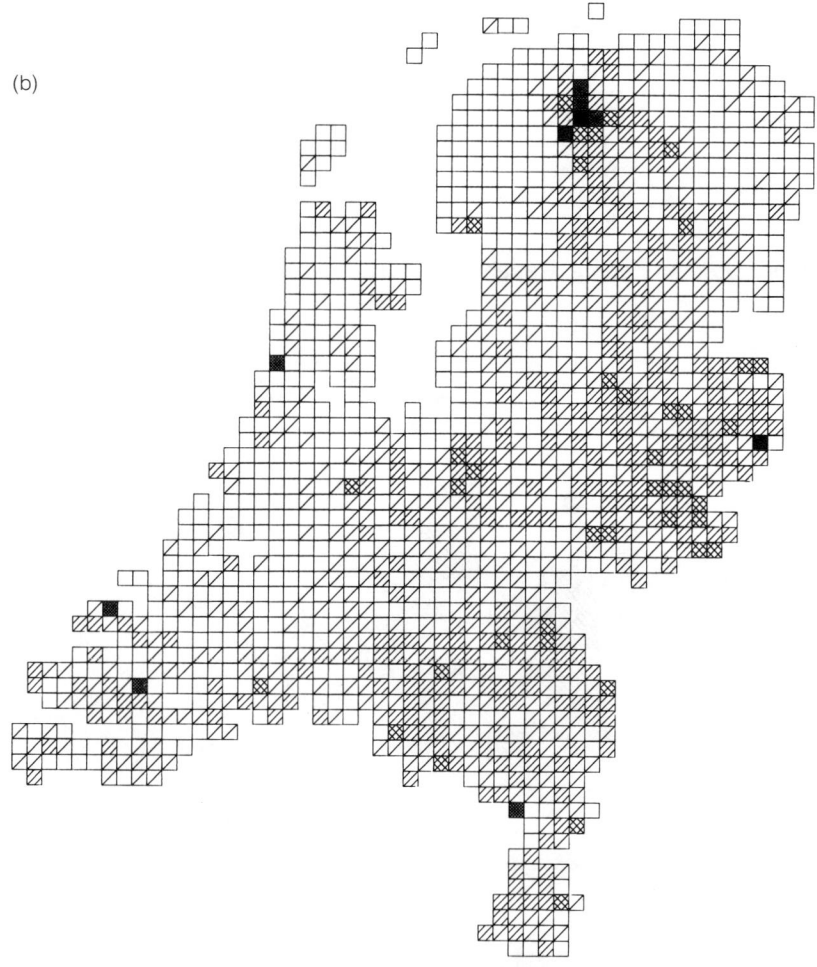

(b)

long distances is a frequently occurring phenomenon. The emission of
SO_2 and NO_x by industry, power plants and car traffic is spread over
large distances in Western Europe. For instance, 46% of the deposition
in The Netherlands comes from surrounding countries (Table 1.4).
Pollution can even influence systems on a worldwide scale, causing
problems like the enforced greenhouse effect and the depletion of the
ozone layer (Arnfield, 1987).

The spatial distribution by air also operates on regional and local
scales (Berendse *et al.*, 1993, this volume). The emission of ammonia in
The Netherlands, for instance, is mainly restricted to the sandy regions
where intensive animal husbandry is concentrated, resulting in acidifi-

Table 1.4 Area of origin of the total potential acid deposition ($2SO_x + NO_y + NH_x$) in The Netherlands for 1980 and 1990 (Heij and Schneider, 1991)

Area of Origin	1980	1989
The Netherlands	48 %	54 %
Germany[a]	13 %	10 %
Belgium	13 %	10 %
UK and Ireland	8 %	9 %
France	10 %	8 %
Eastern Europe	6 %	8 %

[a] Former German Federal Republic territory.

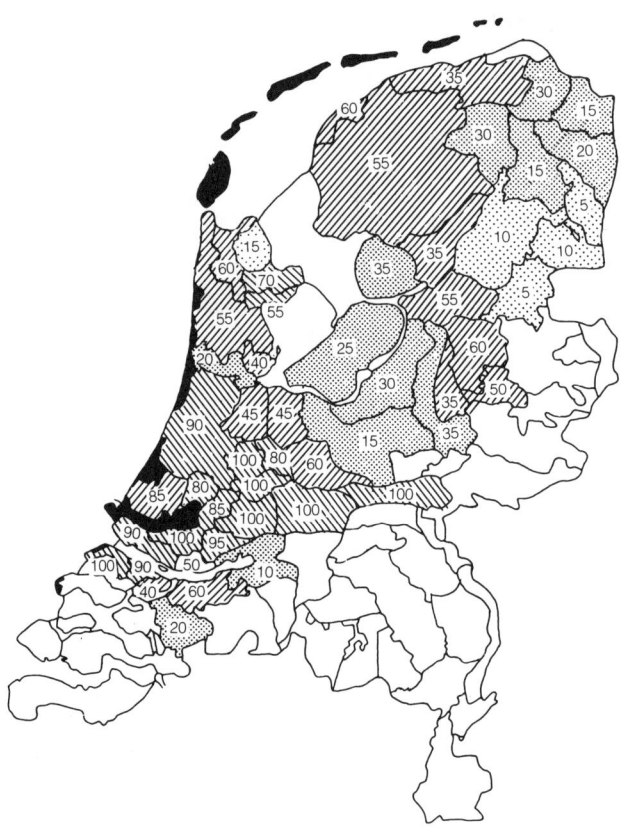

Figure 1.5 The spreading of polluted Rhine water in an extremely dry year (Duel *et al.*, 1989). The shading represents different water management districts.

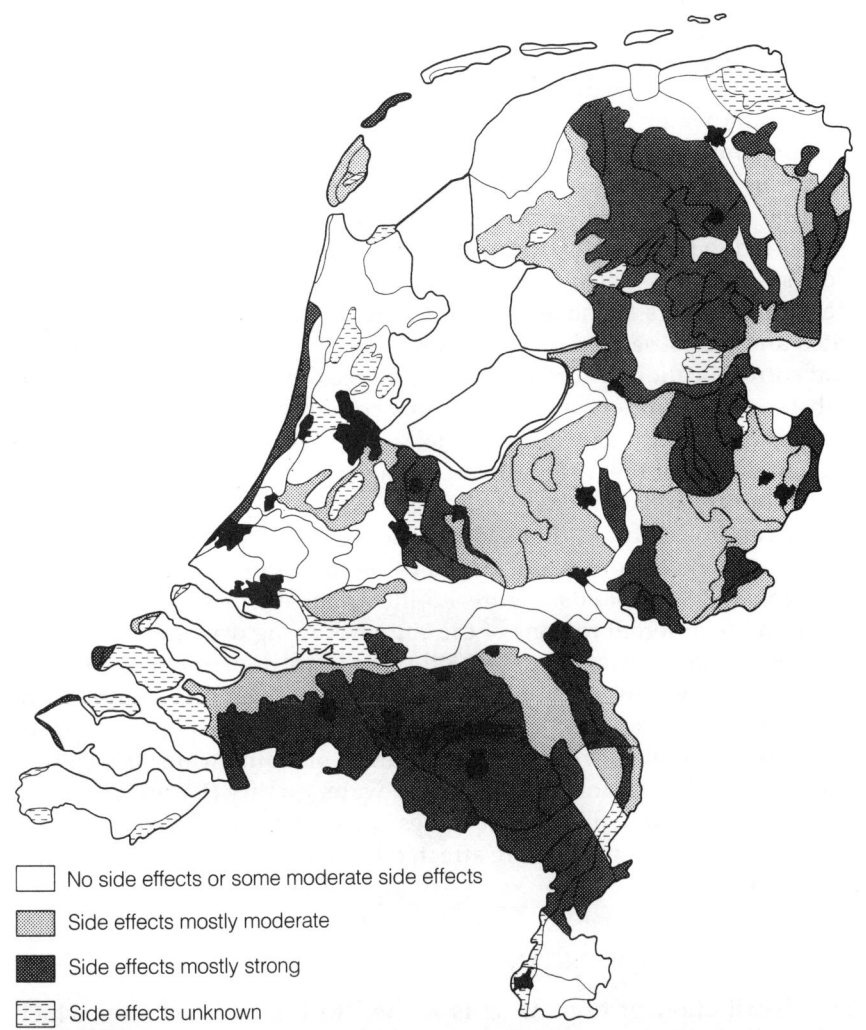

□ No side effects or some moderate side effects

▦ Side effects mostly moderate

■ Side effects mostly strong

▨ Side effects unknown

Figure 1.6 The negative impacts of lowering the groundwater table and the compensation by the inlet of eutroficated water on groundwater-dependent ecosystems in the ecohydrological districts (Van Gool *et al.*, 1990).

cation and eutrophication of ecosystems in this region. On a local scale the distribution of ammonia by air is not only determined by the distance to the source but also by the vegetation structure (Berendse *et al.*, 1993, this volume; Melman and Van Strien, 1993, this volume). In this way areas with a primarily agricultural function affect nature areas in the region.

1.4.2 SPATIAL RELATIONS BY WATER

In The Netherlands there are both natural and artificial water systems. Water systems can be defined as spatially related entities of surface water and groundwater. The flow systems are included in a spatial hierarchy, ranging from local, subregional to regional levels (Engelen *et al.*, 1989). Therefore the impacts of water pollution, groundwater extraction, agricultural drainage and leaching of manure and fertilizers are closely connected to the spatial pattern and the hierarchical level of the affected flow system (Grootjans *et al.*, 1993, this volume). The Netherlands is the delta of the rivers Rhine, Meuse and Scheldt which drain heavily populated and industrialized areas and this spatial relation leads to pollution and eutrophication of surface waters and seas and accumulation of toxic substances in underwater soils. Subsequently this water is spread throughout the country on a regional scale (Figure 1.5), mainly by artificial regulation of the (ground)water table of polders in the western part of the country and of the lakes in Friesland (Duel *et al.*, 1993, this volume). Moreover, in the dunes (Bakker and Stuyfzand, 1993, this volume) and the pleistocene regions water has to be let in to satisfy agricultural and drinking water demands. Figure 1.5 illustrates the percentage of Rhine water in the different regions of the country during dry years.

Many natural systems in The Netherlands are dependent on groundwater quantity and quality (Verhoeven *et al.*, 1993, this volume). Increasing water demands have resulted in a general lowering of groundwater tables, while the distribution of Rhine water has caused a levelling out of regional variations in water quality (Barendregt *et al.*, 1993, this volume). Figure 1.6 gives an overview of groundwater-dependent ecosystems that are affected by lowered groundwater tables and eutrophication.

1.4.3 MOVEMENTS OF ORGANISMS

The distribution of organisms is related to human land use, which influences both the spatial distribution of ecosystems and the movements of organisms. The availability of habitat and the reachability of habitat patches play a role on various spatial scales.

Considering animal movements on a European scale the Dutch delta is of international significance for its wetlands, dunes and intertidal areas (Wolff, 1988). It plays an important role for many bird species as hibernation and feeding area during seasonal migration. On a national scale, different biogeographical regions can be distinguished, ranging from the eutrophic swamps and meadows in the west and north to forests and heathlands in the nutrient-poor sandy regions of the east and south part of the country. These regions are intersected and connected by the river systems. The regions function as core areas for the

Table 1.5 Relative change in abundance of higher plant species, classified into ecological groups, between 1930 and 1980 (Logemann, 1989)

Ecological Groups	Decreased (%)	Increased (%)	Unchanged (%)	Species total
Species of heath-lands and bogs	65	4	31	138
Species of dry pastures	42	6	52	187
Disturbance and pioneer species	39	7	54	105
Species of fertilized pastures	35	5	60	101
Weed species	31	16	54	281
Species of water and water banks	30	9	60	194
Species of dunes and salt marshes	28	16	56	71
Species of forest edges and woodlots	25	9	66	135
Forest species	24	8	69	224
Total	35	9	56	1436

system-specific organisms from which dispersal of species, especially along the river systems, takes place (Descamps *et al.*, 1987; Kwak *et al.*, 1988). The original ecological differences between these regions are disappearing due to the levelling effects of modern land use. This has greatly influenced the distribution of species.

On the landscape level organisms are negatively influenced by the disappearance or fragmentation of natural areas. Natural areas have become more isolated due to increasing distances between patches, while barriers in the form of highways and urban areas were raised and natural elements in the agricultural landscape as connecting elements were erased. Due to these processes local extinction of species becomes more frequent, and suitable areas can no longer be reached. Eventually this may result in extinction on a national scale (Opdam *et al.*, 1993; this volume; Verboom *et al.*, 1993, this volume).

1.5 EFFECTS OF MODERN LAND USE ON NATURAL ECOSYSTEMS

The impacts of modern land use on natural ecosystems in terms of the above-mentioned landscape ecological relations are: destruction, disturbance, acidification, eutrophication, desiccation and fragmentation. The impact of these processes on natural ecosystems has led to a dramatic levelling out of the original spatial variation in abiotic conditions. As a consequence the area occupied by (semi-)natural systems as well as biodiversity has been considerably reduced. A decline in quantity and quality is observed in all landscape types, but most distinctly in the nutrient-poor and groundwater-dependent ecosystems (Table 1.5). Heathlands have been replaced by grasses, while little is left of river dune landscapes, chalk grassland and wet meadows. The total swamp area has diminished, while ruderal species have replaced characteristic dune species. In the tidal region natural variation related to salt content is declining. Although species composition of forest ecosystems is not yet greatly affected, nearly 50% of the forest has a reduced vitality (Ministerie van Landbouw, Natuurbeheer en Visserij, 1990b). Apart from this decline of natural ecosystems, the loss of small natural elements makes the modern agricultural landscapes unsuitable for most organisms. The overall degeneration of natural ecosystems is illustrated by the occurrence of threatened species in all groups of organisms (Table 1.6). As far as higher plants are concerned: 34% of the 1415 species that were found in The Netherlands in 1930 have declined considerably. Plant species restricted to now rare abiotic conditions show the greatest loss, whereas the already common nitrophilous species increased by 8% (Weinreich and Musters, 1989). This decline of species diversity is a common phenomenon in highly industrialized countries (Table 1.6).

1.6 NATURE CONSERVATION STRATEGIES

From the 1970s onward a great variety of environmental regulations have been issued to put a halt to the negative environmental effects of human activities. Structural environmental strategies are focused on

Table 1.6 Threatened animal species for some highly industrialized countries, 1989. a: number of species known, b: percentage threatened species (World Resources Institute, 1990)

Country	Mammals		Birds		Reptiles		Amphibians	
	a	b	a	b	a	b	a	b
The Netherlands	60	65%	257	33%	7	86%	15	67%
Germany[a]	94	47%	305	32%	12	75%	19	58%
France	113	52%	342	40%	36	39%	29	62%
Italy	97	13%	419	14%	46	52%	28	46%
UK	77	31%	233	15%	11	45%	14	14%
USSR[b]	357	22%	765	10%	144	26%	34	26%
US	466	11%	1090	7%	368	7%	222	4%
Japan	186	2%	632	6%	85	4%	58	2%
Canada	210	3%	426	3%	42	2%	41	0%

[a] Former German Federal Republic territory.
[b] Former USSR territory.

the prevention of emission of pollutants. This strategy is combined with the undoing of negative effects of pollutants which have already spread into the environment. For instance water purification, sod cutting of heathlands, removal of underwater soils, are all intended to reduce the accumulation of nutrients. The long-term policy plans are aimed at a sustainable use of the support and production functions, eventually leading to a 'basic quality' for air, water and soil without frustrating regulation and information functions (Tjallingii, 1993, this volume). The plans are directed at different spatial scales depending on the range of distribution of the pollutants or disturbances:

1. on a global level regarding pollutants in high air currents;
2. on a continental level regarding pollutants in air, and sea currents;
3. on a fluvial level regarding pollutants in catchment areas and coastal seas;
4. on a regional level regarding pollutants or disturbances in land-scapes or ecosystems;
5. on a local level regarding changes in landscape elements.

Apart from these environmental measures, in recent years the significance of the spatial configuration of different types of land use and therefore of spatial planning is being recognized (Harms et al., 1993, this volume). Developments in dynamic types of land use such as agriculture or urbanization are unpredictable over a period of time and need a certain flexibility in planning procedures. Natural systems on the other hand require stability in space and time. When these types of land use are integrated on too small a scale, natural systems dependent on specific abiotic conditions will not persist, because of the spatial flows of matter.

The former defensive strategy aimed at merely protecting certain areas as nature reserves therefore often proved to be unsuccessful. Even with intensive management measurements to compensate for external influences, such as the locally heightening of groundwater tables or the removal of topsoil layers, the specific ecosystem could not be maintained. A more offensive strategy was born, in which natural areas of increased size were claimed and buffer zones were developed to protect these areas against unwanted disturbances. At the same time species-oriented nature management became more and more questioned. Not so much species composition but rather the occurrence of natural abiotic processes, such as erosion in the dunes or meandering of streams with minimal human interference, was considered to attribute primarily to the natural values.

These new insights are set out in the latest nature policy plan (Ministerie van Landbouw, Natuurbeheer en Visserij, 1990a), in which a separation of land-use functions was introduced in the form of a robust national ecological network of internationally and nationally significant ecosystems. Likewise, on the regional and landscape scales a planning strategy on the basis of the framework concept is being proposed (Van Buren and Kerkstra, 1993, this volume), and special attention is being paid to the development of connecting zones to link nature reserves as a spatial network (Harms *et al.*, 1993, this volume).

REFERENCES

Arnfield, A.J. (1987) Greenhouse effect, in *The Encyclopedia of Climatology*. Encyclopedia of Earth Sciences Series, Vol. 11, (eds J.E. Olivier and R.W. Fairbridge), Van Nostrand Reinhold, New York, pp. 463–4.

Bakker, T.W.M. and Stuyfzand, P.J. (1993) Nature conservation and extraction of drinking water in coastal dunes of The Netherlands; The Meijendel area as a case study, in *Landscape Ecology of a Stressed Environment* (eds C.C. Vos and P. Opdam) IALE-studies in Landscape Ecology Vol. 1, Chapman & Hall, London, pp.244–620.

Barends, S. (1989) *Percelen in Nederland, veranderingen in de percelering tussen 1900 en nu*. Reeks Landschapsstudies 14, PUDOC, Wageningen.

Barendregt, A., Wassen, M.J. and De Smidt, J.T. (1993) Hydroecological modelling in a polder landscape: a tool for wetland management, in *Landscape Ecology of a Stressed Environment* (eds C.C. Vos and P. Opdam) IALE-studies in Landscape Ecology Vol.1, Chapman & Hall, London, pp. 79–100

Berendse, F., Aerts, R. and Bobbink, R. (1993) Atmospheric nitrogen deposition and its impact on terrestrial ecosystems, in *Landscape Ecology of a Stressed Environment* (eds C.C. Vos and P. Opdam) IALE-studies in Landscape Ecology Vol. 1, Chapman & Hall, London, pp. 104–23.

Daan, N., Bromley, P.L., Hislop, J.R.G. and Nielsen, N.A. (1990) Ecology of North Sea fish. *Netherlands Journal of Sea Research*, **26**, 343–86.

Descamps, H., Joachim J. and Lauga, J. (1987) The importance for birds of the riparian woodlands within the alluvial corridor of the river Garonne, S.W. France. *Regulated Rivers: Research and Management*, **1**, 301–316.

Duel, H., Fiselier, J.L., Klijn, F. and Kwakernaak, C. (1989) *Gebiedsvreemd water in Nederland; een verkenning van de problematiek van gebiedsvreemd water en de ruimtelijke oplossingsmogelijkheden*, Studie- en Informatiecentrum TNO voor Milieu-onderzoek, Delft.

Duel, H., During, R. and Kwakernaak, C. (1993) Artificial wetlands: a device for restoring natural wetland values, in *Landscape Ecology of a Stressed Environment* (eds C.C. Vos and P. Opdam) IALE-studies in Landscape Ecology Vol.1, Chapman & Hall, London, pp. 263–80

Duynisveld, W.H.M., Strebel, O. and Bottcher, J. (1988) Are nitrate leaching from arable land and nitrate pollution of groundwater avoidable? in *Ecological Implications of Contemporary Agriculture* (eds H. Eijsackers and A. Quispel), *Ecological Bulletin*, **39** 116–125.

Engelen, G.B., Gieske, J.M.J. and Los, S.O. (1989) *Grondwaterstromings-stelsels in Nederland*, Achtergrondreeks Natuurbeleidsplan No.2, Ministerie van Landbouw, Natuurbeheer en Visserij, SDU, The Hague.

Eijsackers, H. and Quispel, A. (eds) (1988) Ecological implications of contemporary agriculture, *Ecological Bulletin*, **39**.

Grootjans, A.P., Van Diggelen, R., Everts, F.H. *et al* (1993) Linking ecological and hydrological processes on various spatial scales: a case study of small stream valleys, in *Landscape Ecology of a Stressed Environment* (eds C.C. Vos and P. Opdam) IALE-studies in Landscape Ecology Vol.1, Chapman & Hall, London, pp . 60–78.

Harms, B.H, Knaapen, J.P. and Rademakers, J.G. (1992) Landscape planning for nature restoration: comparing regional scenarios, in *Landscape ecology of a stressed landscape* (eds C.C. Vos and P. Opdam) IALE-studies in Landscape Ecology Vol.1, Chapman & Hall, London, pp . ..-.. .

Heij, G.J. and Schneider, T. (1991) *Acidification research in the Netherlands. Final report of the Dutch priority programme on acidification*, Studies in Environmental Science 46, Elsevier, Amsterdam.

Hurle, K. (1988) How to handle weeds? Biological and economic aspects, in *Ecological Implications of Contemporary Agriculture*, (eds H. Eijsackers and A. Quispel), *Ecological Bulletin* **39**, 63–68.

Kwak, R.G.M., Reyrink, L.A.F., Opdam, P.F.M. and Vos, W. (1988) *Broedvogeldistricten van Nederland, een ruimtelijke visie op de Nederlandse avifauna* PUDOC, Wageningen.

Langeweg, F. (ed.) (1989) *Zorgen voor morgen: nationale milieuverkenning 1985-2010*, Samson H.D. Tjeenk Willink, Alphen aan den Rijn.

Logemann, D. (1989) *De achteruitgang van de natuur in Nederland*, Stichting Natuur and Milieu, Utrecht.

Maas, R.J.M. (ed,) (1991) *Nationale milieuverkenning 1990-2010*, Samson H.D. Tjeenk Willink, Alphen aan den Rijn.

Melman, Th.C.P. and Van Strien, A.J. (1993) Ditch banks as a conservation focus in intensively exploited peat farmland, in *Landscape Ecology of a Stressed Environment* (eds C.C. Vos and P. Opdam) IALE-studies in Landscape Ecology Vol.1, Chapman & Hall, London, pp . 122–42.

Ministerie van Landbouw, Natuurbeheer en Visserij (1990a) *Natuurbeleidsplan*, regeringsbeslissing, SDU, The Hague.

Ministerie van Landbouw, Natuurbeheer en Visserij (1990b) *De vitaliteit van het Nederlandse bos 8*, SDU, The Hague.

Ministerie van Landbouw, Natuurbeheer en Visserij (1991) *Meerjarenplan gewasbescherming*, regeringsbeslissing, SDU, The Hague.

Ministerie van Verkeer en Waterstaat (1989) *De derde nota Waterhuishouding*, regeringsbeslissing, SDU, The Hague.

Opdam, P., Van Apeldoorn, R., Schotman, A. and Kalkhoven, J. (1993) Population responses to landscape fragmentation, in *Landscape Ecology of a Stressed Environment* (eds C.C. Vos and P. Opdam) IALE-studies in Landscape Ecology Vol.1, Chapman & Hall, London, pp . 143–71.

Tjallingii, S.P. (1993) Water relations in urban systems: an ecological approach to planning and design, in *Landscape Ecology of a Stressed Environment* (eds C.C. Vos and P. Opdam) IALE-studies in Landscape Ecology Vol.1, Chapman & Hall, London, pp . 281–302.

Van Buuren, M. and Kerkstra, K. (1993) The framework concept: a new perspective in the design of a multifunctional landscape, in *Landscape Ecology of a Stressed Environment* (eds C.C. Vos and P. Opdam) IALE-studies in Landscape Ecology Vol.1, Chapman & Hall, London, pp . 219–43.

Van de Maarel, E. and Dauvellier, P.L. (1978) *Naar een Globaal Ecologisch Model voor de ruimtelijke ontwikkeling van Nederland*, SDU, The Hague.

Van Gool, C.R., Groen, C.L.G., Runhaar, J. and Van Amstel, A.R. (1990) Verdroging van natuur in Nederland; deel I: inventarisatie van de omvang van het probleem. *Landschap*, **7**, 145–63.

Verboom, J., Metz, J.A.J. and Meelis, E. (1993) Metapopulation models for impact assessment of fragmentation, in *Landscape Ecology of a Stressed Environment* (eds C.C. Vos and P. Opdam) IALE-studies in Landscape Ecology, Vol.1, Chapman & Hall, London, pp . 172–92.

Verhoeven, J.T.A., Kemmers, R.M. and Koerselman, W. (1993) Nutrient enrichment of freshwater wetlands, in *Landscape Ecology of a Stressed Environment* (eds C.C. Vos and P. Opdam) IALE-studies in Landscape Ecology Vol.1, Chapman & Hall, London, pp . 33–59.

Weinreich, J.A. and Musters, C.J.M. (1989) *Toestand van de natuur; veranderingen in de Nederlandse natuur.* Achtergrondreeks Natuurbeleidsplan nr. 4. Ministerie van Landbouw, Natuurbeheer en Visserij, SDU, The Hague.

Wolff, W.J. (ed) (1988) De internationale betekenis van de Nederlandse natuur. Achtergrondreeks Natuurbeleidsplan nr. 3. Ministerie van Landbouw, Natuurbeheer en Visserij, SDU, The Hague.

World Resources Institute (1990) *World Resources 1990–1991, a Guide to the Global Environment.* Oxford University Press, New York.

Zonneveld, J.I.S. (1985) *Levend land; de geografie van het Nederlandse landschap,* Bohn, Scheltema and Holkema, Utrecht/Antwerpen.

Zonneveld, J.I.S. (1991) Naturressourcen und Naturrisiken bei der Landschaftsgestaltung in den westlichen und nördlichen Niederlanden. *Nova acta Leopoldina*, **64**, 67–75.

Part One

Spatial Relations by Water Flows

As a result of artificial water flows throughout the country, polluted river water is the dominant water type, replacing the characteristic local water types. This river water has been charged even further by polluted groundwater, precipitation and surface water from agricultural areas. Consequently, management of surface water and groundwater in the remaining natural wetlands is a basic problem for nature conservation. This problem is further complicated by the current shortage of drinking water and water for irrigation. Yet, wetlands and moist grasslands are regarded as belonging to the most important landscape types to be conserved. Current trends in nature conservation policy aim for extending the area of these landscape types and for encouraging the role of natural processes.

Evidently, this situation gives rise to many questions on the role of groundwater and surface water streams and the nutrient load transported in determining the development of local ecosystems. Answering such questions is the domain of ecohydrology, which is a dominant component of landscape ecology in The Netherlands. It is featured in the three chapters of Part One, and in four chapters in Part Four.

Questions considered in this Part are about the effects of nutrient input, about the impacts of water withdrawal, about the relative contributions of various water streams to the local groundwater quality, in relation to the allocation of nature reserves, and about the effects of hydrological buffer zones. There is a need for predictive models, in which the various sources are integrated. Chapters 2 and 3 are attempts to meet this need for models, although we are still a long way from applying mechanistic models in ecohydrology.

What does the present situation in the Dutch landscape look like? Disregarding the considerable local variation, the general picture is that inputs of nitrogen in wetlands and moist grasslands are generally higher, and locally much higher than prior to the 1950's. To a lesser extent, this is also the case for phosphorus (Chapter 2). This has greatly affected species composition of wetlands and moist grasslands. Effects on fauna are poorly studied, except for meadow birds. In this book these effects are disregarded, because they are mainly secondary (related to changes in vegetation and management) and usually do not have a spatial component. In aquatic systems, the effects of pollution on communities have been described extensively, but spatial relations have been largely neglected up to now.

Ecohydrological research combines analysis of water and nutrient flows with detailed work on the local level, up to the root zone of the vegetation. The objective is to reveal how changes in water level and water chemistry affect decomposition, immobilization and mineralization rates, and therefore nutrient availability to plants, and how this influences the species composition of vegetation (Chapter 2).

So the challenge to ecohydrologists is to link processes on the landscape scale with those on the scale of land units (ecotopes) and the scale of processes in the root zone. This causes considerable methodological problems, since hydrological models of various levels of scale can not be linked simply. The local soil pattern may greatly influence the local pattern of groundwater flow, and consequently the chemistry of local groundwater. A combination of hydrological models on a larger spatial scale with interpretation of local hydrological conditions using plant species as indicators, may give an applicable method of impact assessment (Chapter 3). Statistical models are used here to predict the effect of changes in the water management regime (Chapter 4).

In the near future, research will be greatly influenced by questions related to recent plans to enlarge present nature reserves and to restore nature in former agricultural areas (Chapters 9 and 10). Large-scale hydrological patterns will roughly indicate suitable locations for such areas. More detailed analysis will be necessary to find proper locations, to define the dimensions of buffer zones that will be needed, and to determine how natural processes can be given a new chance. Research on aquatic systems will have to be integrated in this kind of approach. In water management plans, more attention will be paid to the storage of local water instead of replacing it by polluted river water during summer (e.g. Chapter 12). Future problems with drinking and irrigation water supplies will further complicate this type of research.

Nutrient enrichment of freshwater wetlands

2

Jos T.A. Verhoeven, Rolf H. Kemmers and Willem Koerselman

2.1 INTRODUCTION

In prehistoric times, freshwater wetlands formed a major part of the Dutch land surface (Pons, 1992). At present, in a landscape shaped by reclamation and urbanization, Dutch wetland reserves occur as small isolated patches in a matrix of agricultural land or as complexes formed by peat dredging, diking of oxbow lakes, etc. Freshwater dune slacks are common in the extensive coastal sand dune areas. Many wetland reserves are highly valued for their botanical and ornithological richness.

The Dutch wetlands have been strongly modified by humans. Not only were many of them reconstructed or even created as a result of the great reclamations and the peat digging works, but in almost all cases their hydrology is also being carefully controlled. Water tables are generally kept within narrow limits for the purpose of optimizing agricultural production in adjacent fields or pastures, primarily by supplying or draining surface water. In other areas, groundwater is being extracted for drinking water preparation.

Over the past 50 years, the Dutch wetlands that used to be meso- to eutrophic, have been subject to considerable enrichment with nutrients. This is the result of high atmospheric N deposition and N and P transports from the heavily fertilized agricultural areas via shallow groundwater and surface water flow. Infiltration of eutrophicated river

Landscape Ecology of a Stressed Environment
Edited by Claire C. Vos and Paul Opdam
Published in 1993 by Chapman and Hall, London. ISBN 0 412 44820 3

water into wetland reserves and sand dune areas has been another source of nutrient enrichment.

This chapter considers the consequences of this nutrient enrichment for the productivity, plant species density and species composition of freshwater wetlands. Only wetlands fed by rain, groundwater and/or surface water are considered and the focus is mainly on fens, moist grasslands and dune slacks. There is no discussion of ombrotrophic bogs or wetlands in the present floodplains of the great rivers.

The fens dealt with are peat-forming ecosystems occurring mostly in sets of rectilinear ponds originating from peat dredging in former centuries. The ponds have terrestrialized and carry species-rich plant communities or alder scrub. Almost all fens are located in polders in the lower northwestern part of The Netherlands under a regime of carefully controlled water tables (Verhoeven *et al.*, 1988; Van Wirdum, 1991). Where they occur in groundwater discharge areas, water management mainly consists of drainage of excess rain water. Fens in areas where groundwater discharge stopped and turned into groundwater recharge because of human interference in the regional hydrology, are generally supplied with surface water. Although the water managers prefer good-quality 'lithotrophic' ('groundwater-like') water, they often have no other source than the strongly polluted water originating from the Rhine river.

Moist grasslands with a high water table are found in the polders in the western part of the country, on peat or clay soils. They also occur in the middle and lower courses of brook valleys in the sandy southeastern part of The Netherlands, in discharge areas of main aquifers. Many of these systems have been subject to fertilization and drainage in the past but have been taken out of intensive agricultural use. A management of annual harvesting is aimed at removing nutrients and restoring the former species richness. Like the fens, these grasslands are subject to artificial infiltration of polluted river water during dry spells.

Besides coastal defence, nature conservation and recreation, water catchment is one of the major present uses of the Dutch coastal sand dunes. From the 1850s onwards, extraction of groundwater for drinking water supply has occurred. With growing demands for drinking water, lowering of water tables in dune slacks became a major environmental problem (Bakker *et al.*, 1979; Van Zadelhoff, 1981; Van der Meulen, 1982; Bakker and Stuyfzand, 1993, this volume). Artificial recharge of the dune aquifers with water from the rivers Rhine and Meuse, a measure to cope with these problems, started in the 1950s. Increased nutrient concentrations together with increased groundwater flow rates occurring near infiltration ponds (Van Dijk, 1984, 1985b) caused a strong increase of nutrient inputs to infiltration ponds and natural dune slacks fed by local groundwater discharge originating from infiltration ponds.

Although none of the systems treated contain vegetation types unique to The Netherlands, fen and dune slack vegetation have become increasingly rare in Western Europe. Further, the landscape of long rectilinear turf ponds separated by baulks of uncut peat and inhabited by terrestrializing fen vegetation are a typically Dutch landscape feature.

This chapter first considers the various sources of nutrient (i.e. nitrogen and phosphorus) enrichment that are important for Dutch freshwater wetlands. Then, the role of hydrological factors, such as water table and water chemistry, in controlling nutrient availability is considered, and the way in which this relates to the position of the wetland in the landscape. A review of studies on the control of plant growth by nutrients, and of the relation between productivity and species richness and species composition is given, resulting in an evaluation of the effects of enrichment. Finally, some implications of the picture generated for the management of wetland reserves are given.

2.2 ENLARGED INPUTS: RAIN, GROUNDWATER, SURFACE WATER

With the exception of some extensive dune valleys, freshwater wetland nature reserves in The Netherlands are usually small and occur scat-

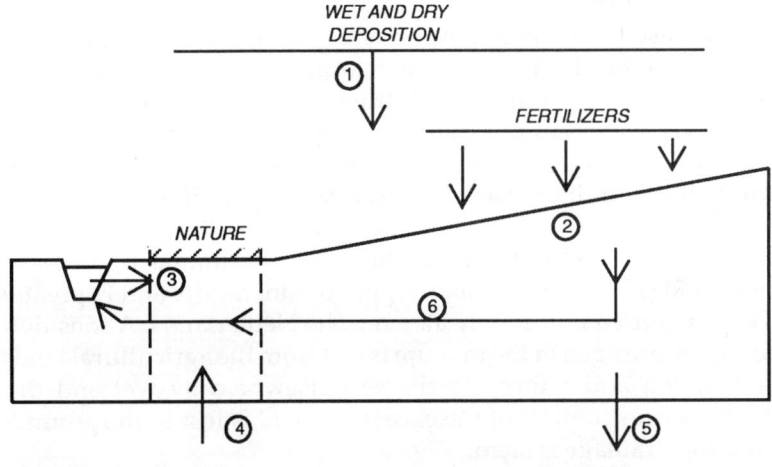

1 atmospheric deposition
2 leaching
3 surface water infiltration
4 discharge from aquifer
5 recharge to aquifer
6 drainage

Figure 2.1 Sources of nutrients for natural ecosystems in catchments with intensively used agricultural land.

tered throughout the landscape where they are bordered by intensively used agricultural land within the same catchment. Most of these (semi-)natural areas are seriously affected by enlarged nutrient inputs from the following sources (Figure 2.1):

1. leaching of nutrients from agricultural land within the same catchment and transport by shallow groundwater flow to the drainage system;
2. wet and dry atmospheric deposition;
3. infiltration of surface water originating from polluted rivers;
4. discharge from aquifers.

The quantitative importance of these sources differs for the various nature reserves. Although the actual inputs to wetland ecosystems have only exceptionally been measured, some general indications can be given of regional differences in the nutrient quantities potentially supplied by the sources mentioned. This general picture is based on an extensive model developed by the Dutch government in the context of the preparation of the third Policy Analysis of the Water Management in The Netherlands (Grashoff et al., 1989; Kroes et al., 1990). In this model, the consequences of eutrophication from the sources mentioned for the nutrient load of surface waters were analysed. All calculations were generalized to the level of catchment areas. Coastal dune areas were not included in the study.

The model used data on atmospheric deposition and application of manure and chemical fertilizers from local and regional data bases administered by the government. Nutrient transports to certain catchments via river water supply during dry spells and via groundwater discharge and recharge, and leaching of nutrients from agricultural areas towards the draining surface waters were quantified by a combination of water quantity and water quality models (Kroes, 1988).

Data on manure and fertilizer applications are summarized by Vos and Zonneveld (1993, this volume). Applied minerals (N and P) greatly exceed crop requirements nearly all over The Netherlands. A considerable part of the nitrogen in the manure is lost from the agricultural lands by volatilization and returns to the soil elsewhere by wet and dry deposition. However, most of the excess nutrient is lost to the groundwater and the drainage system.

A considerable load of nitrogen by atmospheric deposition can be observed all over The Netherlands (Berendse et al., 1993, this volume). Current values average about 45 kgN/ha/year which is 10 times background values (Erisman, 1990). Deposition is fairly light in the coastal areas (20–30 kgN/ha/year) and particularly high in the southeastern part of the country (50–65 kgN/ha/year). Atmospheric depositions of phosphorus and potassium are hardly affected by human activities,

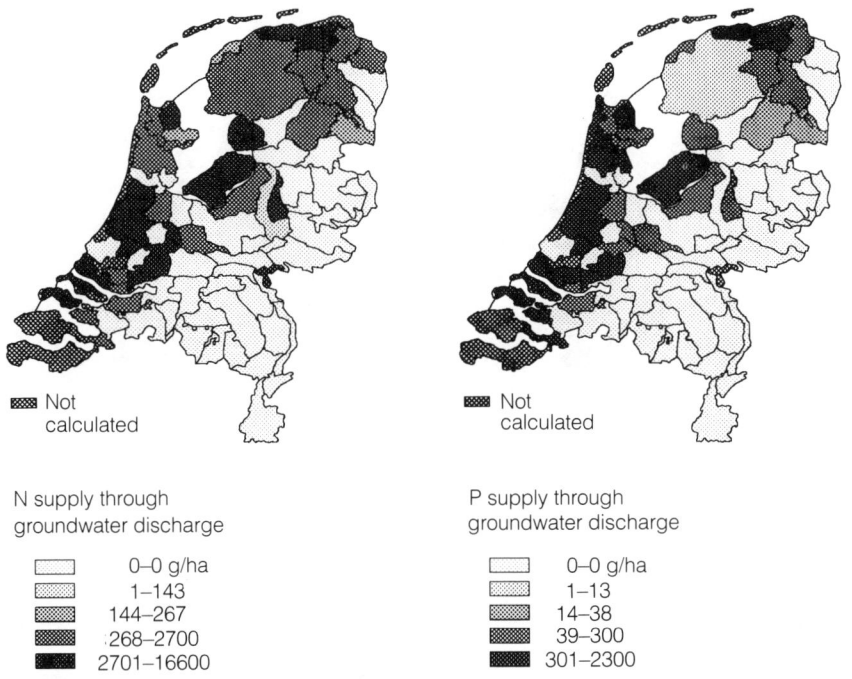

N supply through
groundwater discharge

	0–0 g/ha
	1–143
	144–267
	.268–2700
	2701–16600

P supply through
groundwater discharge

	0–0 g/ha
	1–13
	14–38
	39–300
	301–2300

Figure 2.2 Base load of nitrogen and phosphorus to catchment areas by seepage from aquifers in 1985 (Grashoff *et al.*, 1989; Kroes *et al.*, 1990).

except for some industrial areas near the coast.

Nutrient inputs to catchment areas by the artificial supply of polluted river water during dry spells are of minor importance and only locally exceed 3 kgN/ha/year. In the northwestern part of The Netherlands, the nitrogen load by discharge from aquifers is of some interest. In contrast the phosphorus load from this source is considerable (Figure 2.2) and exceeds all other external inputs for this element. These flows from the aquifer must be considered as a base load originating from numerous peat and clay deposits in the subsoil.

All sources of nutrients were integrated in the models to calculate leaching of N and P to the drainage system by shallow groundwater flow. According to the model calculations, leaching of nitrogen from agricultural land is a quantitatively very important flux (Figure 2.3). Together with the atmospheric N deposition, it may locally be a major cause of eutrophication of freshwater wetland ecosystems.

This rough, regional picture of nutrient enrichment in The Netherlands can only partly be used for evaluating extra nutrient inputs to (semi-)natural ecosystems. Local conditions with respect to atmospheric N deposition and nutrient loads of ground and surface water may be quite different from the average. Further, the inputs depend on

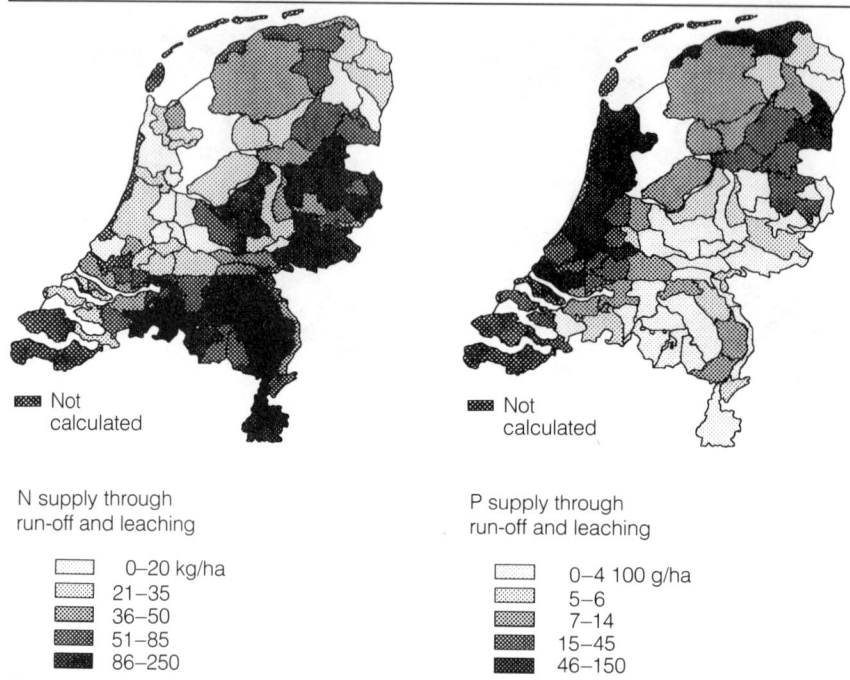

N supply through
run-off and leaching

    ☐   0–20 kg/ha
    ▒   21–35
    ▓   36–50
    ▓   51–85
    ■   86–250

P supply through
run-off and leaching

    ☐   0–4 100 g/ha
    ▒   5–6
    ▓   7–14
    ▓   15–45
    ■   46–150

Figure 2.3 Nitrogen and phosphorus transported to draining surface waters by leaching in the various catchment areas in 1985 (Grashoff *et al.*, 1989; Kroes *et al.*, 1990).

the particular hydrology of the various ecosystems and are, therefore, difficult to generalize.

There are a few examples of nutrient budget studies for wetland ecosystems in which the quantitative importance of various sources was studied. Koerselman *et al.* (1990) calculated for two mesotrophic fens (one in a groundwater discharge area, the other in a groundwater recharge area) that atmospheric deposition provided two to three times more N than shallow groundwater discharge or surface water supply, respectively (Table 2.1). Atmospheric inputs were of the same magnitude as hydrologic inputs for phosphorus. Van Wirdum (1991) found a similar picture for a quaking fen in a groundwater recharge area where considerable amounts of lithotrophic surface water are supplied during the summer months (Table 2.1).

In dune areas where river water is artificially recharged, nutrient inputs on the vegetation along infiltration ponds and in dune slacks fed by discharge of river water are strongly increased. Although there have been no studies of the nutrient budgets of these systems, it is clear that external supplies with river water are by far the most important source of nutrients (Van Dijk, 1984, 1985b, 1989; Van Dijk *et al.*, 1985).

Although the degree to which inputs of nitrogen and phosphorus to

Table 2.1 Nutrient balances of quaking fens in Westbroek (groundwater discharge area). Molenpolder (groundwater recharge area) (Koerselman *et al.*, 1990) and Stobbenribben (groundwater recharge area) (Van Wirdum. 1991). Values as kg/ha/year

Stobbenribben	N In	N Out	P In	P Out	K In	K Out
Precipitation						
Wet	25	-	0.3	-	10	-
Dry	(20)	-	-	-	-	-
Fixation	(20)	-	-	-	-	-
Surface water	11	0	0.4	0.0	62	0
Groundwater	0	8	0.0	1.3	0	74
Harvest	-	57	-	3.2	-	54
Sum total	76	65	0.7	4.5	72	128
Mineralization	115	-	5.2	-	-	-

Westbroek	N In	N Out	P In	P Out	K In	K Out
Precipitation						
Wet	24	-	0.7	-	6	-
Dry	18	-	-	-	-	-
Fixation	13	-	-	-	-	-
Surface water	1	21	0.1	0.7	4	13
Groundwater	20	0	0.5	0.0	6	0
Harvest	-	66	-	5.6	-	44
Sum total	76	88	1.3	6.3	16	58
Mineralization	67	-	3.4	-	-	-

Molenpolder	N In	N Out	P In	P Out	K In	K Out
Precipitation						
Wet	26	-	0.6	-	6	-
Dry	18	-	-	-	-	-
Fixation	2	-	-	-	-	-
Surface water	7	9	0.5	1.0	19	22
Groundwater	0	1	-	0.1	-	2
Harvest	-	38	-	3.9	-	32
Sum total	53	49	1.1	5.0	25	56
Mineralization	315	-	43.8	-	-	-

wetlands and moist grasslands are enlarged may be difficult to indicate for any particular site, it can be concluded that inputs of nitrogen, and to a lesser extent of phosphorus, are generally higher, and locally much higher than they were before the 1950s. In order to evaluate the effects of enlarged inputs of nutrients, it is necessary to consider the internal nutrient cycling processes within the ecosystem. Nutrients may be stored in soil pools not accessible to plants or may leave the ecosystem through enlarged output flows. On the other hand, increased nutrient inputs may accelerate decomposition and mineralization rates, and thus have a positive feedback on nutrient availability to plants. Such effects may also be induced by changes in water chemistry (pH, composition of major ions).

The next section indicates how the internal nutrient cycling in wetland ecosystems is indirectly controlled by soil moisture and water chemistry. Both these factors are primarily dependent on the local hydrology of the wetland. Further, it will be pointed out that because of these controls, the position of the wetland in the landscape is of great importance for its nutrient dynamics.

2.3 HYDROLOGY AND SOIL CONDITIONS AS FACTORS CONTROLLING NUTRIENT AVAILABILITY

2.3.1 CONTROL OF NUTRIENT AVAILABILITY

Nutrient dynamics in freshwater wetland ecosystems are influenced by hydrology in both a direct and an indirect way. The direct influence is associated with the role played by water as an agent for mass flow. Water flows as carriers for inputs and outputs of nutrients have been briefly dealt with in the previous section. Both qualitative and quantitative aspects of the water regime (i.e. water chemistry and soil moisture, respectively) have an indirect influence on the nutrient cycling of any particular site. This section indicates the ways in which these factors control nutrient availability and then investigates how their impact is dependent on the position of a site in the landscape.

Water chemistry distinctly influences mineralization and humification processes of soil organic matter which is the main internal source of nutrients. A high calcium content of the interstitial water (lithotrophic water, Van Wirdum, 1991) favours decomposition of litter which leads to low C/N ratios and generally also to mobilization of nitrogen from organic matter (Kemmers and Jansen, 1985). A study of decomposition and mineralization in Dutch fens and bogs (Verhoeven et al., 1990) showed faster decomposition and lower C/N ratios in the fens with a supply of base-rich water. However, N mineralization was not correlated with C/N ratio and was significantly higher in mires with atmotrophic than in those with lithotrophic water. The combination of high N mineralization and low decomposition in atmotrophic mires is probably due to the absence of bacterial immobilization.

In contrast to the availability of nitrogen that is mainly determined by bacterial processes as decomposition and mineralization, the availability of phosphorus is also dependent on physicochemical processes as adsorption/desorption and precipitation/dissolution (Nichols, 1983; Richardson and Marshall, 1986). The presence of calcium, aluminium and iron in the interstitial soil water and in the soil particles is of great importance for these processes. Soil pH and redox potential play a distinct conditioning role: orthophosphate is adsorbed to iron and aluminium oxides at pH 4.5–6.5 and to calcium hydroxides at pH >6.5 (Stumm and Morgan, 1981). As orthophosphate is more effectively bound to Fe(III) oxides than to Fe(II) oxides (Patrick and Khalid, 1974), a high redox potential will result in stronger adsorption of orthophosphate.

Soil moisture, as controlled by the water table, is another parameter of great importance because of its role in conditioning soil temperature, aeration, redox status and soil acidity. All these factors directly control the rates of decomposition and mineralization of soil organic matter and, therefore, the cycling of nitrogen and phosphorus. Soil acidity and redox status, as explained above, also control the adsorption/desorption and precipitation/dissolution of inorganic phosphates.

2.3.2 POSITION IN THE LANDSCAPE

The relief as an expression of the topographic factor *sensu* Jenny (1941) can be considered as a basic factor underlying hydraulic gradients which generate groundwater flow through the landscape. In most landscapes, a major distinction can be made between relatively dry groundwater recharge areas and relatively wet groundwater discharge areas (Figure 2.4a, b; Kemmers, 1986a). The two hydrological factors that indirectly control nutrient availability, i.e. water chemistry and soil moisture, are strongly dependent on the position of a particular site in the landscape (Figure 2.4a).

Groundwater flow through aquifers is accompanied by the dissolution of calcite deposits by natural weathering processes, resulting in an increase of Ca^{2+} and HCO_3^- ions with growing residence time (Kemmers, 1986b). Consequently, each site has distinct hydrochemical features depending on its position in the hydraulic gradient in the landscape. Discharge areas are enriched with Ca^{2+} ions which creates sites with a high base state of the soil interstitial water. In contrast, recharge areas have a low base state because of leaching of cations due to the continuous infiltration of the precipitation surplus to the aquifer.

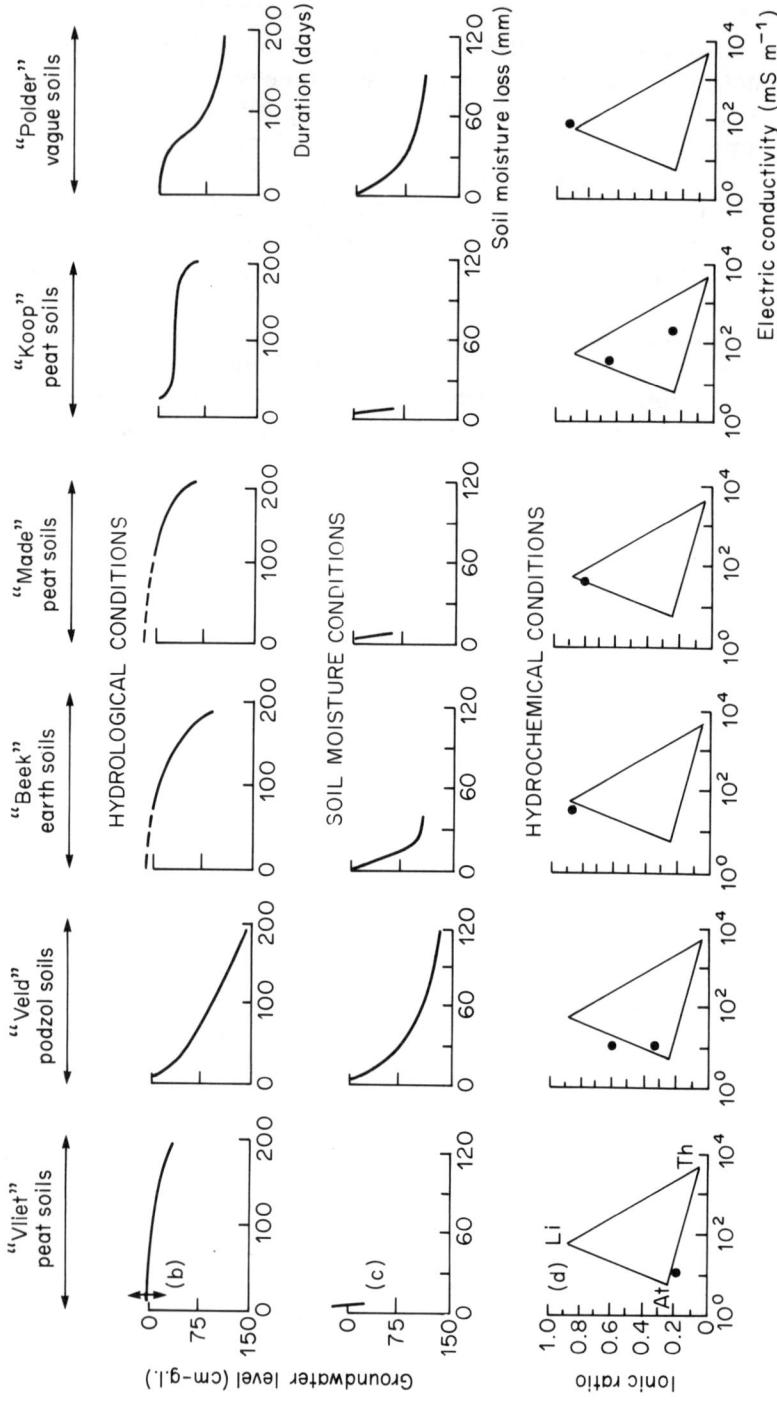

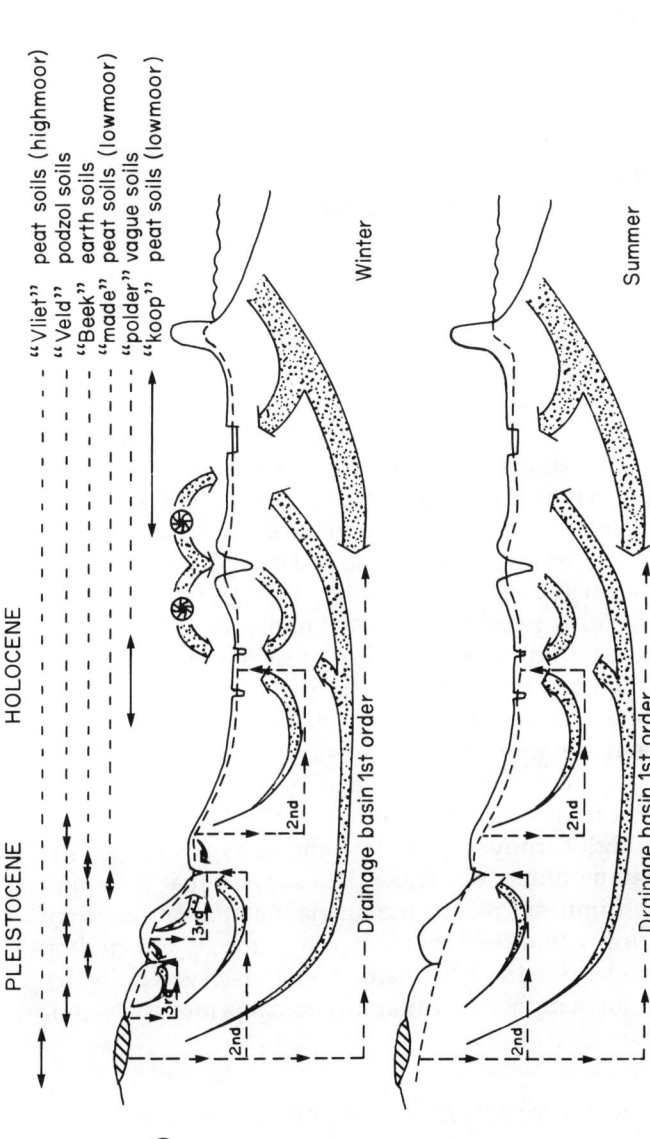

Figure 2.4 Position of wetland soils in the landscape as related to groundwater flow systems in The Netherlands. Groundwater discharge to drainage areas has a permanent character but may be periodical in catchments of a lower order. (a) Main site conditions occurring in wetland soils are controlled by the hydrological position in the landscape. (b) Hydrological conditions as depth – duration – frequency curves of groundwater levels. (c) Soil moisture conditions as moisture loss from the soil related to the water table. (d) Hydrochemical conditions as the ionic ratio (IR) versus electrical conductivity (EC). The vertices (At, Li, Th) represent reference values of atmotrophic, lithotropic and thalassotrophic groundwater types. (After Kemmers, 1986a.)

Recharge areas with artificial infiltration of river water similar in chemistry to groundwater have a base state comparable with that in discharge areas (Van Wirdum, 1991).

Van Wirdum (1980) suggested to use the ionic ratio (IR; $Ca^{2+}/Ca^{2+} + Cl^-$ in equivalent concentrations of charge) and the electric conductivity (EC) as a simple parameter set to characterize water chemistry. He introduced the IR–EC diagram with reference points for the chemical composition of the main water types in the hydrological cycle: rainwater (At), groundwater (Li), and seawater (Th). Currently in Dutch research the hydrological position of sites is deduced from IR–EC diagrams (Figure 2.4d), IR and EC both being low for rainwater-fed and high for groundwater-fed areas. Surface water-fed sites show intermediate IR in combination with high EC values.

Soil moisture, as controlled by the water table, is also strongly related to the position of the ecosystem in the landscape. Soil moisture is generally lower in recharge areas than in discharge areas (Figure 2.4c).

In conclusion, the position of any particular wetland in the landscape is of great importance for the availability of nutrients to the vegetation. Discharge of groundwater rich in Ca or Fe will keep orthophosphate availability low by continuously adding fresh adsorption sites and by controlling the pH. Several recent studies have provided statistical evidence that the phosphorus concentration in the soil interstitial water will be buffered at a low level as long as the soil complex maintains a high calcium saturation (Boyer and Wheeler, 1989; Kemmers, 1990; Wassen, 1990). On the other hand, the high soil moisture and high base status in wetlands in groundwater discharge areas are favourable for bacterial decomposition and recycling of nitrogen.

Hence, groundwater discharge sites are characterized by a high turnover rate of organic matter with ample supply of nitrogen but a restricted availability of phosphorus. On the other hand, recharge sites with a low calcium saturation establish both high C/N and C/P ratios with a surprisingly high turnover rate of N and P, as a result of the absence of bacterial immobilization. These data suggest that phosphorus might play a key (limiting) role in the nutrient status of discharge sites, which would imply that they would not be strongly affected by enrichment with nitrogen. On the other hand, wetlands in groundwater recharge sites would have higher P availability and a primarily N-controlled plant growth.

2.3.3 'INTERNAL' EUTROPHICATION AS A RESULT OF HYDROLOGICAL CHANGES

Apart from the direct nutrient enrichment due to increased inputs as described in section 2.2, many wetlands in The Netherlands have become enriched because of increased availability of nutrients previously stored in inaccessible pools. This phenomenon is here referred to as 'internal eutrophication' defined as an acceleration of nutrient cycling in the soil associated with a change in environmental conditions, i.e. a water table draw-down or a change in water chemistry.

As a result of modifications of the water regime to improve agricultural production, water tables in many areas have been lowered substantially. Over the period 1950–1987, water tables were significantly lowered in 77 Dutch nature reserves, most commonly by 30 cm (Garritsen et al., 1990, Vos and Zonneveld, 1993, this volume). Experimental evidence suggests that this must have resulted in a strong increase of decomposition and N mineralization rates (Kemmers and Jansen, 1985; Kemmers, 1990). Grootjans (1985) found enhanced N mineralization in grasslands as a result of water table draw-down. The effects on P availability are more difficult to predict, as P mineralized by microbes may be readily adsorbed by newly formed iron oxyhydroxides under conditions of soil aeration and a pH < 6.5 (Patrick and Khalid, 1974).

Another impact affecting nutrient cycling is the groundwater extraction that has turned many groundwater discharge areas into groundwater recharge areas. This impact causes a major alteration of the water chemistry in wetland areas. P availability is particularly affected by changes in soil pH and base saturation (Stumm and Morgan, 1981). Ca-P oxyhydroxides that are thought to be important P sinks in groundwater-fed wetlands (section 2.3.2), disintegrate when the pH drops below 6.5, resulting in an increase of P availability. There are also indications that an increased concentration of sulphate enhances P availability (Caraco et al., 1989).

Although effects of 'internal eutrophication' have rarely been studied in situ, the process must be at least partially responsible for the nutrient enrichment of Dutch wetland ecosystems.

2.4 CONTROL OF PLANT GROWTH BY NUTRIENT AVAILABILITY

2.4.1 N AND P AS ELEMENTS LIMITING PLANT GROWTH

For an evaluation of the effects of nutrient enrichment of ecosystems on the species composition of the vegetation, it is essential to know which of the major plant nutrients, nitrogen, phosphorus and potassium, is in short supply and, therefore, controls plant growth and productivity. The best way to identify the limiting nutrient is to measure the biomass increase in response to nutrient additions in a fertilization experiment in the field (Chapin et al., 1986). N, P and K are added separately and in various combinations. Measurements of nutrient uptake may give further clues on nutrient availability in the various treatments.

A number of such 'fertilization' studies in various parts of the world have indicated that nitrogen availability most commonly controls primary production in terrestrial ecosystems, e.g. in wetlands (Graneli, 1983; Bowden, 1987; Morris, 1991), tundra (Shaver and Chapin, 1980), and abandoned fields (Tilman, 1984). Willis (1963) found N–limited plant growth in British dune slacks. A simultaneous control by N and P, indicating low availability for both of these elements has been found in floating *Carex* and *Equisetum* stands in Norway (Solander, 1983) and in wetlands in Alaska (Sanville, 1988). Examples in which P or K is the primary factor limiting growth are mostly non-fertilized hayfields

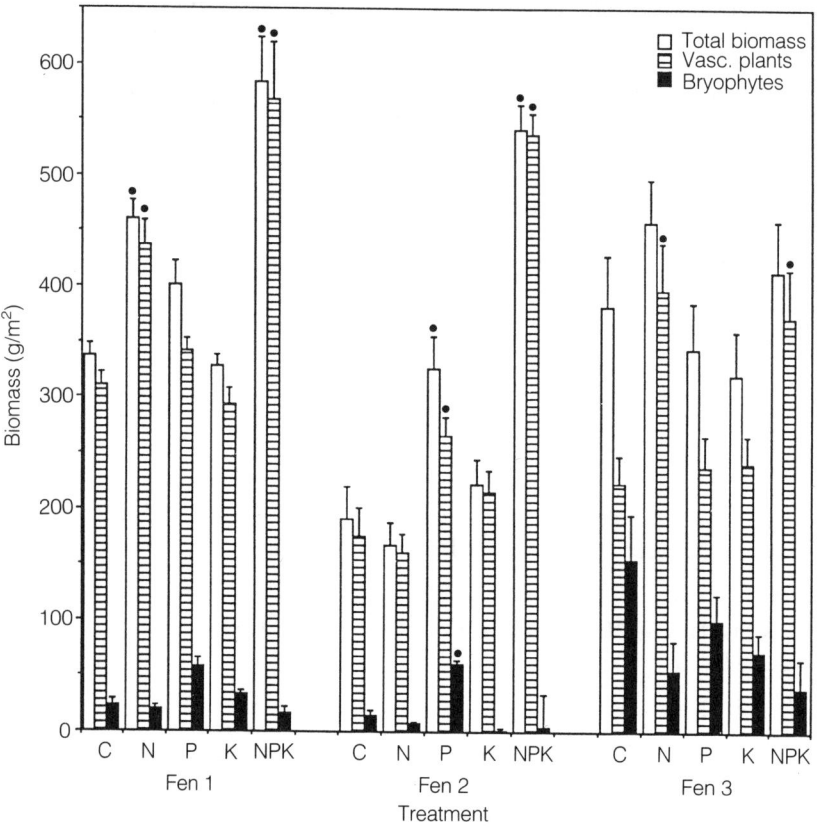

Figure 2.5 Response of aboveground biomass (harvested in August) to fertilization with N, P, K and all three of these elements compared to a control in three fens (after Verhoeven and Schmitz, 1991). Fen 1: fen in groundwater discharge area terrestrialized in the 1940s and harvested since the 1960s; fen 2: fen in groundwater discharge area terrestrialized about 1900 and harvested since the 1930s; fen 3: as fen 1 but in groundwater recharge area.

with a summer-mowing regime (e.g. Egloff, 1986; Hayati and Proctor, 1991).

In view of the very high atmospheric nitrogen deposition on all terrestrial ecosystems in The Netherlands for several decades now, one would expect N availability in these systems to increase with the possible effect that its role in controlling plant growth is taken over by P or K. However, a comparison of N mineralization rates found in Dutch fens and bogs with values for similar systems in other areas of the world did not suggest a higher N mineralization in The Netherlands, so that N enrichment seems not to have affected N cycling (yet) (Verhoeven et al., 1990).

Further, fertilization experiments carried out in fens and moist grasslands (Vermeer, 1986; Verhoeven and Schmitz, 1991) showed N-controlled plant growth in most systems studied. Only in fens in which prolonged annual harvesting of the vegetation had caused a depletion of the soil in P and K, was plant growth controlled by the availability of P (Figure 2.5). A study of nutrient budgets has shown that the management of annual harvesting that is commonly practised to prevent trees from establishing in these species-rich herbaceous stands leads to near-equilibrium situations for N but high net exports of P and K from the system (Table 2.1; Koerselman et al., 1990).

Similar results have been obtained by H. Olff (personal communication, 1991) for moist 'abandoned'[†] grasslands in a brook valley under a regime of annual hay-making: plant growth proved to be distinctly N-limited in grasslands 5 years after abandonment, whereas addition of P as well as K resulted in increased biomass production in grasslands 44 years after abandonment.

T. Oomes (personal communication, 1991) carried out fertilization experiments in grasslands with different water tables. He found that plant growth responded to N fertilization in the situation with the low water table (30–70 cm below the soil surface in summer) and to K fertilization in the situation with the high water table (10–50 cm below soil surface in summer). Fertilization experiments with N, P and K in wet heathlands showed no response of biomass production, although strong shifts in species dominances occurred, particularly upon P fertilization (Aerts and Berendse, 1988; section 2.5).

No fertilization experiments have been conducted in the Dutch dune slacks that have been impacted by falling water tables as a result of drinking water extraction and nutrient enrichment due to infiltration of polluted river water. In the absence of such studies, an indication of nutrient limitation may be obtained from the nutrient concentrations

---

[†]'Abandonment' here means a switch from commercial, agricultural management to nature management: cessation of grazing and fertilization, continuation of annual harvesting.

in the aboveground plant material. If N or P concentrations are lower than the so-called 'critical value' (after De Wit *et al.*, 1963), the nutrient is considered limiting plant growth. We used the ratio between N and P in plant biomass as an indication for growth limitation by these elements in dune slack vegetation. Under circumstances of low supply of both N and P, the N/P ratio in the plant is assumed to approximate 20 (De Wit's critical N value divided by the critical P value). N/P ratios well over 20 indicate P limitation, whereas N/P ratios less than 20 are indicative of situations where plant growth is N-controlled (Figure 2.6). In pristine dune slacks, the vegetation N/P ratio is close to 20, and the vegetation may be either P or N limited, or limited by both elements together (co-limitation). Fertilization studies by Willis (1963) in pristine British dune slacks have shown that nitrogen was the primary factor limiting plant growth there. This is also probably the case for pristine dune slacks in The Netherlands. The vegetation in dune slacks in dune

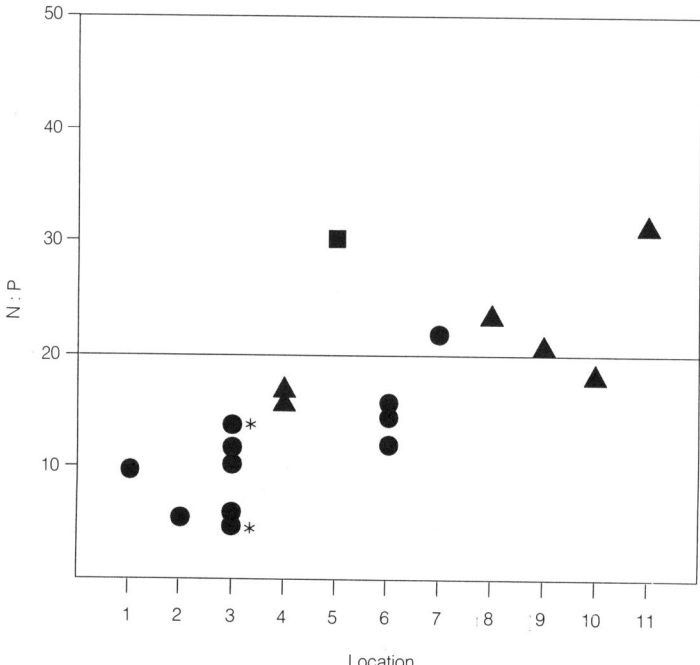

Figure 2.6 N/P ratio of vegetation in dune slacks and along dune ponds. ●, Sites fed by infiltration water rich in P; ■, sites fed by infiltration water low in P due to dephosphatization; ▲, natural sites outside water catchment areas. Locations: 1, 2: Berkheide, 3: Meijendel, 4, 5: North-Holland Dune Reserve, 6, 7: Amsterdam Municipal Waterwork Dunes, 8: Terschelling, 9: Zwanewater, 10: Voorne's Dune, 11: South-Kennemerland Dunes. Source: location 1, 2, 3*, 5, 7–11, Van Dijk (1985a); location 3, 4, 6, Koerselman (1991). Modified after Koerselman (1991).

areas with artificial river water infiltration shows generally low N/P ratios, indicating N-limited plant growth (Figure 2.6, location 5). This river water is eutrophic and particularly rich in phosphates; in the case where dephosphatized water was infiltrated, the vegetation showed much higher N/P ratios indicating P-limited growth.

The information reviewed in this section suggests that either nitrogen or phosphorus may act as the key factor controlling plant growth in fens, moist grasslands and dune slacks in The Netherlands. Early-succession fens, recently abandoned grasslands and pristine dune slacks show a nitrogen-limited plant growth, whereas P-limited growth occurs in systems that have been under a regime of annual summer mowing and hay-making for more than 30 years and in dune slacks with infiltration of dephosphatized river water.

2.4.2 IMPACT OF NUTRIENT ENRICHMENT ON PRIMARY PRODUCTIVITY

Does the increased atmospheric nitrogen input result in increased bio-mass production in systems with N-limited plant growth? Several investigations have shown that such a biomass increase does not necessarily take place in cases with a regime of annual mowing. Many of the grasslands have been in agricultural use in the past and have built up large nutrient stocks in the soil (Bakker, 1989). The export of nutrients due to annual harvesting is substantial and exceeds annual atmospheric inputs, in particular during the first 10 years after cessation of fertilization (Table 2.2; Bakker, 1989, the values for atmospheric inputs measured for fens give an approximation of such inputs for the grass-

Table 2.2 Exports of nutrients (kg/ha/year) as a result of annual harvesting in moist grasslands and fens in The Netherlands. The grasslands have been in agricultural use in the past. Annual inputs for the fens are total values composed of atmospheric deposition, groundwater flow and N fixation.

g/f	Soil	Exports			y	Inputs			Reference
		N	P	K		N	P	K	
g	peat	156	66	91	3				Bakker (1989)
g	peat	176	30	93	11				Bakker (1989)
g	sand	100	14	22	11				Oomes and Mooi (1985)
fd	kragge	66	6	44		56	1.0	16	Koerselman et al. (1990)
fr	kragge	38	4	32		40	0.7	25	Koerselman et al. (1990)

y, number of years of harvesting without fertilizer applications.
g, grassland.
fd, fen in groundwater discharge area.
fr, fen in groundwater recharge area.
kragge, floating mat with > 80% organic matter

lands). Under the present circumstances of high nitrogen deposition, depletion of the P and K stocks in the soil is probably relatively more rapid than that of the N stock, as is indicated by the response to P and K addition in the grasslands where fertilization has been stopped more than 40 years ago. In grasslands taken out of agricultural production, therefore, prolonged annual harvesting is expected to result in a decrease in productivity primarily caused by decreasing nitrogen availability, and eventually in a stable level of productivity due to limitation by P and/or K, possibly in combination with N.

In the Dutch terrestrializing fens that have never been fertilized but have built up their nutrient stocks by accumulation of organic matter in their floating mats since the 1930s (Verhoeven, 1986; Verhoeven et al., 1988), plant growth is N–limited in the earlier successional stages. In these relatively young systems, the high N inputs due to atmospheric deposition and groundwater enrichment may result in a slowly increasing biomass production. Prolonged annual harvesting will, however, result in a shift from N-controlled to P-controlled plant growth and stabilization or even decrease in productivity.

Distinct increases in productivity have been documented for dune slack vegetation; here, the increase has been primarily attributed to the high nutrient load of the river water artificially recharged in the dune areas for storage and later dinking water preparation (Van Dijk, 1984, 1985a). However, dune slacks not influenced by artificial river water recharge have also commonly shown an increase in biomass production. Heil et al. (1990) have attributed these increases to high atmospheric N deposition.

In conclusion, evidence for increases of plant biomass production as a result of the high atmospheric nitrogen deposition exists only for wetland systems without annual harvesting of hay, such as dune slacks. In 'abandoned' grasslands and fens under a prolonged regime of annual mowing, productivity does not show a rising trend. Also, there are no indications of a general shift from limitation by N towards limitation by P or K as a result of high atmospheric N deposition. Such a shift has taken place only in response to certain management procedures: prolonged annual harvesting of fens and grasslands depletes the soil relatively stronger in P and K than in N, and supply of water rich in N and poor in P may eventually lead to P-limited plant growth in dune slacks[†]. Hence, the successional stage and historical management regime of individual wetlands are key factors in determining N and P

[†]In certain dune slacks, plant growth may be controlled by grazing by rabbits. A high grazing pressure prevents the development of a tall sward, thereby creating suitable circumstances for the establishment of characteristic dune slack species of low stature even under eutrophic conditions (Heil et al., 1990). High grazing pressure may 'mask' effects of eutrophication.

availability to the vegetation.

It is remarkable that this conclusion is different from that reached in section 2.3, in which the control of nutrient availability was primarily attributed to water chemistry and soil moisture, as dependent on groundwater hydrology. The low availability of phosphates in wetlands in groundwater discharge areas has been contradicted by several fertilization experiments in which nitrogen was shown to limit plant growth in groundwater-fed fens.

This paradox may be explained by assuming that easily available orthophosphate is scarce in such wetlands but that the many characteristic plant species inhabiting them have special adaptations for dissolving/desorbing the large store of inorganically bound P that has accumulated in these systems. In this way, these particular species would be able to overcome the P limitation in their habitat and would become N-limited. If, however, the orthophosphate availability in such systems increases as a result of changes in water chemistry bringing about dissolution of the inorganic bound P, the 'lithotrophic' species would lose their advantage and be replaced.

2.5 CONSEQUENCES OF NUTRIENT ENRICHMENT FOR SPECIES COMPOSITION

In herbaceous vegetation, species richness and species composition are related to productivity and, hence, to nutrient availability. Grime (1973, 1979) presented a relation between species richness per area and annual maximum biomass. The relation had the shape of an optimum curve with low species richness at low productivity, high species richness at intermediate productivity, and low species richness at high productivity. The low species richness at the lower end of the biomass range was attributed to the fact that only few species have adapted to extreme environmental conditions. It was hypothesized that the high species richness at the intermediate biomass values is associated with ample possibilities for niche differentiation in the case of competition for nutrients. Much more limited niche differentiation was thought to occur in the case of competition for light (higher end of the biomass range).

Grime used a range of different vegetation types (grasslands, forest understorey, wetlands) not randomly distributed over the biomass range. Later studies confirming Grime's theory were similar in that they compared different vegetation types (e.g. Vermeer and Berendse, 1983; Wheeler and Giller, 1983; Day et al., 1988). Moore and Keddy (1989) studied 15 wetland vegetation types and concluded that the relation proposed by Grime does exist when comparing different vegetation types. However, within each type studied, there was no correlation between species richness and biomass production. Similarly, Vermeer and Verhoeven (1987) did not find such a correlation for a data set originating from one major vegetation type, i.e. quaking fens.

These results make any prediction of changes in species richness as a result of a very gradual productivity increase precarious. Only if the species composition has changed to the extent that vegetation structure has been strongly affected (e.g. a transition from a stand dominated by small sedges to one dominated by tall reeds), can species diversity be expected to change according to the relation presented by Grime.

Effects of nutrient enrichment on the vegetation are most pronounced where there is no management of annual harvesting. Prolonged nitrogen enrichment from atmospheric deposition has resulted in drastic changes in wet heathlands: the dominant dwarf shrub *Erica tetralix* has been almost totally replaced by the grass *Molinia caerulea* (Aerts and Berendse, 1988; Berendse *et al.*, 1993, this volume). Rare species such as *Gentiana pneumonanthe*, *Drosera intermedia* and *Lycopodium inundatum* have disappeared together with *Erica*.

Changes in the vegetation composition following eutrophication of dune slacks and dune ponds are well documented (Bakker *et al.*, 1979; Van Zadelhoff, 1981; Van Dijk, 1984; 1989; Van Dijk *et al.*, 1985; Meltzer and van Dijk, 1986). Species that have benefited from the increased nutrient supply are nitrophilous species such as *Epilobium hirsutum*, *Eupatorium cannabinum*, *Urtica dioica*, *Mentha aquatica*, *Lycopus europaeus*, *Cirsium arvense* and *Calamagrostis epigejos*. Fifty plant species indicative of oligotrophic or mesotrophic conditions have declined strongly in artificial recharge areas (Van Dijk, 1989). Similar changes were observed in dune slacks not affected by artificial recharge. Characteristic dune slack species have often been replaced by the tall grass *Calamagrostis epigejos*. This process has changed the character of dune slack vegetation dramatically: low-productive, species-rich dune slack communities have been replaced by highly productive, almost monospecific stands.

Actual rises in productivity and accompanying changes in species richness over time have not been documented for the Dutch moist grassland and fen ecosystems in which plant biomass is being harvested annually. There are data for the reverse process, a decrease in productivity accompanied by a rise in species richness in grasslands taken out of agricultural use and harvested annually to reduce nutrient availability (H. Olff, personal communication, 1991). As stated above, increases in productivity may not occur at all in systems harvested annually in the summer, even in the strongly enriched Dutch landscape.

Changes in species composition as a result of modified nutrient availability may not always be associated with a change in productivity. The shifts in dominance described above for wet heathlands occurred

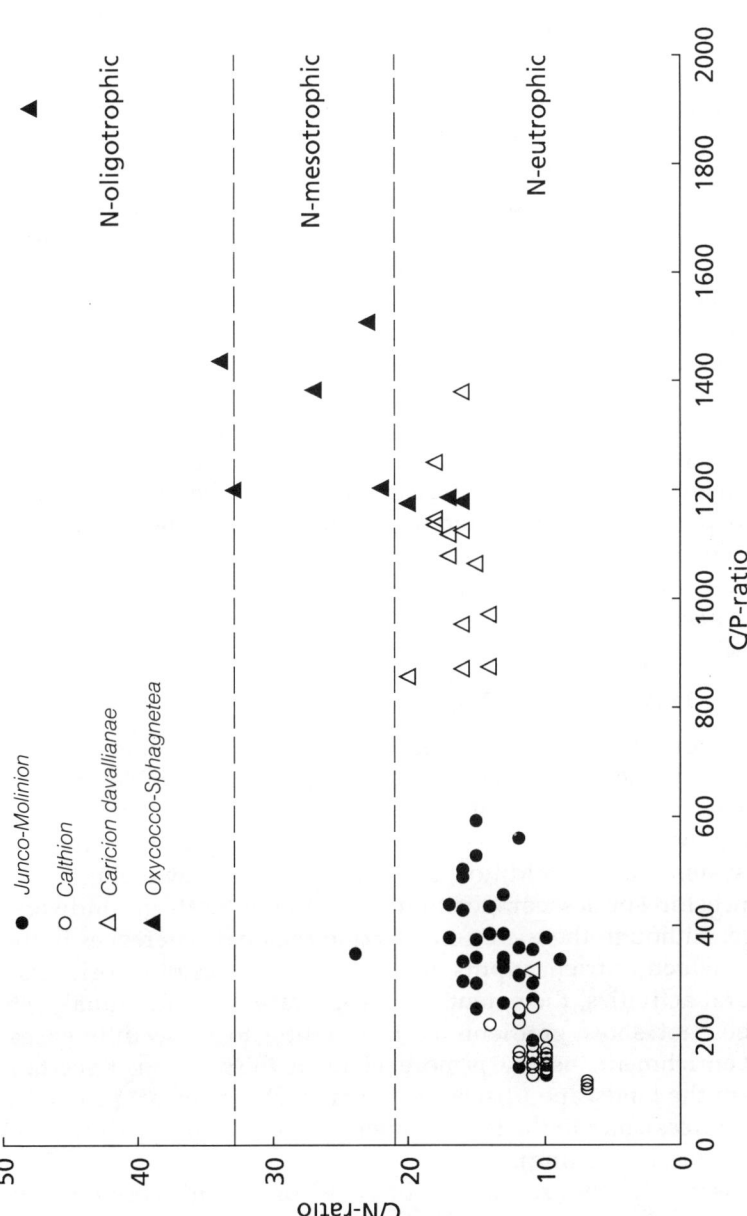

Figure 2.7 Trophic status of fens and bogs indicated by C/N ratios versus C/P ratios of the top soil. N-eutrophic sites (C/N ratio <21 according to Succow, 1988) are inhabited by mesotraphent plant communities of the Junco-Molinion and Caricion davallianae, suggesting that C/P ratios are a better indicator of the trophic status of these wetlands.

without a distinct increase in productivity. As physiological adaptations in plants to N-poor conditions differ from those to P-poor conditions, species composition of N-limited stands will differ from that of P-limited stands, even in the same biomass range. A survey of the nutrient status of several mires in The Netherlands and Germany revealed that C/P and C/N ratios of the soil correlate with species composition, suggesting that plant communities in N-poor mires are different from those under P-poor conditions (Figure 2.7). Egloff (1986) and Willis (1963) have shown that an increase in P availability led to shifts in the species composition without affecting biomass production in moist grasslands and dune slacks, respectively. As indicated in section 2.4, the characteristic, species-rich vegetation in fens and moist grasslands with groundwater discharge probably contains many species with special adaptations for growing in soils low in easily available orthophosphate.

It can be concluded that increases in productivity will only result in changes in species composition if they are drastic and lead to distinct changes in the vegetation structure. Changes in the species composition may also occur without an increase in productivity when there is a shift in the element that limits plant growth. Species-rich communities with many rare plant species characteristically occur in wetlands in groundwater discharge areas under conditions of poor P supply. It is assumed that this poverty in P is not always discovered in a fertilization experiment (see also section 2.4.).

2.6 IMPLICATIONS FOR MANAGEMENT

In the previous sections, it has been shown that freshwater wetland ecosystems in The Netherlands are subject to considerable nutrient enrichment, primarily with nitrogen, in the case of some dune slacks also with phosphorus. N is the factor controlling plant growth in many of these systems, and in addition, low orthophosphate availability is of importance for species composition in wetlands with groundwater discharge. Although there are considerable regional differences in the degree to which nutrient inputs are enlarged as a result of primarily agricultural activities, the extent of eutrophication of individual wetlands does not show gradients corresponding to these differences. Nutrient enrichment, instead, is more related to the position of a certain wetland in the landscape (discharge versus recharge areas, proximity to point sources), and to the management of the wetland (hydrological manipulations, harvesting).

For conservation and restoration of plant communities occurring in mesotrophic wetlands in The Netherlands, it is necessary to manipulate nutrient availability in a well-designed way. Measures should basically aim at removing nutrients from the system or at making them unavailable in certain soil storage pools. Knowledge of the factor that controls plant growth is essential. The most common types of management aimed at nutrient removal are harvesting the aboveground biomass and modifying the hydrological fluxes.

Annual harvesting of aboveground biomass has been shown to be very effective in removing nutrients from the system. Total removal of N inputs from atmospheric deposition is almost guaranteed with this management procedure; P and K are depleted from the system even more drastically, as their inputs are relatively lower (Den Hoed *et al.*,1987; Koerselman *et al.*, 1990). The studies by Bakker (1989), Oomes and Mooi (1985) and Olff (personal communication) have shown that for restoration of previously fertilized moist grasslands, annual harvesting in the summer is the most effective management.

A hydrological measure aimed at reducing N availability is the maintenance of a high water table to reduce soil N mineralization rates. In dune slacks and moist heathlands with a very high total N pool in the soil, even removal of the top soil may be required. Manipulation of biogeochemical processes that act as P stores can be used to reduce P availability, e.g. by hydrological measures that stimulate discharge of Fe- and Ca-rich groundwater in the wetland. It is interesting to note that in weakly acid and circum-neutral soils a high water table may decrease N availability (by its effect on mineralization) but is expected to increase P availability, as orthophosphate is more effectively bound to Fe(III) oxides than to Fe(II) oxides (Patrick and Khalid, 1974).

It is very important to maintain the inflow of groundwater in discharge wetlands, even if they show N-limited plant growth, in order to prevent changes in the species composition by keeping orthophosphate availability low. Infiltration of river water, particularly when little polluted with N and P and having a 'lithotrophic' water chemistry, may provide a successful alternative for seepage from aquifers (Vermeer and Joosten 1992, Van Wirdum, 1991). Although water from the river Rhine has been shown to contain too high concentrations of NaCl to be a suitable water source, water from the Meuse does have an acceptable water chemistry.

Prepurification of river water in dune infiltration areas, and harvest operations may induce a shift from N limitation towards P limitation. In a country where precipitation has been severely polluted with nitrogen, nutrient-poor vegetation types may only survive under conditions of P limitation. Thus, long-term strategies may specifically address the availability of P.

REFERENCES

Aerts, R. and Berendse, F. (1988) The effect of increased nutrient availability on vegetation dynamics in wet heathlands. *Vegetatio*, **76**; 63–70.

Bakker, J.P. (1989) *Nature Management by Grazing and Cutting*. Geobotany 14. Kluwer, Dordrecht, pp. 1–400.

Bakker, T.W.M., Klijn, J.A. and Van Zadelhoff, F.J. (1979) *Duinen en duinvalleien*. Pudoc, Wageningen.

Bakker, T.W.M. and Stuyfzand, P.J. (1993) Nature conservation and extraction of drinking water in coastal dunes of The Netherlands: the Meijendel area as a case study, in *Landscape Ecology of a Stressed Environment* (eds C.C. Vos and P. Opdam), IALE-Studies in Landscape Ecology vol. 1, Chapman & Hall, London, p.p. 244–260.

Berendse, F. Aerts, R. and Bobbink, R. (1993) Atmospheric nitrogen deposition and its impact on terrestrial ecosystems, in *Landscape Ecology of a Stressed Environment* (eds C.C. Vos and P. Opdam) IALE-Studies in Landscape Ecology Vol. 1, Chapman and Hall, London, pp. 104–21.

Bowden, W.B. (1987) The biogeochemistry of nitrogen in freshwater wetlands. *Biogeochemistry* **4**, 313–48.

Boyer, M.H.L. and Wheeler, B.D. (1989) Vegetation patterns in spring-fed calcareous fens: calcite precipitation and constraints on fertility. *Journal of Ecology* **77**, 597–609.

Caraco, N.F., Cole, J.J. and Likens, G.E. (1989) Evidence for sulphate-controlled phosphorus release from sediments of aquatic systems. *Nature*, **341**, 316–18.

Chapin, F.S. III, Vitousek, P.M . and Van Cleve, K. (1986) The nature of nutrient limitation in plant communities. *American Naturalist* **127**, 48–58.

Day, R.T., Keddy, P.A., MacNeill, J.M. and Carleton, T.J. (1988) Fertility and disturbance gradients: a summary model for riverine marsh vegetation. *Ecology* **69**, 1044–54.

Den Hoed, M.A., Moberts, F.M.L. and Stuyfzand, P.J. (1987) *Mogelijkheden voor een verschralend maaibeheer in twee (ver)natte duinterreinen; een landschapsecologische evaluatie van een natuurbeheersmaatregel*. KIWA NV Nieuwegein, SWE 86.007.

De Wit, C.T., Dijkshoorn, W. and Noggle, J.C. (1963) Ionic balance and growth of plants. *Verslagen Landbouwkundig Onderzoek* **69**, 15.

Egloff, T.B. (1986) *Auswirkungen und Beseitigung von Dungungseinflussen auf*

Streuwiesen. Veroffentl. des Geobot. Inst. der ETH Stiftung, Rubel, Zurich, Heft 89.

Erisman, J.W. (1990) *Acid Deposition in The Netherlands.* RIVM Report nr. 723001002.

Garritsen, A.C., Van Amstel, A.R. and Rolf, H.L.M. (1990) Verdroging van natuur in Nederland. II. Hydrologische aspecten van de inventarisatie. *Landschap,* **7,** 165–81.

Graneli, W. (1983) *Biomass Response after Nutrient Addition to Natural Stands of Reed, Phragmites australis.* XXII SIL. Congress, Lyon, France, 1–8.

Grashoff, P.S., Menke, M.A., Van Belois, C.H. and Ruygh, E.F.W. (1989) *PAWN-vermesting. Verzamelen en berekenen van invoergegevens ten behoeve van waterkwaliteitsberekeningen.* Delft, Waterloopkundig Laboratorium, Rapport 1420.

Grime, J.P. (1973) Control of species density in herbaceous vegetation. *Journal of Environmental Management* **1,** 151–67.

Grime, J.P. (1979) *Plant Strategies and Vegetation Processes.* Wiley, Chichester.

Grootjans, A.P. (1985) *Changes of groundwater regime in wet meadows,* Thesis, University Groningen.

Hayati, A.A. and Proctor, M.C.F. (1991) Limiting nutrients in acid-mire vegetation: peat and plant analyses and experiments on plant responses to added nutrients. *Journal of Ecology,* **79,** 75–96.

Heil, G.W., Van der Meulen, F. and Ten Harkel, M.J. (1990) Invloed van atmosferische depositie op de ontwikkeling van droge duingrasland vegetaties. *Geografisch Tijdschrift,* **24,** 427–32.

Jenny, H. (1941) *Factors of Soil Formation.* MacGraw-Hill, London.

Kemmers, R.H. (1986a) Perspectives in the modelling of processes in the root zone of spontaneous vegetation of wet and damp sites in relation to regional water management. *Proc. and Inform. CHO-TNO* **34**: 91-116, The Hague.

Kemmers, R.H. (1986b) Calcium as a hydrological characteristic for ecological states. *Ecology (CSSR)* **5,** 271–82.

Kemmers, R.H. (1990) De stikstof- en fosforhuishouding van mesotrofe standplaatsen in relatie tot mogelijkheden van aanvoer van gebiedsvreemd water. Utrecht. *The Utrecht Plant Ecology News Report,* **10,** 7–23.

Kemmers, R.H. and Jansen, P.C. (1985) *Stikstofmineralisatie in onbemeste half-natuurlijke graslanden* Inst. Cultuurtechniek en Waterhuishouding, Wageningen. Rapport, **14,** 1–20.

Koerselman, W. (1991) *Verruiging van (ver)natte duinvalleien; een literatuuronderzoek naar de relatie tussen de grondwaterstand, de beschikbaarheid van voedingsstoffen en de vegetatiestruktuur.* KIWA N.V. Nieuwegein, SWE 91–006.

Koerselman, W., Bakker, S.A. and Blom, M. (1990) Nitrogen, phosphorus and potassium budgets for two small fens surrounded by heavily fertilized pastures. *Journal of Ecology* **78,** 428–42.

Kroes, J.G. (1988) *Animo Version 2, Users guide.* Wageningen, ICW. Nota 1848.

Kroes, J.G., Roest, C.W.J., Rijtema, P.E. and Locht, L.J. (1990) *De invloed van enige bemestingsscenario's op de afvoer van stikstof en fosfor naar het oppervlaktewater in Nederland.* Staring Centrum Wageningen, Rapport nr. 55.

Meltzer, J.A. and Van Dijk, H.W.J. (1986) The effects of dissolved macronutrients on the herbaceous vegetation around dune pools. *Vegetatio* **65,** 53–61.

Moore, D.R.J. and Keddy, P.A. (1989) The relationship between species richness and standing crop in wetlands: the importance of scale. *Vegetatio*, **79**, 99–106.

Morris, J.T. (1991) Effects of nitrogen loading on wetland ecosystems with particular reference to atmospheric deposition. *Annual Review of Ecology and Systematics*, **22**, 257–80.

Nichols, D.S. (1983) Capacity of natural wetlands to remove nutrients from wastewater. *Journal of Water Pollution Control*, **55**, 495–505.

Oomes, M.J.M. and Mooi, H. (1985) The effect of management on succession and production of formerly agricultural grassland after stopping fertilization, in *Sukzession auf Gruenlandbrachen* (ed. K.F. Schreiber), Muenstersche Geogr. Arb. **20**, 59–67.

Patrick, W.H. Jr. and Khalid, R.A. (1974) Phosphate release and sorption by soils and sediments: effect of aerobic and anaerobic conditions. *Science*, **186**, 53–55.

Pons, L.J. (1992) Holocene peat formation in the lower parts of The Netherlands. In: *Fens and Bogs in The Netherlands: Vegetation, History, Nutrient Dynamics and Conservation* (ed J.T.A. Kerhoeven), Kluwer Academic, Dordrecht, in press.

Richardson, C.J. and Marshall, P.E. (1986) Processes controlling movement, storage and export of phosphorus in a fen peatland. *Ecological Monographs* **56**(4), 279–302.

Sanville, W. (1988) Response of an Alaskan wetland to nutrient enrichment. *Aquatic Botany*, **30**, 231–243.

Shaver, G.R. and Chapin, F.S.III. (1980) Response to fertilization by various plant growth forms in an Alaskan tundra: nutrient accumulation and growth. *Ecology* **61**, 662–675.

Solander, D. (1983) Biomass and shoot production of *Carex rostrata* and *Equisetum fluviatile* in unfertilized and fertilized subarctic lakes. *Aquatic Botany*, **15**, 349–366.

Stumm, W. and Morgan, J.J. (1981) *Aquatic Chemistry*. Wiley, New York.

Stuyfzand, P.J. (1983) *Kwaliteitsveranderingen van voorgezuiverd Lekwater bij kunstmatige infiltratie in het duingebied ten westen van Castricum* . KIWA N.V. Nieuwegein, SWE 366.

Succow, M. (1988) *Landschaftoekologische Moorkunde*. Gustav Fischer, Jena, Germany.

Tilman, D. (1984) Plant dominance along an experimental nutrient gradient. *Ecology*, **65**, 1445–1453.

Van der Meulen, F. (1982) Vegetation changes and water catchment in a Dutch coastal dune area. *Biological Conservation*, **24**, 305–316.

Van Dijk, H.W.J. (1984) *Invloeden van oppervlakte-infiltratie t.b.v. duinwaterwinning op kruidachtige vegetaties*. Thesis, RUL. 240 pp.

Van Dijk, H.W.J. (1985a) Bodemchemische processen bij herstel van bodem en vegetatie, in *Terugkeer vochtige duinvalleien* (ed Stichting Duinbehoud), pp. 9–15

Van Dijk, H.W.J. (1985b) The impact of artificial dune infiltration on the nutrient content of ground and surface water. *Biological Conservation*, **34**, 149–67.

Van Dijk, H.W.J.(1989) Ecological impact of drinking water production in Dutch coastal dunes, in *Perspectives in Coastal Dune Management* (eds F. Van der Meulen, P.D. Jungerius and J.H. Visser) SPB Academic Publishers. The Hague, pp. 163–82

Van Dijk, H.W.J., Noordervliet, M.A.W. and De Groot, W.T. (1985) Nutrient supply of herbaceous bank vegetations in Dutch coastal dunes; the importance of nutrient mobilization in relation to artificial infiltration. *Acta Botanica Neerlandica*, **34**, 301–21.

Van Wirdum, G. (1980) *Eenvoudige beschrijving van de waterkwaliteitsverandering gedurende de hydrologische kringloop*. CHO-TNO, rapporten en nota's 5,Den Haag, pp 118–43.

Van Wirdum, G. (1991) *Vegetation and Hydrology of Floating Rich-fens*. Datawyse, Maastricht.

Van Zadelhoff, F.J. (1981) *Nederlandse kustduinen*; Geobotanie. Pudoc, Wageningen.

Verhoeven, J.T.A. (1986) Nutrient dynamics in minerotrophic peat mires. *Aquatic Botany*, **25**, 117–38.

Verhoeven, J.T.A., Koerselman, W and Beltman, B. (1988) The vegetation of fens in relation to their hydrology and nutrient dynamics: a case study, in *The Vegetation of Inland Waters* (ed. J.J. Symoens), Kluwer Academic, Dordrecht. *Handbook of Vegetation Science*, vol. 15, pp. 249–82.

Verhoeven, J.T.A., Maltby, E. and Schmitz, M.B. (1990) Nitrogen and phosphorus mineralization in fens and bogs. *Journal of Ecology*, **78**, 713–26.

Verhoeven, J.T.A. and Schmitz, M.B. (1991) Control of plant growth by nitrogen and phosphorus in mesotrophic fens. *Biogeochemistry*, **12**, 135–48.

Vermeer, J.G. (1986) The effect of nutrients on shoot biomass and species composition of wetland and hayfield communities. *Oecologia Plantarum*, 7(21); 31–41.

Vermeer, J.G. and Berendse, F. (1983) The relationship between nutrient availability, shoot biomass and species richness in grassland and wetland communities, *Vegetatio* **53** , 121–6.

Vermeer, J.G. and Joosten, J.H.J. (1992) Conservation and management of bog and fen reserves in The Netherlands, in *Fens and Bogs in The Netherlands: Vegetation, History, Nutrient Dynamics and Conservation* (ed: J.T.A. Verhoeven), Kluwer Academic Publishers, Dordrecht, pp. 433–78.

Vermeer, J.G. and Verhoeven, J.T.A. (1987) Species composition and biomass production of mesotrophic fens in relation to the nutrient status of the organic soil. *Acta Oecolgia Oecologia Plantarum*, 8(22), 321–30.

Vos, C.C. and Zonneveld, J.I.S. (1993) Patterns and processes in a landscape under stress: introduction of the study area, in *Landscape Ecology of a Stressed Environment* (eds C.C. Vos and P. Opdam) IALE-studies in Landscape Ecology Vol. 1, Chapman Hall, London, pp. 1–28.

Wassen, M.J. (1990) *Water Flow as a Major Landscape Ecological Factor in Fen Development*. University of Utrecht, PhD Thesis.

Wheeler, B.D. and Giller, K.E. (1983) Species richness of herbaceous vegetation in Broadland, Norfolk, in relation to the quantity of above-ground plant material. *Journal of Ecology*, **70**, 179–200.

Willis, A.J. (1963) Braunton Burrows: the effects on the vegetation of the addition of mineral nutrients to the dune soils. *Journal of Ecology*, **51**, 353–74.

Linking ecological patterns to hydrological conditions on various spatial scales:

3

a case study of small stream valleys

Ab P. Grootjans, Rudy van Diggelen, Henk F. Everts, Piet C. Schipper, Jan Streefkerk, Nico P.J. de Vries and Azing Wierda

3.1 INTRODUCTION

Stream valley systems can be considered as a series of gradient situations (ecoclines) that are connected by a stream. The properties of a stream valley depend on the properties of the total landscape system bordering the stream. Modelling stream valley systems means that we should be able to simulate both quantitative and qualitative aspects of

Landscape Ecology of a Stressed Environment
Edited by Claire C. Vos and Paul Opdam
Published in 1993 by Chapman and Hall, London. ISBN 0 412 44820 3

the hydrology of all the land(scape) units (Zonneveld, 1989) involved. A consequence of applying regional hydrological models is that statistically reliable results can only be obtained with respect to the influence of large-scale interferences with the hydrology. These include abstraction of groundwater from aquifers and the effects of large-scale agricultural drainage systems. However, the impact of these changes on a local scale, e.g. on the species composition of stream valley meadows, can only be understood when not only changes in water levels are known but also the changes in fine-scale pattern of the shallow groundwater flows and related differences in groundwater chemistry are revealed (Wilcox *et al.*, 1986; Siegel and Glaser, 1987; Grootjans *et al.*, 1988). The next problem would be to couple models describing the saturated zone to models that simulate chemical processes in the unsaturated zone (Kemmers, 1986). Although there have been studies which have solved most of these problems on a local scale (Jørgenson *et al.*, 1988), it is practically impossible to refine the whole regional model in such a way that fine-scale ecological questions can be answered. Finally, hydrological research on a regional scale can hardly be translated to ecological processes on a local scale. In our view this is not really a problem. To

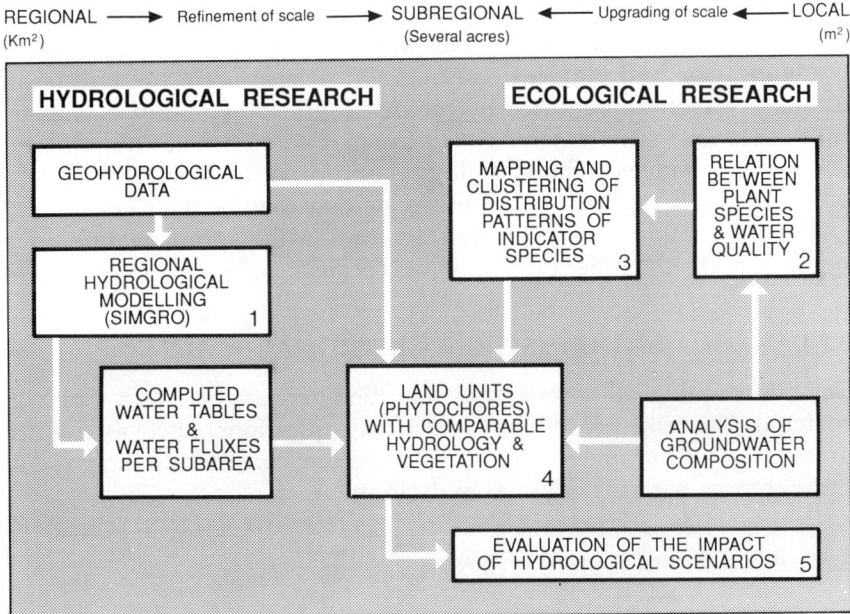

REGIONAL ⟶ Refinement of scale ⟶ SUBREGIONAL ⟵ Upgrading of scale ⟵ LOCAL
(Km²)   (Several acres)   (m²)

Figure 3.1 Outline of a hydroecological approach to evaluate the influence of different hydrological scenarios and to predict potentials for marsh and meadow development.

make reliable statements on hydrological conditions on a local scale, the hydrologist should first model regional conditions, because interferences in a regional scale can have considerable effects on conditions at a local scale. In order to gain an understanding of the functioning of the regional hydrological system, ecological research should be performed on that same scale in order to detect ecological processes on that scale.

This chapter presents a practical solution for the coupling of hydrological modelling to ecological research on both regional and local scales. On a regional scale we will use indicator species for hydrological conditions such as groundwater composition (Jeglum, 1971; Pietch, 1982)(Figure 3.1). Distribution patterns of indicator species may then be used to refine the scale of research by 'translating' the impact of large-scale hydrological interferences to the level of individual land units. The delineation of land units is based on combinations of distribution patterns of indicator species (phytochores). They reflect qualitative and quantitative aspects of the hydrology. The indication of hydrological conditions, thus obtained, may then be tested in the field by measuring the groundwater chemistry and compared to the detailed results of the hydrological model. This procedure allows an evaluation of the ecological impact of computed hydrological scenarios.

Distribution patterns of species are also used on a local scale as guidelines for hydrological research. On a local scale the effects of drainage are often identified by ecological interpretation of (a change in) vegetation pattern, which may be used to formulate hypotheses concerning changes in the hydrology. Appropriate hydrological models which simulate water flows on this scale may then be selected.

Two case studies are presented illustrating the approach; one on a regional scale; the other on a very local scale.

3.2 LANDSCAPE ANALYSIS ON A REGIONAL SCALE

The hydroecological approach was applied here as a research tool in a regional planning procedure considering questions, such as: which areas are most suitable for nature regeneration, considering the present interferences with the regional hydrology?

3.2.1 DESCRIPTION OF THE STUDY AREA

The study area is situated in the sandy area (Fig.1.2, Vos and Zonneveld, 1993, this volume). It consists of the catchment area of the Drentse Aa (30 000 ha) (Figure 3.2). Parts of the adjoining catchment of the Peizer- and Lieverense Diep will also be discussed. The valleys are situated in a landscape with little relief; the highest point of the Drenthian Plateau

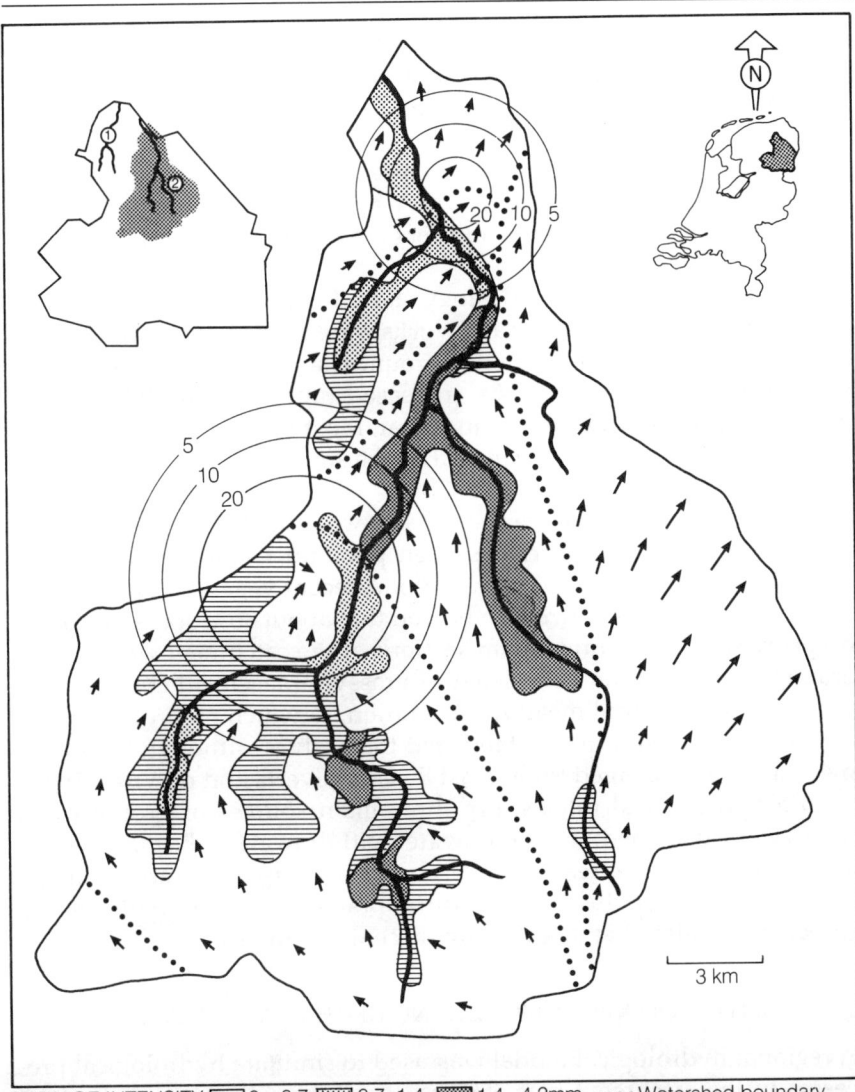

SEEPAGE INTENSITY ▤ 0 – 0.7 ▨ 0.7–1.4 ▰ 1.4– 4.2mm •••• Watershed boundary
⁄5⁄10⁄ Groundwater draw down in cm.

Figure 3.2 Some results of the hydrological model SIMGRO (Drentse Aa catchment), showing exfiltration areas with various seepage intensities, and the influence of groundwater withdrawal. The dotted lines divide the different groundwater systems. The arrows indicate flow direction of groundwater in the second aquifer. The length of the arrow is proportional to the flow velocity. 1 = Nature reserve Mensinge, 2 = Drentse Aa catchment.

is 21 m above sea level. A series of subregional hydrological systems create flow patterns which mostly end in the stream valleys (Engelen, 1981). Semipervious loamy layers of glacial origin (Saale ice-age) add to the differentiation in water flows. Shallow groundwater flows that pass through the leached cover sands are generally poor in dissolved minerals. Groundwater that originates from deeper aquifers has passed through non-leached deposits and is, therefore, rich in calcium and hydrocarbonate. The hydrological basis is formed by marine clay layers, which are locally pushed upwards by salt domes. Other 'obstacles' for groundwater flows in the aquifers are large clay bodies, formed in pre-Saalien lakes.

Interferences with the hydrology started mainly in the 1960s, when large-scale agricultural drainage works were carried out, mainly in the upstream areas. The abstraction of groundwater from the main aquifers was modest in the early 1950s, increased rapidly in the 1970s to the present situation of 12 million m^3/year. Most water-winning facilities are situated in the lower basin and in the southwest part of the catchment.

Most of the present nature reserve area was established between 1970 and 1990, and much effort has been put into restoring the original species-rich grasslands (Bakker, 1989). A recent vegetation map of the 3226 ha reserve area, however, showed that about 30% of the meadows on peaty soils in the midstream sections of the catchment were degenerating. The vegetation consisted of grassy replacement communities of former seminatural meadow vegetation, in which no protected species had remained or had established themselves, although the meadows had been managed for at least 8 years (Everts and de Vries, 1991).

In a hydrological pilot study three main sources of groundwater losses were identified: (1) groundwater withdrawal for the public water supply; (2) reduced stream flow discharge from upstream areas, and (3) increased drainage due to agricultural practices. Further hydrological research was aimed at quantifying these water losses.

3.2.2 THE HYDROLOGICAL MODELLING OF THE CATCHMENT

A regional hydrological model was used to simulate hydrological processes in the catchment area. The reasons for using a regional model were twofold. First, natural differences in the hydrology of the stream valleys are mainly determined by large infiltration areas of the Drenthian Plateau and elevated sandy ridges. Also, the influences of several groundwater abstraction facilities are so large that they influence the regional hydrological system. Second, the regional distribution of clay layers plays a key role in the assessment of infiltration and exfiltration areas.

To simulate mathematically groundwater flow in the catchment and to validate the geohydrological parameters, such as vertical flow resistance (C) and (horizontal) transmissivity (Kd), the model SIMGRO was used. With this model fluctuation patterns of groundwater tables can be simulated, given hydrological conditions in the catchment, and information on the management of surface waters (Querner, 1989).

The model area is 30 000 ha, divided into 135 hydrological homogeneous subareas, which were delineated using abiotic criteria, such as soil type, surface water regime and height above sea level. The model comprised 1000 elements. The smallest elements are situated in the stream valley itself and cover an area of about 10 ha. The necessary input data were gathered for each separate model element: C-values, thickness of clay layers, height above sea level, drainage characteristics and natural groundwater recharge (only available per subarea). The model calculates water tables in several aquifers and water balances of each separate element, including seepage intensities.

Several measured and calculated parameters (water tables, discharge from five subcatchment areas) were used to calibrate the model. Various hydrological scenarios were calculated. Each scenario varied with respect to groundwater withdrawal, changes in stream flow discharge, drainage intensity and afforestation.

Five groundwater systems could be distinguished on the basis of flow pattern of the groundwater and infiltration and exfiltration characteristics (Figure 3.2). Only two groundwater systems in the middle basin are still more or less natural, whereas the two in the lower basin are purely artificial, due to the impact of large groundwater-winning facilities. The fifth system in the eastern part of the study area belongs to another catchment area; the adjoining and low-lying Hunze valley drains the groundwater infiltrating in the sandy ridges along the Drentse Aa.

The hydrological system of the lower basin is affected by pumping stations. The seepage intensity is very low here and large parts have even turned into infiltration areas. The groundwater systems of the middle reaches are least affected by the large-scale influence of pumping stations. The stream valleys in this area are mainly fed by recharge from nearby elevated sandy ridges. The seepage intensity is still high (1.4–4.2 mm/day), although local drainage channels may have a relatively large effect on adjoining meadow complexes. The most affected areas are situated in the upper reaches. An extensive agricultural drainage system, groundwater abstraction and the presence of many semipervious layers, reduce the groundwater flow down to the stream valley meadow to a minimum (0–0.7 mm/day).

The water levels in the streams themselves are also unnatural in most of the southern half of the catchment area. A reduction of the highest level of 100 cm was calculated in one of the main streams here. The mean level had decreased by 20 cm. Under such conditions the stream causes much additional drainage in the meadows.

Indic. group	Phytochore type	1	2	3	4	5	6	7	8	9	10	11	12	13
1	*Eriophorum angustifolium*	-	-	-	-	-	o	o	-	-	o	-	o	-
2	*Hydrocotyle vulgaris*	o	o	o	-	■	-	o	o	o	o	o	o	o
2	*Viola palustris*	o	o	-	o	o	o	■	o	o	o	o	o	o
2	*Carex curta*	o	o	o	o	-	o	o	o	o	o	-	o	-
2	*Juncus acutiflorus*	o	o	o	o	■	o	o	o	o	o	o	o	o
2	*Montia fontana*	-	o	o	■	o	o	o	o	o	-	-	-	-
2	*Ranunculus aquatilis/pelt.*	-	-	-	-	o	o	o	■	o	-	■	o	o
3	*Hottonia palustris*	-	-	o	o	o	o	-	o	o	o	■■	■	■■
3	*Potamogeton polygonifolius*	-	o	o	-	o	-	-	-	o	o	-	-	-
3	*Eleocharis acicularis*	-	-	-	o	o	o	-	o	o	-	o	o	-
3	*Carex rostrata*	-	o	o	o	■	■	■	■	■	■	■	■	■
3	*Potentilla palustris*	-	o	-	o	■	o	o	-	o	o	o	o	o
3	*Menyanthes trifoliata*	-	-	-	-	o	o	-	-	o	o	o	-	o
3	*Scirpus sylvaticus*	-	-	-	-	-	o	o	o	■	o	-	-	-
4	*Caltha palustris*	-	-	-	■	o	■	■	■	■■	■	■	o	o
4	*Berula erecta*	-	-	-	-	-	o	■	■	■■	o	o	-	■
4	*Cardamine amara*	-	o	o	o	-	o	o	■	■	-	-	-	-
4	*Veronica beccabunga*	-	-	o	o	o	o	o	■	■	o	-	-	-
5	*Carex acutiformis*	-	-	-	-	o	■	o	■	■	-	-	-	-
4	*Carex aquatilis*	-	-	-	o	o	o	o	■	■	■	■	■	■
4	*Carex paniculata*	-	-	-	-	o	o	o	o	o	o	o	o	o
4	*Ranunculus lingua*	-	-	-	-	-	o	o	■	o	o	-	o	o
7	*Equisetum fluviatile*	-	■	■	■	■	■	■	■■	■■	■■	■	■	■
7	*Glyceria maxima*	-	-	■	-	■	■	■	■■	■■	■■	■■	■■	■■
6	*Potamogeton natans*	-	o	o	o	o	■	o	o	-	o	■■	o	o
6	*Potamogeton pusillus*	-	o	o	-	o	o	-	-	-	o	■	■	o
4	*Sagittaria sagittifolia*	-	-	-	-	o	o	-	-	-	o	■	■	■
x	*Nuphar lutea*	-	-	o	-	-	■	-	-	-	o	■	o	■
6	*Potamogeton trichoides*	-	-	-	-	-	-	-	-	-	-	-	■	o
4	*Carex acuta*	-	-	-	-	-	o	-	-	-	o	■	■	■
?	*Senecio aquaticus*	-	-	-	-	o	o	-	-	-	■	o	o	o
6	*Potamogeton crispus*	-	-	-	-	-	-	-	-	-	o	o	o	-
?	*Carex pesudocyperus*	-	-	-	-	o	o	-	-	-	-	-	■	o
6	*Potamogeton pectinatus*	-	-	-	-	o	o	o	-	o	■	o	o	o
6	*Potamogeton perfoliatus*	-	-	-	-	o	-	-	-	-	o	o	-	o
6	*Butomus umbellatus*	-	-	-	-	-	-	-	-	-	-	o	o	■
5	*Carex disticha*	-	-	-	-	o	o	-	o	-	o	■	-	-
6	*Ranunculus circinatus*	-	-	-	-	-	-	-	-	-	-	o	o	o
6	*Carex riparia*	-	-	-	-	-	-	-	-	-	-	-	o	o
6	*Acorus calamus*	-	-	-	-	-	o	-	-	-	-	o	o	■
3	*Potamogeton acutifolius*	-	-	-	-	-	o	-	-	-	-	-	o	o
6	*Sium latifolium*	-	-	-	-	-	-	-	-	-	-	o	o	o
?	*Cicuta virosa*	-	-	-	-	-	-	-	-	-	-	o	o	o
6	*Hippuris vulgaris*	-	-	-	-	-	-	-	-	o	-	■	-	-
6	*Chara vulgaris*	-	-	-	-	-	-	-	-	o	-	o	o	o

o – rare/occasional   ■ – common   ■■ – abundant

Figure 3.3 Distinguished land units (phytochores) in the catchment of the Drentse Aa (after Everts and de Vries, 1991). The phytochore types (numbered in map) are characterized by the occurrence of species groups that indicate differences in groundwater composition. Indicator groups: 1 = indicators of soft (calcium-poor) groundwater; 3 = indicators of moderately soft groundwater; 5 = indicators of hard (calcium-rich) groundwater; 6 = indicators of groundwater which is rich in magnesium and chloride

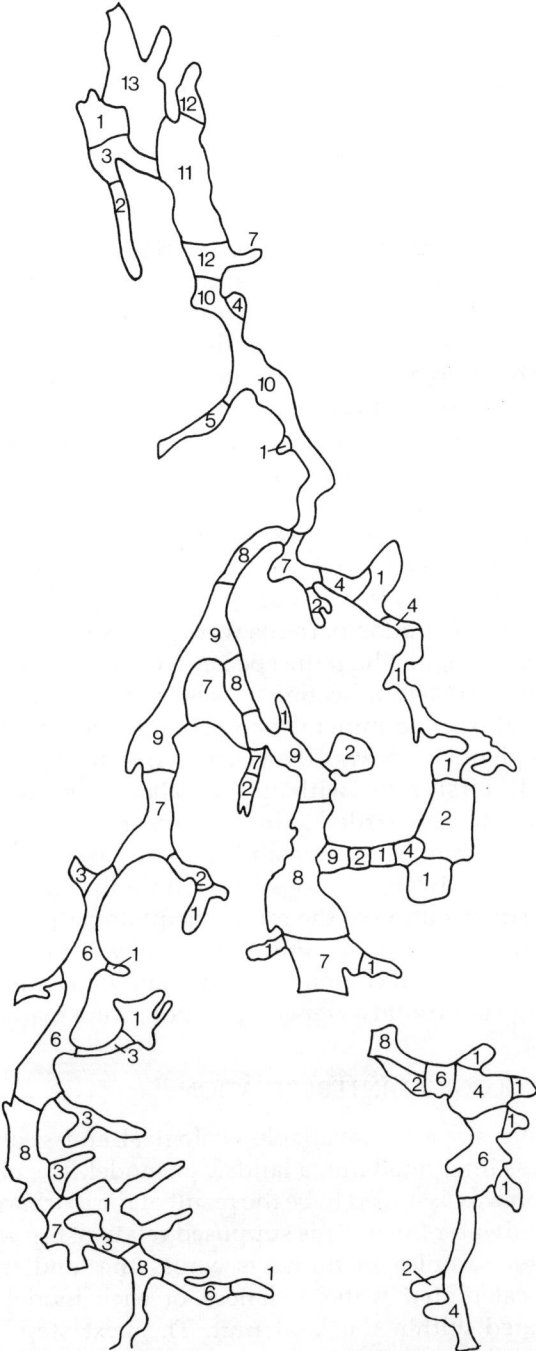

.**Figure 3.3** continued   7 = not indicative for groundwater composition in the study area.
The indicator groups 2 and 4 are intermediate between two species groups.

We concluded that most of the calculated boundaries of the hydro-logical systems are unnatural and relatively recent. It is unlikely that distribution patterns of species have adapted to this new hydrological situation.

3.2.3 DELINEATION OF PHYTOCHORES

Ecological criteria were, therefore, used to refine the scale of research and delineated smaller land units based on distribution patterns of indicator species (phytochores). An advantage is that the indicator species can be expected to reflect both qualitative and quantitative aspects of the hydrology, whereas the quantitative hydrological model only presents groundwater levels and water balances.

The delineation was carried out as follows. The first step was a survey of some 50 marsh and water plant species, which were not rare and were expected to indicate differences in water quality. The indica-tor values were obtained from the literature (Ellenberg, 1978; Pietch, 1982; De Lyon and Roelofs, 1986; Grootjans et al., 1988). These results were re-examined in the research area (Everts et al., 1988; Everts and de Vries, 1991). The distribution patterns were grouped using cluster tech-niques (Fresco, 1991), in which the species were arranged in ecological groups according to their indication values with respect to their prefer-ence for mineral-poor to mineral-rich groundwater (Figure 3.3). The delineation of phytochores was then carried out in a hierarchical man-ner (Figure 3.4). First, areas without *Equisetum fluviatile* were deline-ated. These areas were regarded as infiltration areas, despite the results of the hydrological model. The remaining species were used to identify differences in groundwater composition within the supposed exfiltra-tion areas. The distribution of the more calciphilous species were used for the delineation of the smaller units. Eventually, 13 phytochore types, characterized by different combinations of plant indicators and reflect-ing a wide range of groundwater composition, were mapped.

3.2.4 HYDROECOLOGICAL INTERPRETATION

In the following stage the available ecological and geohydrological information was integrated into a landscape model. The observed eco-logical zonation was assumed to be the result of the discharge of several different groundwater flows. This supposed relationship was tested by analysing water samples in transects across the land units and by studying the calculated water balances of each model element of SIMGRO situated within the land unit. The next step consisted of analysing discrepancies between the ecological interpretations and the results of the hydrological modelling. The discussions with geologists

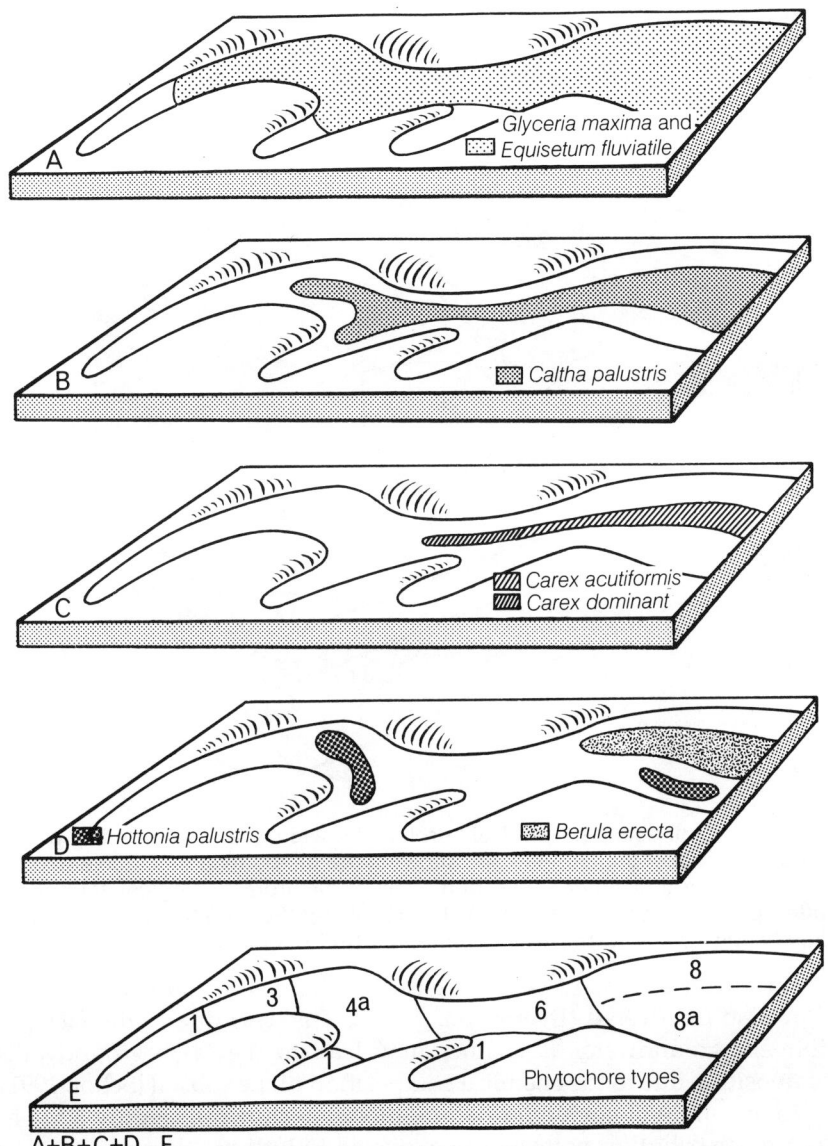

Figure 3.4 Procedure followed in delineating phytochores (after Everts and de Vries, 1991). For explanation see text.

and hydrologists led to various adjustments in the hydrological model. On one occasion it was possible to detect errors in the results of the hydrological modelling, which were based on incorrect extrapolation of the geological data. On the other hand incorrect interpretations of

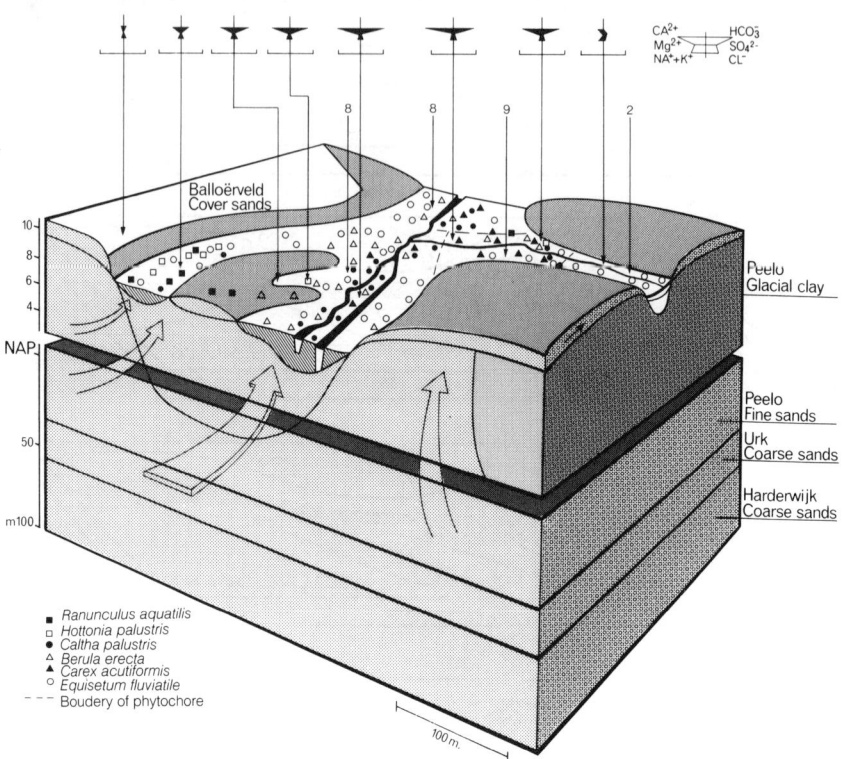

Figure 3.5 Landscape model of a relatively wide scale land unit in a mid-stream section of the Drentse Aa, showing the distribution of species and the interpreted groundwater flows and their origin from geological formations (after: Everts and de Vries, 1991). The results of the groundwater analysis of the shallow groundwater are presented as Stiff-diagrams.

the occurrence of plant species to the discharge of groundwater flows were also discovered (Everts *et al.*, 1988). This was due to the fact that different groundwater flow lines may lead to the same groundwater composition due to the chemical composition of the subsoil (Schot, 1991).

An example of such a landscape model is shown in Figure 3.5. The species distribution patterns are expected to indicate the presence of a gradient in groundwater types, which must be sustained by different groundwater flows. On the left hand side of the valley, a species such as *Hottonia palustris* indicates the presence of a relatively calcium-poor subsurface flow (De Lyon and Roelofs, 1986) whereas the relatively calciphilous species in the centre, such as *Carex acutiformis* and *Berula erecta*, indicate the discharge of calcium-rich groundwater from a deep aquifer (Everts and de Vries, 1991).

Analysis of the water samples from the shallow groundwater confirmed this hypothesis on groundwater composition. The calcium-rich groundwater types were found in the centre of the valley, whereas calcium-poor groundwater was found at the valley flanks. The hypothesis was also consistent with the results of the geohydrological modelling by SIMGRO. The seepage intensity is highest in the centre, providing calcium-rich groundwater from the deep aquifer, whereas the larger infiltration area on the left can be expected to provide calcium-poor groundwater from the leached cover sands.

3.3 LANDSCAPE ANALYSIS ON A LOCAL SCALE

In the following, another type of application will be discussed, namely a case study aimed at a regeneration project on a local scale.

Most nature reserves in stream valleys are so small that they easily fit in one or two elements of a large-scale hydrological model such as SIMGRO. An analysis of hydrological problems on a regional scale, therefore, disregards the influence of local circumstances on the hydrology of the reserves. Technically, it is quite possible to refine the model in certain subareas and focus it on very local problems, but this is only practical if a clear hypothesis about the impact of local hydrological conditions on the vegetation is available. An illustration is given as to how a hydroecological pilot study on a local scale can provide such a hypothesis. The research started with a vegetation survey, which was interpreted to supposed groundwater flow using indicator species. Detailed geological and geochemical research followed and the results led to a landscape model, which was used as a basis for further hydrological research.

3.3.1 DESCRIPTION OF THE STUDY AREA

The meadow reserve 'Mensinge' is situated along the Lieverense Diep (53°08'N and 6°35'E). The area consists of 46 ha meadows and *Phragmites australis* reed with adjoining wet heath (13 ha) and forest (70 ha) on the valley flanks. Some parts of the meadows are very rich in species. Apart from the more common species of *Caltha palustris* meadows, *Polygonum bistorta*, *Valeriana dioica* and *Dactylorhiza incarnata* have large populations here. Other parts of the reserve are degenerating (Grootjans *et al.*, 1988); in the course of the last 10 years all characteristic marsh and meadow species have shown a marked decline in these parts of the reserve.

The occurrence of many calciphilous species seem to indicate that calcium-rich groundwater is discharging in the centre of the valley. The reserve is situated at a site where the original river(s) broke through a cover sand ridge. The thin clay layer, indicated on the geological map, is not present but has been eroded away in this process, leaving the

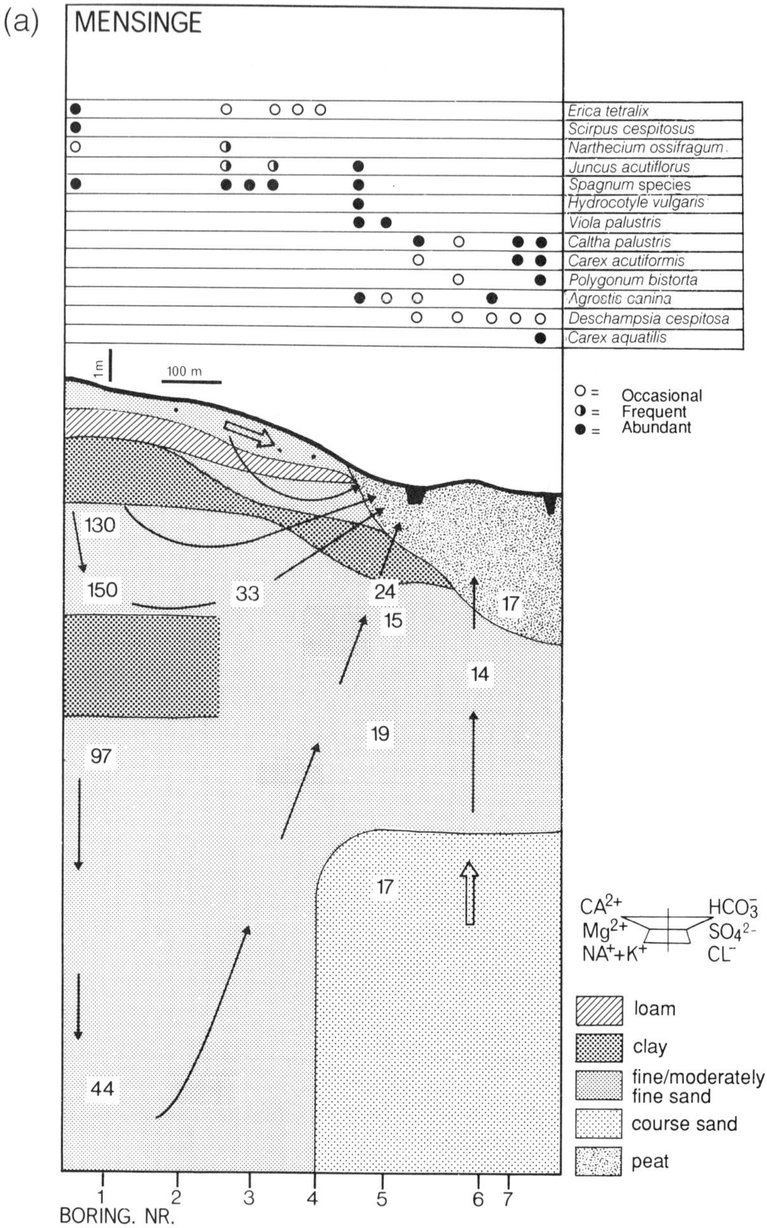

Figure 3.6 (a) Distribution of Cl⁻ concentrations (mg/l) and interpreted flow pattern of groundwater in a geological profile of the valley reserve Mensinge. The occurrence of some plant species along the transect is also indicated in relation to the calcium content of the shallow groundwater. (b) Macro-ionic composition of the groundwater, sampled from minifilters at metre intervals in boring 1 in (a). All values are expressed in meq/l, except for NO_3^- which is expressed in mg/l.

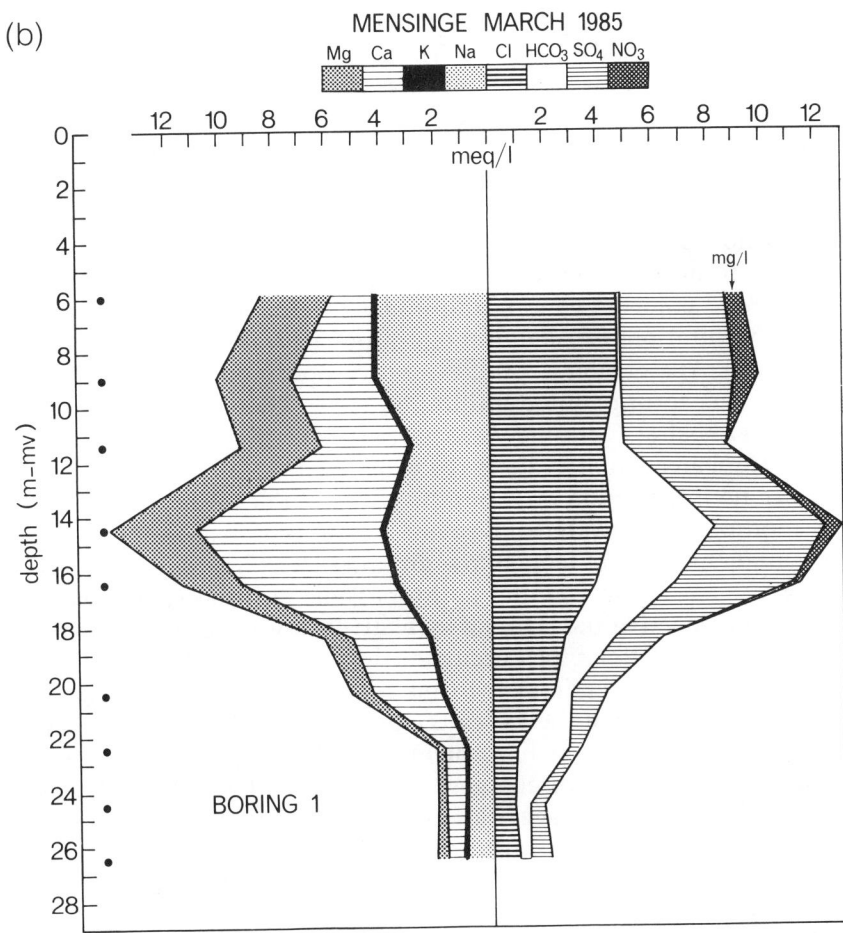

aquifer in open connection with the phreatic groundwater (Figure 3.6). The presence here of acidophilous species, such as *Hydrocotyle vulgaris* and *Viola palustris*, on the valley flanks suggests the discharge of a subsurface flow with very calcium-poor groundwater.

3.3.2 GROUNDWATER ANALYSIS

In order to test these interpretations, geological borings up to 28 m were carried out along a transect across the valley (Figure 3.6a). The results showed a picture more complex than that based on species occurrence Several (thin) clay layers were present in the valley flank instead of only one. They were separated by sandy layers, in which groundwater flow is possible. Groundwater tubes were placed in the bore-holes and filters were installed at various depths. The analysis of the shallow groundwater discharging along the valley flanks confirmed the hypothesis of mineral-poor groundwater here. The interpretation of the occurrence of indicator species was only partly successful. The landscape model obtained could explain the presence of calciphilous species in the reserve, but not the observed decline of these species in part of the study area. Also the results of the chemical analysis of groundwater from deeper layers on the valley flank were inconsistent with the distribution of indicator species. Figure 3.6(b) shows the results of the groundwater analyses from nine minifilters. Very high concentrations of sulphate, sodium and chloride were found under the first clay layer, whereas high concentrations of calcium and bicarbonate were found underneath the second. At a greater depth the ionic composition resembles that of groundwater in unpolluted infiltration areas. These findings were interpreted as being the result of excessive fertilization in a neighbouring agricultural area situated along the watershed.

Since no information on groundwater flow lines was available, the direction of groundwater movement was interpreted on the basis of measured chloride contents in the groundwater (Figure 3.6). Chloride can be used as a tracer ion for anthropogenic influences, because it is biologically inactive and is found in relatively low concentrations in the natural groundwater types occurring in this region. It appears that polluted groundwater from the first aquifer discharges at the valley flanks. In the centre of the valley the groundwater has low concentrations of chloride but is rich in calcium. So this is where groundwater from the main aquifer discharges. The pollution from the valley side does not yet affect the vegetation in the meadows. But the low water levels in the stream and the drainage ditch stimulate the replacement of calcium-rich groundwater by infiltration water, creating acidic conditions in the rooting zone.

A further simulation of the supposed flow of groundwater has up to now failed, due to inadequate information on the resistance to water flow of the clay layers and insufficient data on the influence of a nearby groundwater abstraction.

3.4 EVALUATION OF THE HYDROECOLOGICAL APPROACH

The use of indicator species for groundwater composition proved to be a relatively simple and helpful tool in this process of interpreting the results of hydrological modelling in terms of ecological consequences. An exact delineation of the land unit itself is not crucial in the approach, because the aim of the research is to gain an understanding of the hydrological functioning of the land units in question. The use of indicator species may help to improve the results of the hydrological modelling.

Landscape models, designed for each land unit, were the key for a fruitful interpretation of the results of geological, hydrological and ecological research. A critical review of all the data is, however, necessary; geohydrological data, for instance, are mostly collected on a very wide scale, local differences may be overlooked, but can have marked consequences for hydrological modelling.

The interpretation of indication values of plant species may also be incorrect, when applied outside the area where they were established (Niemann, 1973). *Glyceria maxima* and *Carex disticha*, for instance, are associated with exfiltration areas with a discharge of relatively calcium-rich groundwater in the pleistocene part of the catchment, but in the holocene part with clay layers the relationship with groundwater composition is, however, unclear. *Hydrocotyle vulgaris* shows a distinct preference for calcium poor groundwater in the pleistocene stream valleys, but not on the Dutch Wadden islands, where it is also found in calcareous dune slacks. The use of indicator species is, therefore, linked to numerous fundamental and practical problems. It is like prospecting for valuable minerals with the help of plant species (Cole, 1980). As long as it works no-one will challenge the method, but problems may arise if the indication is only valid within a restricted landscape ecological setting. And this is practically always the case with indicator species (Zonneveld, 1988). It is, therefore, essential to re-examine the relationship between the occurrence of plant species and parameters of the hydrology in every type of landscape with different geology and climate (Niemann, 1973). Recent changes in the hydrology may obscure the indication value of a plant species. Misinterpretations, due to retarded responses of soils or species (Van Diggelen *et al.*, 1992) or due to stratification of water types (Everts *et al.*, 1988), can be minimized by not concentrating on the ecology of single species, but by considering also the indicator value of the species combinations (local vegetation types).

The coupling of a regional model, such as SIMGRO (Van Diggelen *et al.*, 1991), to responses of vegetation types or species is seemingly less complicated in hydrologically more homogeneous areas. The computed water levels can be used as input in an ecological prediction procedure in which references are stored regarding responses of species to water-table fluctuations or moisture conditions (Gremmen *et al.*, 1990). However, many species with the same preference for water table fluctuations may react differently to a change in groundwater composition (Van Wirdum, 1981; Wassen *et al.*, 1989; Barendregt *et al.*, 1993, this volume) and associated changes in nutrient conditions (Verhoeven *et al.*, 1993, this volume). It is unlikely that regional hydrological models capable of coping with such problems will be operational in the near future. Until that time the use of indicator species or any other expert knowledge on the relationship between plant species and hydrological conditions will be needed in environmental impact studies.

A perspective of hydroecological approaches, as the one discussed here, is that each discipline spends less time in perfecting its own separate model, but concentrates on gaps in knowledge that are essential in understanding the functioning of hydrological systems and its consequences for ecological processes.

ACKNOWLEDGEMENTS

Jacque Peerboom and M. Kortleve carried out parts of the hydrological modelling. We are grateful to Brian Wheeler and Dick Boeye for their constructive comments on the manuscript. Evert Leewinga, Frits Everts and Dick Visser prepared the figures. Joan Purdell-Lewis corrected the English text. Their help is gratefully acknowledged.

REFERENCES

Bakker, J.P. (1989) *Nature Management by Grazing and Cutting*, Geobotany 14, Kluwer, Dordrecht.

Barendregt, A., Wassen, M.J. and de Smidt, J.T. (1993) Hydroecological modelling in a polder landscape; a tool for wetland management, in *Landscape Ecology of a Stressed Environment* (eds. C.C. Vos and P. Opdam), Chapman & Hall, London, pp. 79–100.

Cole, M.M. (1980) Geobotanical expression of ore-bodies. *Transactions of the Institute of Mineralogy and Metereology*, **89B**, 73–91.

De Lyon, M.J.H. and Roelofs, J.G.M. (1986) *Waterplanten in relatie tot waterkwaliteit en de bodemgesteldheid*. Report, Department of Aquatic Ecology, University of Nijmegen.

Ellenberg, H. (1978) *Vegetation Mitteleuropas mit den Alpen*. Ulmer Verlag.

Engelen, G.B. (1981) A system approach to groundwater quality, in *Studies in Environmental Science*, vol **17** (eds. R.W. van Duijvenbooden *et al.*), Elsevier, Amsterdam pp. 1–25.

Everts, F.H. and de Vries, N.P.J. (1991) *The vegetation development in streamvalley systems; a landscape ecological study of Drenthian stream valleys*. Thesis, University of Groningen, Historische Uitgeverij Groningen (in Dutch with English summary).

Everts, F.H., Grootjans, A.P. and de Vries N.P.J. (1988) Distribution of marshplants as guidelines for hydrological research. *Colloques Phytosociologiques*, **16**, 271–92.

Fresco, L.F.M. (1991) *VEGROW, Processing of Vegetation Data*. Department of Plant Ecology, University of Groningen, Haren.

Gremmen, N.J.M., Reijnen, M.J.S.M., Wiertz, J. and Van Wirdum, G. (1990) A model to predict and assess the effects of groundwater withdrawal on the vegetation in the pleistocene areas of the Netherlands. *Journal of Environmental Management* **31**, 143–55.

Grootjans, A.P., van Diggelen, R., Wassen, M.J. and Wiersinga, W.A. (1988) The effects of drainage on groundwater quality and plant species distribution in stream valley meadows. *Vegetatio*, **75**, 37–48.

Jeglum, J.K. (1971) Plant indicators of pH and water level in peatlands at Candle Lake, Saskatchewan. *Canadian Journal of Botany*, **49**, 1661–76.

Jørgenson, S.E., Hoffmann, C.C. and Mitch, W.J. (1988) Modelling nutrient retention by a reedswamp and wet meadow in Denmark, in *Wetland Modelling* (eds. W.J. Mitch, M. Straskraba and S.E. Jørgensen) Developments in

78 Linking ecological patterns

Environmental Modelling **12**, 133–151, Elsevier, Amsterdam.

Kemmers, R.H. (1986) Perspectives in modelling of processes in the root zone of spontaneous vegetation at wet and damp sites in relation to regional water management. Committee for Hydrological Research TNO. *Proc. and Inf.*, **34**, 91–116. The Hague.

Neimann, E. (1973) Grundwasser und Vegetationsefüge. *Nova Acta Leopoldina*, **6**, 147 pp.

Pietch, W. (1982) Ökochemische Beschaffenheit der Gewässer, in *Ausgewählte Methoden der Wasseruntersuchungen.* **Bd II** (Biologische, microbiologische und toxicologische methoden) (eds. G. Breiting and W. von Tümpling), Gustav Fischer Verlag, Jena, pp. 67–88.

Querner, E.P. (1989) Description of a regional groundwater flow model SIMGRO and some applications. *Agricultural Water Management*, **14**, 209–18.

Schot, P.P. (1991) *Solute transport by groundwater flow to wetland ecosystems.* Thesis University of Utrecht.

Siegel, D.I. and Glaser, H. (1987) Groundwater flow in a bog-fen complex, Lost River Peatland, Northern Minnesota. *Journal of Ecology*, **75**, 743–754.

Van Diggelen, R., Grootjans, A.P. Wierda, A. *et al.* (1991) Prediction of vegetation responses to various hydrological scenarios, in *Hydrological Basis of Ecologically Sound Management of Soil and Water*, (Proceedings Symposium IAHS, Vienna, 11–24 August 1991). IAHS Publ. no 202, 1991, pp. 71–80.

Van Diggelen, R., Molenaar, W., Casparie, W.A. and Grootjans, A.P. (1992) Paläeoökologische Untersuchungen als Hilfe in der Landschaftsanalyse im Gorecht-gebiet. *Telma*, 21, 57–73.

Van Wirdum, G. (1981) Linking up the natec subsystem in models for water management. Committee for Hydrological Research TNO, *Proc. and Inf.*, **27**, 108–28,. The Hague.

Verhoeven, J., Kemmers, R.M. and Koerselman, W. (1993) Nutrient enrichment of freshwater wetlands, in *Landscape Ecology of a Stressed Environment* (eds. C.C. Vos and P. Opdam), Chapman & Hall, London, pp. 33–59.

Vos, C.C. amd J.I.S. Zonneveld (1993) Patterns and processes in a landscape under stress: introduction of the study area, in *Landscape Ecology of a Stressed Environment* (eds. C.C. Vos and P. Opdam), Chapman & Hall, London, pp. 1–28.

Wassen, M.J., Barendregt, A., Bootsma, M.C. and Schot, P.P. (1989) Groundwater chemistry and vegetation gradients from rich fen to poor fen in the Naardermeer (The Netherlands). *Vegetatio*, **79**, 117–32

Wilcox, D.A., Shedlock, R.J. and Hendrickson, W.H. (1986) Hydrology, water chemistry and ecological relations in a raised mound of Cowles Bog. *Journal of Ecology*, **74**, 1103–17.

Zonneveld, J.I.S. (1988) Environmental indication, in *Vegetation Mapping*, (eds. A.W. Küchler and J.I.S. Zonneveld), Handbook of Vegetation Science vol. **10**, 491–498, Kluwer, Dordrecht.

Zonneveld, J.I.S. (1989) The land unit, a fundamental concept in landscape ecology and its application. *Landscape Ecology*, **3**, 67–86.

Hydroecological modelling in a polder landscape: a tool for wetland management

4

Aat Barendregt, Martin J. Wassen and Jacques T. de Smidt

4.1 INTRODUCTION

Dutch polder areas are losing much of their characteristic biotic diversity because many wetland species are decreasing or have disappeared. As a consequence communities impoverish and ecosystems desintegrate. The origin of this problem is complex: eutrophication, acidiphication, inversion of groundwater or surface water flows. The inversion of groundwater flows, for instance caused by groundwater abstraction, leads to the infiltration of polluted surface water instead of the former exfiltration of clean seepage water. Due to the inversion of surface water flows polluted river water is imported, instead of the former export of autochtonous surplus water. The need to supply extra river water is partly to compensate for the cease of seepage water, partly caused by pumping out water surpluses in winter, to facilitate agriculture. Policy makers and responsible authorities wish to develop water management in a way that gives new opportunities for the re-establishment of lost biotic values. If a number of management alternatives exist, a tool is needed to assess the ecological impact of the alternatives. For that reason a model is developed that predicts the responses of wetland

Landscape Ecology of a Stressed Environment
Edited by Claire C. Vos and Paul Opdam
Published in 1993 by Chapman and Hall, London. ISBN 0 412 44820 3

species to different water management scenarios.

Modelling efforts in regard to freshwater marshes are just beginning (Mitsch *et al.*, 1988). Most attention in wetland modelling is directed to shallow lakes (Jorgensen, 1988; Straskraba and Gnauck, 1983) and to tidal marshes and estuaries (Hopkinson *et al.*, 1988). Important topics in wetland models are eutrophication (or nutrient availability) and succession in relation to hydrology. The biotic parameter is mostly the dominant species or the primary production (e.g. Hopkinson and Day, 1980; Reddy *et al.*, 1990; Mitsch and Reeder, 1991; Mitsch *et al.*, 1991).

Many wetland models incorporate the master factor hydrology only by its quantity, using parameters for water levels and fluctuations. If water quality is incorporated, the parameter is mostly the nutrient availability. However, in our opinion the concentrations of all ions in the water compartment of the system are important (cf. Moss, 1988). Directly or indirectly the water chemistry affects ecologically important properties of the system, such as pH and nutrient availability (Stumm and Morgan, 1981). Another important parameter that influences the biotic parameters is the relation between salinity and osmotic value in plants and animals (e.g. Beadle, 1943; McKee and Mendelssohn, 1989); the ratio of different major ions also appears to be important (Van Wirdum, 1991). However these parameters work on different scales and causal relations are largely unknown. This means that a causal model cannot be constructed. When little or insufficient knowledge is available for a mechanistic model, descriptive statistical methods offer an alternative. If morphology of the water area, water level, water chemistry and characteristics of the soil and mudlayer are included, the number of variables becomes too large to handle. In that case a common procedure in modelling is to reduce the number of parameters. On the other hand Caswell (1988) has stated 'anything not explicitly included in a model is implicitly assumed to be unimportant in nature'.

To detect the pattern that explains the distribution of the biotic components in a particular landscape with a large number of parameters, we applied a multiple regression method. This method requires large datasets with values of independent variables collected in representative units on the landscape level. The presence of a biotic component in this multidimensional space is described in terms of optimum and tolerance levels. Before the actual modelling of the aquatic ecosystems in a polder landscape could start, according to Jorgensen (1990) three actions have to be taken. First the problem must be described (see above). Second, important ecosystem characteristics are chosen on the basis of a landscape ecological analysis. The third condition, the availability of resources to construct a model, was in the present case met by the collecting of data (section 4.3). After processing the data to a model (section 4.4), the application of the model in the practice of water management is explained (section 4.5).

4.2 LANDSCAPE ECOLOGICAL CHARACTERISTICS OF POLDER AREAS

The western part of The Netherlands mainly consists of polders, that lie below sea level (Vos and Zonneveld, 1993, this volume). In polders the surface water levels of the ditches and canals are regulated artificially. Water surpluses are drained away by pumps and discharged into rivers or principal canals called 'boezem'. This drainage is necessary to prevent the polders flooding since there is a net precipitation surplus during most of the year. In summer, when evapotranspiration exceeds rainfall, water may be in short supply. This is remedied by a reverse action: supply water from the river or the 'boezem' is pumped into the polder. The water balance of a polder is also influenced by non-technical hydrology: seepage leads to an extra surplus and infiltration causes extra water losses.

Much peatland was turned into polders in the period between 1000 and 1500 AD. The wetlands were drained by digging ditches and milling the water outside the surrounding dykes. In a number of polders the peat was dug away to supply the expanding cities with fuel. The remaining strips of peatland were eroded by wind, which led to the formation of lakes. Most of these lakes have been reclaimed again since the nineteenth century. Thus, between the polders where the peat layer is still present, deeper polders are found with a soil of clay or sand. Drainage in peat polders has led to settling and mineralization of the peat, which explains why many of them have sunk below sea level. During these changes rivers and 'boezems' were maintained at their original level (about sea level), which led to an inversion of the landscape morphology. In the natural draining system of the original peat landscape the rivers had the lowest water level. At present the polders comprise the lowest parts of the landscape and consequently surface water has to be pumped upward from the polders into the rivers.

The three main types of groundwater flow in the total polder area of the western Netherlands are: (1) a flow from elevated sandy areas (dunes, hill ridges) into the polders; (2) a flow from the sea or from rivers into the low-lying polders; (3) a flow from polders with a relatively high level to polders with a lower level. These groundwater flows differ in chemistry. In the first flow type, the groundwater is fresh and rich in calcium. The second type is salty or eutrophicated, depending on the recharge source (respectively the sea or a river). In the third type, the groundwater is normally a mixture of rain and river water. Salinity, trophic level and acidity show a large variation even within the same polder.

The main landscape ecological characteristics of polder areas are connected to the morphology and the hydrology. The morphology of a polder reflects the natural and (agri)cultural history of the last millenium. Old polders have a coarse network of wide ditches in wet peatland (mostly mediaeval landscape with an elevation just below sea level). Recent polders in reclaimed lakes have a denser network of narrow ditches in mineral soil (down to 6m below sea level). In the latter type the water in the ditches is deeper below the soil surface. The discharge of groundwater in these former lakes is more intense due to their low position in the landscape.

The hydrological relation between polders causes considerable transport of solutes. This transport, the water balance and chemical composition of the water sources largely determine the chemical types of surface water in a polder. The hydrology of many polder areas is complex (e.g. Higler, 1989). The quality of the inflowing water is a result of interaction of regional and subregional groundwater systems with systems on a local scale, leading to complex hydrochemical gradients in the fens (Schot et al., 1988; Schot, 1989; Koerselman, 1989). Structure and species composition of plant communities are closely related to these gradients (Gorham, 1956; Van Wirdum, 1979; Waughman, 1980; Wheeler, 1980a, b, c; Grootjans, 1985; Wilcox et al., 1986, Succow, 1988; Verhoeven et al., 1988, 1993, this volume; Wassen et al., 1989, 1990; Wassen, 1990; Van Diggelen et al., 1991). The cause of these gradients is a diversity in water level and buffer-capacity of the systems. These chemical balances are explained by Stumm and Morgan (1981). The calcium concentration is concluded to be an important variable in ecosystems (Wheeler, 1980b; Boyer and Wheeler, 1989; Wassen, 1990), next to the acidity (Giller and Wheeler, 1988). Aeration, and thus water level, induces mineralization of the organic soil and in this way raises the nutrient levels. Much attention is paid by hydroecologists to the availability of phosphorus and nitrogen and their biotic impact (e.g. Waughman and Bellamy, 1980; Wilson and Fitter, 1984; Grootjans et al., 1985, 1986; Richardson and Marshall, 1986; Verhoeven and Arts, 1987; Kemmers and Janssen, 1988; Verhoeven et al., 1988). A complex of variables, including at least water level, soil, pH and concentration of calcium and iron, determines concentration and availability of nutrients. Apart from these factors, other chemical variables influence the ecological processes in the system (e.g. salinity; McKee and Mendelssohn, 1989).

Gradients in this complex of variables are the effect of hydrological processes. They can be found within one polder. At one end of the gradient rain water may infiltrate and cause acidification. Such places may suffer from water deficit in summer. The opposite end may receive upward seepage providing stable chemical and hydrological conditions throughout the year. Storage of rain water leads in extreme circumstances to acidiphication of fen ecosystems and development of *Sphagnum* vegetation; supply with calcium-rich groundwater leads to a low productive, species rich *Carex* vegetation. Supply with eutrophicated water from the rivers leads to a high productive, species-poor vegetation. This supply may create more complex gradients, but in most cases it means the disappearance of previous diversity.

4.2.1 THE MAIN LANDSCAPE ECOLOGICAL PARAMETERS

The diversity in non-biotic parameters is demonstrated with a Principal Components Analysis (Figure 4.1) based on a dataset including variables on water chemistry, hydrology and dimensions of the systems, describing a wide range of aquatic polder systems. The first axis of variation is related to the salinity of the samples. The second axis corresponds with the morphological differences between seepage areas and wide and deep aquatic systems. The third and fourth axes correlate with nutrients and pH.

A transect through the Vecht river valley (Figure 4.2) shows the

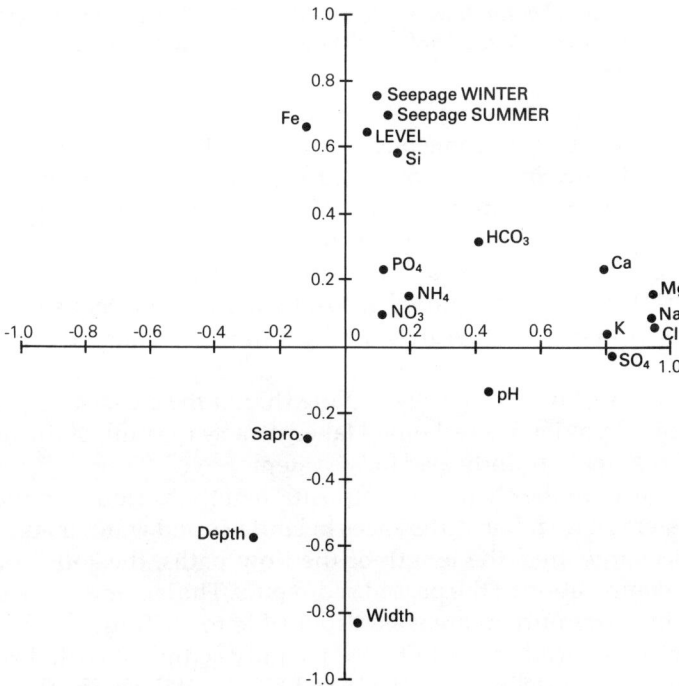

Figure 4.1 Principal Components Analysis of the dataset (n = 770) with variables on water chemistry, hydrology and dimensions of the aquatic system.

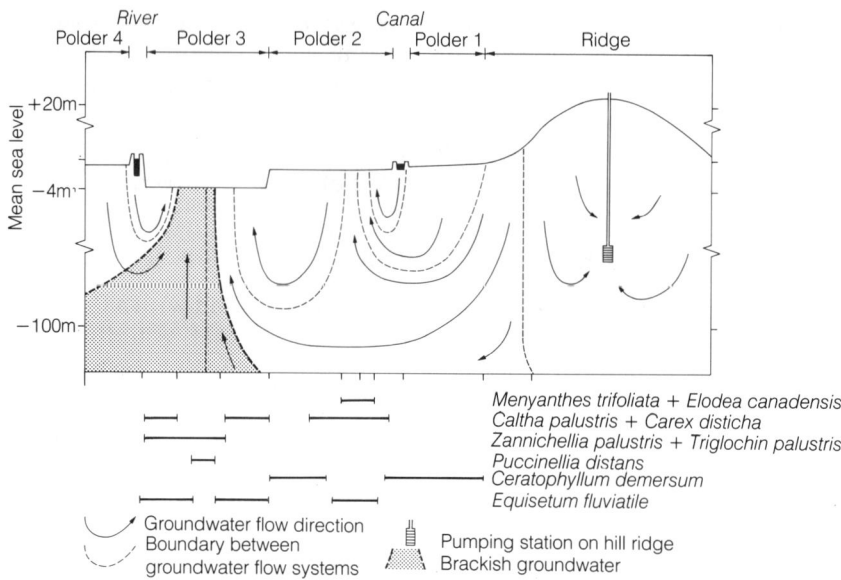

Figure 4.2 Groundwater flow systems of the Vecht river plain (schematic cross-section, after Schot, 1989, 1991) and schematic distribution of some plant species.

various groundwater flow systems of different origin and hierarchy, and the dependency of some plant species on these groundwater flows. The groundwater in the ice-pushed ridge that borders the river plain, is recharged by precipitation water. Most of it is abstracted by drinking water supply companies. Only a minor part of the ridge water still flows westward and comes at the surface in the polders of the plain (Witmer, 1989). In this regional groundwater flow system several sub-regional systems are hierarchically arranged, recharged by surface water from a polder, a canal or the river. They discharge to adjacent polders with a lower water level. Note that in the cross-section shown in Figure 4.2 polder 3, a reclaimed lake, attracts the bulk of the regional and subregional groundwater flow systems.

Differences in the chemical composition of the various groundwater flow systems depend on differences in land use and water management in the recharge area, the length of the flow paths, the soil conditions and the composition of the passed sediments. The leaking of the sewage system in certain urban areas is responsible for pollution of the ridge water at some locations (Schot, 1991). Sandy sediment with dissoluble calcite leads to enrichment with Ca and HCO_3^- of the ridge flow which discharges in deep polders (Figure 4.2) (Wassen *et al.*, 1988). However, in some places the passage of marine sediments results in a brackish water type. The supply with eutrophic river water that is pumped in during dry periods eutrophicates the subregional systems (Roijackers and Verstraelen, 1988).

Mesotraphent freshwater species (*Menyanthes trifoliata, Eleodea canadensis*) are confined to the area in polder 2 where ridge water flows upward and fills the ditches. These species are absent where the water in the ditches is a mixture of brackish and fresh water (polder 3) (Wassen *et al.*, 1988). The range of *Caltha palustris* and *Carex disticha* is not restricted to the mesotrophic conditions. *Zannichellia palustris, Triglochin palustris* and *Puccinellia distans* are related to more or less brackish water. *Ceratophyllum demersum*, indicating eutrophic conditions, is in greatest abundance where artificially supplied eutrophic water infiltrates. *Equisetum fluviatile* indicates seepage.

4.3 THE DATA: VARIABLES AND SAMPLING PROCEDURE

A limited number (25) of non-biotic parameters were incorporated in the dataset. These variables are in the literature generally accepted to be of importance for the presence of species (section 4.2). Included are water chemistry (pH, concentration of macro-ions and nutrients), groundwater flow (calculated seepage or infiltration, in summer and in winter), surface water level (in relation to the surface of the land), width and depth of the aquatic system, morphology of the banks, the soiltype of the bordering terrestrial surface and the subaquatic system. Not included is the chemistry of the subaquatic bottom.

The biotic component could, in principle, be represented by the vegetation types. There is, however, no applicable classification system available for these aquatic polder systems. If such classification were present, problems would still arise in dealing with mixtures of different vegetation types. In fact the only biotic parameter that can be recorded consistently is the species composition of the vegetation. Recording was restricted to the hydrophytes and phreatophytes (Londo, 1988), because these species depend on accessible surface water or groundwater.

A proper description of relations between biotic and non-biotic components requires a sufficiently large number of samples of all variable combinations. The dataset should not be contaminated by including many variables that do not meet the requirement of independence. The correlation between these variables will make the results hard to interpret and will undermine the validity of the explanation. Moreover, variables that are not functionally interrelated but only show similar spatial distribution should not predominate in the dataset. For this reason a great diversity of polders with different combinations of characteristics were sampled in an area of 100 by 50 km in the provinces of Noord-Holland and Utrecht.

The sample areas (in different versions respectively 770, 193 and 745) are homogeneous aquatic parts of the landscape with a size of approximately 0.5–2 ha. The first step was to select the locations by means of published data. Sample locations should be homogeneous with respect to all parameters. For this reason spatial differences within the sample area were not accepted with respect to soil type, electric conductivity of surface water (maximal 15%), hydrology (water levels up to 10 cm, seepage or infiltration in summer or winter), vegetation (structure and floristic composition) and morphology (maximal 20% in width and depth). The second step was to test this homogeneity in the field. Sample locations were, if necessary, subdivided until all parameters were within the acceptable range. Local inhomogeneous sites were avoided. Areas where the hydrology or the management had changed in recent years were not included, to avoid recording species which were not yet in balance with the environmental conditions.

The field work was restricted to a few months time and locations were only sampled once. Characteristic (absence of) changes in parameters during the year (Henderson-Sellers and Archer, 1982; Stumm and Morgan, 1981) were avoided, for practical reasons. The incorporation of a dynamic equilibrium is possible but would have required five to ten times more data. As a consequence the modelling was restricted to a stable situation, and the occurrence of seasonal fluctuations was avoided by collecting data in summer only.

4.4 PROCESSING THE DATA TO A MODEL

For each plant species a model was constructed that predicts its presence or absence at a given set of non-biotic parameters. The following model criteria were designed.

1. Only parameters that explain the presence or absence of the species were incorporated.
2. If necessary more than one parameter was incorporated.
3. Both discrete parameters (at a nominal scale, e.g. soil types) and continuous parameters (at a ratio scale, e.g. concentration) could be modelled.
4. The presence of the species could be described on the basis of a bell-shaped optimum curve (for continuous parameters).
5. The resulting model had to be easy to apply and facilitate impact assessment of changes in hydrology in biotic terms.

These criteria led to the use of multiple logistic regression analysis (Hosmer and Lemeshow, 1989), with a regression model of the group of the generalized linear models (Nelder and Wetherburn, 1974). The values of the response variable are transformed to presence/absence values, to obtain a binomial distribution, characterized by a parameter, p, in this case representing the probability of the presence of the species. By including continuous parameters, both linearly and quadratic, bell-shaped curves are obtained that describe the presence of a species in terms of optimum and tolerance levels with respect to these parameters (Ter Braak, 1987). Discrete parameters can also be added (Jongman *et al.*, 1987).

Complications that arose in processing the dataset were the skewed frequency distribution of parameters and the correlation between parameters. The former complication was solved by logarithmic transformation and the latter by a forward selection procedure. Our selection procedure was more general than the standard one. In each step the linear and quadratic terms of a continuous variable were considered jointly for inclusion in the model. In each step, the continuous or discrete variable that decreased the deviance most, was added. The procedure was stopped when the deviance decreased below 3% of the starting deviance or when seven variables (out of 25) were added. Finally, each quadratic term that lowered the mean deviance was removed, which means that a linear combination (resulting in a sigmoid curve) describes the presence better than the quadratic of the Gaussian curve. The procedure was implemented in the GENSTAT macro language (Alvey, 1977) and carried out only for species that occurred more than 25 times in the dataset (to avoid over-fitting in the model).

4.4.1 THE MODEL ICHORS

The resulting species models with the multiple regression were incorporated in a computer program ICHORS (= Influence of Chemical and Hydrological variables on the Response of plant Species; Barendregt *et al.*, 1986). This program, written in Fortran, allows application of all species models at the same time. In different versions of ICHORS, respectively 163, 76 and 135 species models were installed. Defining a special constellation of all incorporated parameters (a description of a special environment in non-biotic variables), the program predicts the response of the species to that set of variables. The effect of a planned hydrological change can now be expressed in ecological parameters. This is done by comparing the responses of the species to one set of non-biotic variables (representing the present conditions) with the responses to a new set (representing the conditions after the change). The response to the new conditions can be compared with particular management aims, to decide if a planned hydrological change will be successful.

Most of the selected variables in the three versions are mentioned in Table 4.1. No master factor can be detected. Both morphology, soil,

Table 4.1 Incorporated parameters in three versions of ICHORS, with the frequency of the parameters expressed in percentage of the total number of regression equations

ICHORS-version	3.0	3.1	3.2
Number of regression equations	163	76	135
Average number of used parameters	4.8	4.0	3.8
Level surface water	33	25	13
Width of surface water	41	61	23
Depth of surface water	16	17	11
Thickness of sapropel/mud layer	8	14	5
Turbidity of surface water	12	14	6
Soil under surface water	13	16	22
Soil at the shore line	69		45
Seepage or infiltration in summer	9	17	13
Seepage or infiltration in winter	9	14	15
Chemistry of water			
pH	17	25	17
HCO_3	8	13	7
Cl	20	14	8
SO_4	9	8	12
Na	10	9	9
Mg	34	8	26
Ca	19	12	10
K	14	24	15
PO_4	15	5	21
NO_3	11	20	14
NH_4	14	12	10
N-mineral-total	9	13	6
Si	16	22	16
Fe	4	11	13
Index $(Cl + SO_4) / (Cl + SO_4 + HCO_3)$	11	9	8
Index $(Cl) / (Cl + SO_4)$	18	12	12
Index $(Ca + Mg) / (Ca + Mg + Na + K)$	4	5	2
Index $(Ca) / (Ca + Mg + Na + K)$	4	3	9
Summary			
Morphology etc.	98	117	52
Turbidity	12	14	6
Soil	82	16	67
Hydrology (seepage or infiltration)	18	31	28
pH	17	25	17
Anions (HCO_3, Cl, SO_4)	37	35	27
Cations (Na, Mg, Ca)	63	29	45
Nutrients (P, N, K)	63	74	66
Si + Fe	20	33	29
Water typology (index)	37	29	31

Table 4.2 Three tests of ICHORS with independent datasets. The non-biotic parameters in these datasets are the input of ICHORS. The obtained predicted values of the plant species are divided into eight (seven) classes with increasing values. For these classes the mean observed presence of the plant species in the datasets is calculated.

Class number	1	2	3	4	5	6	7	8
Interpolation ICHORS version 3.0 at 50 locations								
$n =$	1917	1365	1245	812	839	779	705	488
Mean predicted response	0.0	0.4	2.7	7.4	14.5	27.0	46.3	76.3
Mean observed presence	4.2	11.9	20.9	33.9	41.7	56.6	66.3	71.7
Extrapolation ICHORS version 3.0 at 62 locations								
$n =$	2052	1409	1363	868	1182	1070	1163	999
Mean predicted response	0.0	0.4	2.7	7.3	14.6	16.8	46.7	78.2
Mean observed presence	0.9	3.6	4.6	11.6	16.5	24.2	36.6	57.1
Extrapolation ICHORS version 3.2 at 95 locations								
$n =$	1261	1587	2108	1520	2648	2166	1535	
Mean predicted response	0.0	0.5	2.7	7.2	17.1	36.5	64.8	
Mean observed presence	1.6	7.6	10.8	11.8	20.9	30.6	44.9	

hydrology and water chemistry (including nutrients and water typology) are needed to describe the presence or absence of the species. The mean explained deviance in the versions is respectively 35% (standard deviation, SD=18%), 29% (SD=14%) and 20% (SD=10%). The linear regression fitted better than the logistic regression in respectively 12%, 21% and 17% of the cases.

It is not possible to calibrate this model, while there are no independent parameters that can be improved with new information. Only a validation can be applied with independent datasets that contain all incorporated non-biotic variables as well as the presence–absence data of all incorporated species. In this region only one dataset ($n=50$) was available for this purpose. As it is possible that specific ecological relations are characteristic only for the selected area, it is in general not allowed to use the model for an extrapolation outside the region. With two datasets ($n=62$ and $n=95$) of bordering areas with comparatively the same ranges in parameters, the extrapolation of the model will be tested.

Another problem that had to be tackled, was the difference between the field data (presence or absence of a species) and the model-output (the probability of encountering a species with a value between 0 and 1). Therefore the lists of calculated responses for all localities in the test-sets were combined with the presence–absence data. Each test with x locations and y species models obtained in this way $x \times y$ pairs of values. These pairs of values were sorted on the calculated responses from ICHORS and divided in classes with increasing values (Table 4.2). The average response can be compared with the average real presence of species. If the model functions in a proper way, the constructed classes with increasing predictions should be in agreement with the observed data from the area. The model gives a reliable prediction of the responses on an ordinal scale, both in interpolation as in restricted extrapolation. This makes the model a useful tool to compare different options in water management on their relative effectivity to restore degenerated wetland ecosystems.

4.5 APPLICATION OF THE HYDROECOLOGICAL MODEL

ICHORS is applied in the wetland area of Molenpolder (Figure 4.3). This polder is located north of the city of Utrecht on a gradient from a hill ridge to the river Vecht. From the foot of the hills, seepage water flows through ditches toward the infiltration area near the river, where Molenpolder is located. The polder with a peat soil has a great number of artificial turf ponds, where peat has been excavated. Vegetation in these ponds shows a characteristic succession with mainly reed (*Phragmites australis*) and sedges (*Carex sp.*) In recent decades the hydrology of Molenpolder has been affected in two ways. First, the amount of seepage water has decreased due to groundwater abstraction in the hill ridge, which had gradually increased since 1920 (Witmer, 1989). Second, a submerged sandpit just south of Molenpolder was constructed in 1960, leaving a lake of 70 ha, and 30 m deep. This lake has strongly affected the flows of groundwater in the area. These two changes have caused a water shortage in the Molenpolder of 500 000–600 000 m^3/year. As this is being compensated by the inlet of Vecht river water, the ecosystems of Molenpolder receive river water instead of seepage water deriving from the hills. This hydrological change has worked out negatively because the river water is eutrophicated. As a consequence the great diversity of succession phases with, for example, quaking fens disappeared and was replaced by eutraphent water vegetation with *Ceratophyllum* and *Lemna* species.

Today management authorities aim at the return of the mesotrophic situation with its numerous succession phases. For that reason the ecological effects of the options for the available supply water are assessed with ICHORS by calculating responses of species on these

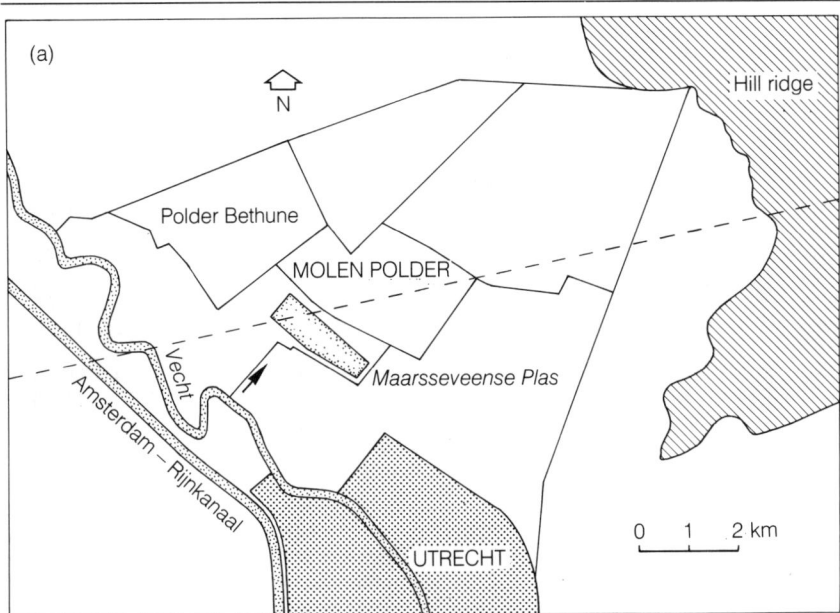

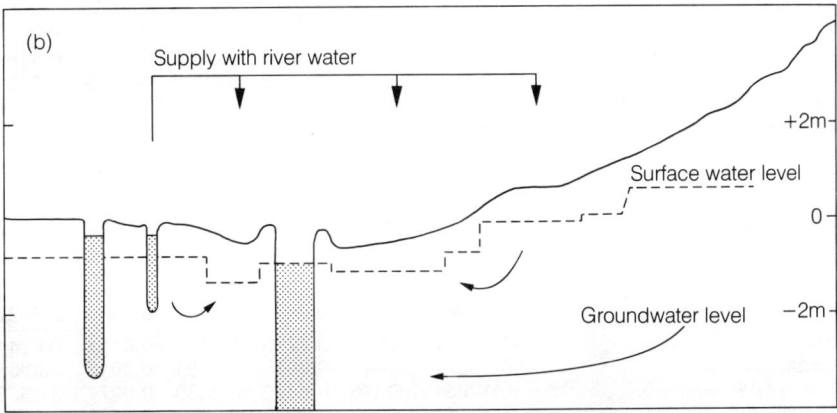

Figure 4.3 (a) The location of Molenpolder, with the supply of water from River Vecht to Molenpolder indicated by an arrow. (b) Cross-section with the hydrological situation at the location of the broken line in (a).

water types (Table 4.3). In this application of the model a combination of variables is present in the dataset, so an interpolation is performed (an extrapolation of a response curve is doubtful). The first option from Table 4.3 is the present water type (column 1). This is a mixture of seepage water from the hill ridge, with supply water from the River Vecht and rain water. Five options for the supply of surface water were

Table 4.3 Defined chemistry and calculated responses (probability of encountering) in an application of ICHORS version 3.1 in Molenpolder with six options for surface water supply. 1 = present situation; 2 = water from River Vecht; 3 = as 2, dephosphated; 4 = dephosphated water from canal Amsterdam-Rijnkanaal; 5 = water from Maarseveense Plas; 6 = water from Bethune Polder

Input to ICHORS 3.1		*1*	*2*	*3*	*4*	*5*	*6*
pH		7.37	7.48	7.98	7.25	7.37	8.30
HCO$_3$	(mg/l)	163.55	168.10	168.09	169.00	250.00	177.00
CL	(mg/l)	110.25	117.45	137.45	154.00	44.00	36.00
SO$_4$	(mg/l)	39.54	73.28	73.28	68.00	8.80	7.20
PO$_4$	(mg/l)	0.16	3.10	0.42	0.05	0.07	0.05
NO$_3$	(mg/l)	0.46	4.30	4.30	4.30	0.25	0.25
NH$_4$	(mg/l)	0.21	2.60	2.60	0.67	1.89	0.05
Na	(mg/l)	69.59	77.58	92.58	95.00	25.70	19.00
Mg	(mg/l)	7.29	8.31	8.31	11.50	7.00	5.50
Ca	(mg/l)	55.29	65.39	65.39	70.00	70.00	55.00
K	(mg/l)	7.54	10.50	10.50	6.70	2.60	1.30
Si	(mg/l)	0.49	3.33	3.33	1.69	5.30	4.00
Fe	(mg/l)	0.11	0.15	0.15	0.50	2.35	0.10

Output of ICHORS 3.1:	*1*	*2*	*3*	*4*	*5*	*6*
Ceratophyllum demersum	- 0.7363	0.7225	0.7059	0.4648	0.0010	0.1134
Stachys palustris	- 0.6709	0.6505	0.6463	0.8002	0.1725	0.1116
Lysimachia thyrsiflora	- 0.2131	0.2823	0.3779	0.3852	0.0000	0.0344
Alisma plantago-aquatica	- 0.2156	0.3666	0.3147	0.2621	0.1349	0.1349
Lemna trisulca	- 0.1814	0.2662	0.1912	0.0596	0.0112	0.0692
Hydrocharis morsus-ranae	- 0.5088	0.9397	0.8855	0.4112	0.1076	0.4964
Frangula alnus	- 0.0501	0.4276	0.4462	0.0204	0.0000	0.0165
Carex paniculata	- 0.6666	0.9177	0.9177	0.9177	0.6257	0.6257
Alnus glutinosa	- 0.5361	0.9781	0.9781	0.8094	0.6035	0.4405
Equisetum fluviatile	- 0.0219	0.0000	0.0000	0.0000	0.3861	0.6169
Epilobium palustre	- 0.0038	0.0000	0.0000	0.0002	0.1909	0.3101
Utricularia vulgaris	- 0.0168	0.0000	0.0000	0.1226	0.4242	0.6638
Sparganium erectum	- 0.3036	0.3353	0.3172	0.3983	0.6109	0.5392
Cicuta virosa	- 0.2734	0.0163	0.0163	0.1974	0.5611	0.4806
Mentha aquatica	- 0.1623	0.0682	0.0164	0.2209	0.4679	0.4566
Glyceria maxima	- 0.1280	0.1518	0.1298	0.1817	0.3551	0.3280
Potentilla palustris	- 0.0033	0.0000	0.0000	0.0084	0.0968	0.0546
Nuphar lutea	- 0.5558	0.2450	0.2450	0.2861	0.8106	0.8202
Nymphaea alba	- 0.4043	0.1142	0.1868	0.1267	0.1144	0.4096
Berula erecta	- 0.2906	0.0671	0.0671	0.1056	0.0600	0.2627
Solanum dulcamare	- 0.3820	0.2052	0.1273	0.1782	0.0151	0.0244
Stratiotes aloides	- 0.2952	0.0041	0.0075	0.0751	0.0013	0.0058
Phragmites australis	- 0.1265	0.0376	0.0376	0.0585	0.0277	0.0333
Lysimachia vulgaris	- 0.1014	0.0232	0.0139	0.0134	0.0118	0.0423
Nasturtium microphyllum	- 0.0017	0.0383	0.0003	0.0000	0.0001	0.0000
Potamogeton lucens	- 0.0015	0.0057	0.0000	0.6196	0.0119	0.0045
Hydrocotyle vulgaris	- 0.0476	0.0050	0.0021	0.1236	0.0007	0.0022
Typha latifolia	- 0.2891	0.2665	0.2665	0.2547	0.8336	0.3155
Calamagrostis canescens	- 0.3530	0.5473	0.6078	0.6826	0.8328	0.5977
Carex pseudocyperus	- 0.3250	0.2526	0.2076	0.3341	0.7188	0.1200
Caltha palustris	- 0.0717	0.0977	0.0720	0.3010	0.7165	0.0305
Scirpus lacustris lacustris	- 0.3251	0.2866	0.1915	0.3615	0.4659	0.1963
Potamogeton obtusifolius	- 0.0012	0.0000	0.0000	0.0000	0.0000	0.1038
Hottonia palustris	- 0.0103	0.0000	0.0002	0.0000	0.0032	0.0586
Sium latifolium	- 0.0242	0.0000	0.0000	0.0331	0.1757	0.8153

assessed; assuming that surface water in Molenpolder is fully replaced by water of the new source. Three options are based on water from a river or canal: the present supply from the River Vecht (column 2), the same water but dephosphated technically (column 3) and dephosphated water from the Amsterdam–Rijnkanaal (column 4). The remaining options are based on the surplus of surface water (due to seepage) from neighbouring polders: Maarsseveense Plas (column 5) and Bethune Polder (column 6). The chemistry of the water varies with each option (Table 4.3; average values). Other variables, such as soil type, dimension and hydrology, are modelled as stationary.

Different strategies are possible to process the calculated responses of the species. First, the assessment procedure can stress the effects for rare and endangered species or for dominating species. The results can also be arranged according to an ecological sequence (e.g. eutrophication or salinity) or to a phytosociological distribution (species per vegetation type). In the example of the Molenpolder (Table 4.3), the responses of a number of species are arranged in groups with increasing responses. The effects of the different options are in this way expressed in a list of species that are stimulated.

The scenarios with water from the river or canal (columns 1–4) show stimulation of species with quite different ecology (mostly eutraphent: *Ceratophyllum* to *Alnus*) than the scenarios with water from the surrounding polders (columns 5 and 6; mostly mesotraphent species, *Equisetum* to *Nuphar*). Supply with untreated river water (column 2) differs very little in effect from supply with dephosphated water from the River Vecht (column 3). Also supply with surface water from the canal (column 4) produces about the same responses. No striking differences in origin or manipulation are evident among the plant species in columns 2–4, except for the stimulation of some aquatic species like *Utricularia vulgaris* and *Potamogeton lucens* in column 4. The options with supply of polder water (columns 5 and 6) show mutually only slight differences, and both stimulate the requested mesotraphent species, e.g. *Equisetum fluviatile* and *Potentilla palustris*. The characteristic species for advanced succession phases, e.g. *Cituta virosa* and *Sparganium erectum*, are optimal, especially in column 5 with *Typha latifolia*, *Carex pseudocyperus* and *Scirpus lacustris*. Column 6 represents the desired mesotraphent aquatic vegetation with, for example, the species *Potamogeton obtusifolius, Hottonia palustris, Hydrocharis morsus-ranae* and *Utricularia vulgaris*. The present situation (column 1) is intermediate to all options, with respect to the hydrological and chemical properties. The high response of *Stratiotes aloides* in this column is surprising. In the river plain this species inhabits the contact zone between groundwater and river water.

The advice to the management is to use the surpluses of seepage water from the surrounding polders. Most of the species that are stimulated by these hydrological changes achieve their optimum in the species-rich succession phases, characteristic for mesotrophic turf ponds. Because of the low responses of many eutraphent species and the high responses of mesotraphent species, the aim of returning to the former diversity will be achieved with this management option. An alternative source of (dephosphated) surface water, from the canal or from the river, is no solution to the problem.

4.6 DISCUSSION AND CONCLUSIONS

A landscape is too complex to be processed in one mechanistic model and it is questionable if all relations can be analysed empirically. We restricted the modelling to the landscape scale, focusing on the relation between hydrology and plant species. On the landscape level the scales of the hydrological characteristics (regional and local) have to be combined with those of the soil and morphology (regional and local) and those of the nutrients (local). However, the descriptive method yields reliable results (Table 4.2) and the possibilities in management are illustrated (section 4.5).

Logistic regression is earlier applied in ecological research (e.g. Austin et al., 1984; Ter Braak, 1987; Murtaugh, 1988). The addition of the stepwise approach clearly facilitates the modelling of a complex of interacting variables, when the functional relations between the parameters are unknown.

An important assumption in ICHORS is that the relation between plant species and the non-biotic parameters is detectable. Other assumptions are that these parameters are stable in time, and that single sampling of surface water at the plots (in summer, by day) is sufficient to indicate the chemistry. Two important restrictions have to be mentioned. The model is constructed for a particular area. Interpolation within the area and within the ranges of the model is allowed. Some extrapolations might be correct but have to be tested, for the actual relations between non-biotic parameters and plant species are involved in this method. Some of these relations are characteristics of this particular dataset. If for instance all clay soils in a dataset have high salinity, the relation clay–salinity is modelled in a complex represented by one variable. However, in an area not represented in the dataset where clay and salinity are not correlated, the model will predict uncertain responses. The second restriction is caused by the required input data of the model, for it is impossible to apply ICHORS without accurate descriptions of the non-biotic variables. For correct application, the hydrology in its hierarchical arrangement and the local differences in landscape (soil, morphology, chemistry) have to be known in present and have to be predicted in future. This is a restriction to the general application of the model, but at the same time it is a stimulation for the research of pattern and processes in the landscape. The required formation of datasets from large areas will provide excellent opportunities to detect landscape ecological relations.

In summary, the important advantages of ICHORS are (1) a complex of variables with different scales is modelled; (2) the included variables of the model explain on the landscape level; (3) the output contains the responses of a great number of species, so that the assessment procedure can be aimed at different information (e.g. endangered species, salinity, eutrophication); (4) the output provides a simple basis for direct advice to policy makers and managers, and gives insight as to which management option is best to reach the desired environmental aim.

REFERENCES

Alvey, N.G. (1977) *GENSTAT: a general statistical program*. Rothamsted Experimental Station, Harpenden, UK.

Austin, M.P., Cunningham, R.B. and Fleming, P.B. (1984) New approaches to direct gradient analysis using environmental scalars and statistical curve-fitting procedures. *Vegetatio*, **55**, 11–27.

Barendregt, A., De Smidt, J.T. and Wassen, M.J. (1986) The impact of groundwater flow on wetland communities, in *Vegetation et Geomorphologie* (ed J.M. Gehu), Colloques Phytosociologiques XIII, Kramer, Berlin, pp. 603–12.

Beadle, L.C. (1943) Osmotic regulation and the fauna of inland waters. *Biological Reviews*, **18**, 172–83.

Boyer, M.L.H. and Wheeler, B.D. (1989) Vegetation patterns in spring-fed calcareous fens: calcite precipitation and constraints on fertility. *Journal of Ecology*, **77**, 597–609.

Caswell, H. (1988) Theory and models in ecology: a different perspective. *Ecological modelling*, **43**, 33–44.

Giller, K.E. and Wheeler, B.D. (1988) Acidification and succession in a floodplain mire in Norfolk Broadland, U.K. *Journal of Ecology*, **76**, 849–66.

Gorham, E. (1956) The ionic composition of some bog and fen waters in the English Lake District. *Journal of Ecology*, **44**, 142–52.

Grootjans, A.P. (1985) *Changes of groundwater regime in wet meadows*. Thesis, University Groningen.

Grootjans, A.P., Schipper, P.C. and Van der Windt, H.J. (1985) Influence of drainage on N mineralisation and vegetation response in wet meadows. I. *Calthion palustris* stands. *Oecologia Plantarum*, **6**, 403–17.

Grootjans, A.P., Schipper, P.C. and Van der Windt, H.J. (1986) Influence of drainage on N mineralisation and vegetation response in wet meadows. II. *Cirsio-Molinietum* stands. *Oecologia Plantarum*, **7**, 3–14.

Henderson-Sellers, B. and Archer, P.B.R. (1982) Models of annual cycles in lentic water bodies. *Hydrobiologia*, **88**, 89–91.

Higler, L.W.G. (1989) Hydrobiological research in peat polder ditches. *Hydrobiological Bulletin*, **23**, 105–9.

Hopkinson, C.S. Jr and Day, J.W. Jr (1980) Modeling hydrology and eutrophication in a Louisiana swamp forest ecosystem. *Environmental Management*, **4**, 325–35.

Hopkinson, C.S.,Wetzel, R.L. and Day, J.W. Jr (1988) Simulation models of coastal wetland and estuarine systems: realisation of goals, in *Wetland Modelling, Developments in Environmental Modelling* 12 (eds. J.M. Mitsch, M. Straskraba and S.E. Jørgensen), Elsevier, Amsterdam, pp. 67–98.

Hosmer, D.W. and Lemeshow, S. (1989) *Applied Logistic Regression.* Wiley, New York.

Jongman, R.H.G., Ter Braak, C.J.F. and Van Tongeren, O.F.R. (eds) (1987) *Data Analysis in Community and Landscape Ecology.* Pudoc, Wageningen.

Jørgensen, S.E. (1988) Modelling eutrophication of shallow lakes, in *Wetland Modelling, Developments in Environmental Modelling* 12 (eds. J.M. Mitsch, M. Straskraba and S.E. Jørgensen), Elsevier, Amsterdam, pp. 177–88.

Jørgensen, S.E. (1990) Ecosystem theory, ecological buffer capacity, uncertainty and complexity. *Ecological Modelling,* **52**, 125–33.

Kemmers, R.H. and Jansen, P.C. (1988) Hydrochemistry of rich fen and water management. *Agricultural Water Management,* **14**, 399–412.

Koerselman, W. (1989) *Hydrology and nutrient budgets of fens in an agricultural landscape.* Thesis, University Utrecht.

Londo, G. (1988) *Nederlandse freatofyten.* Pudoc, Wageningen.

McKee, K.L. and Mendelssohn, I.A. (1989) Response of a freshwater marsh plant community to increased salinity and increased water level. *Aquatic Botany,* **34**, 301–16.

Mitsch, W.J. and Reeder, B.C. (1991) Modelling nutrient retention of a freshwater coastal wetland: estimating the roles of primary productivity, sedimentation, resuspension and hydrology. *Ecological Modelling,* **54**, 151–87.

Mitsch, J.M., Straskraba, M. and Jorgensen, S.E. (1988) Summary and state of the art of wetland modelling, in *Wetland Modelling, Developments in Environmental Modelling* 12 (eds J.M. Mitsch, M. Straskraba and S.E. Jørgensen) Elsevier, Amsterdam, pp. 217–22.

Mitsch, W.J.,Taylor, J.R. and Benson, K.B. (1991) Estimating primary productivity of forested wetland communities in different hydrologic landscapes. *Landscape Ecology,* **5**, 5–92.

Moss, B. (1988) *Ecology of Fresh Waters – Man and Medium.* Blackwell Scientific Publications, Oxford.

Murtaugh, P.A. (1988) Use of logistic regression in modelling prey selection by *Neomysis mercedis. Ecological Modelling,* **43**, 225–33.

Nelder, J.A. and Wetherburn, R.W.M. (1974) Generalized linear models. *Journal of the Royal Statistical Society. A.,* **135**, 370–84.

Reddy, K.R., Rao, P.S.C. and Jessup, R.E. (1990) Transformation and transport of ammonium nitrogen in a flooded organic soil. *Ecological Modelling,* **51**, 205–16.

Richardson, C.J. and Marshall, P.E. (1986) Processes controlling movement, storage, and export of phosphorus in a fen peatland. *Ecological Monographs,* **56**, 279–302.

Roijackers, R.M.M. and Verstraelen, P.J.T. (1988) Ecological investigations in three shallow lakes of different trophic level in The Netherlands. *Verhandlungen internationalen Vereinigung fuer theoretische und angewandte Limnologie,* **23**, 489–95.

Schot, P.P. (1989) Groundwater systems analysis of the Naardermeer wetland, The Netherlands. In Selected papers from the 28th International Geological Congress, Washington. *IAH Selected papers on Hydrogeology* 1 (eds. E.S. Simpson and J.M. Sharp) Hannover, pp. 257–69.

Schot, P.P. (1991) *Solute transport by groundwater flow to wetland ecosystems*. Thesis, University Utrecht.

Schot, P.P., Barendregt, A. and Wassen, M.J. (1988) Hydrology of the Naardermeer; influence of the surrounding area and impact on vegetation. *Agricultural Water Management*, **14**, 459–70.

Straskraba, M. and Gnauck, A. (1983) *Aquatische Oekosysteme – Modellerung und Simulation*. Fischer, Stuttgart.

Stumm, W. and Morgan, J.J. (1981) *Aquatic Chemistry*. Wiley, New York.

Succow, M. (1988) *Landschafsökologische Moorkunde*. Gebr. Borntraeger, Berlin-Stuttgart.

Ter Braak, C.J.F. (1987) *Unimodal models to relate species to environment*. Thesis, University Wageningen.

Van Diggelen, R., Grootjans, A.P., Kemmers, R.H. *et al.* (1991) Hydro-ecological analysis of the fen system Lieper Posse, eastern Germany. *Journal of Vegetation Science*, **2**, 465–76.

Van Wirdum, G. (1979) Dynamic aspects of trophic gradients in a mire complex. *Proceedings and Information 25, CHO-TNO*, The Hague, pp. 66–82.

Van Wirdum, G. (1991) *Vegetation and hydrology of floating rich-fens*. Thesis, University of Amsterdam.

Verhoeven, J.T.A. and Arts, H.H.M. (1987) Nutrient dynamics in small mesotrophic fens surrounded by cultivated land. II. N and P accumulation in plant biomass in relation to the release of inorganic N and P in the peat soil. *Oecologia* (Berlin), **72**, 577–61.

Verhoeven, J.T.A., Kemmers R.H. and Koerselman, W. (1993) Nutrient enrichment of freshwater wetlands, in *Landscape Ecology of a Stressed Environment* (eds C.C. Vos and Opdam, P.), Chapman & Hall, London, pp. 33–56.

Verhoeven, J.T.A., Kooijman A.M. and Van Wirdum, G. (1988) Mineralization of N and P along a trophic gradient in a freshwater mire. *Biogeochemistry*, **6**, 31–43.

Vos, C.C. and Zonneveld, J.I.S. (1993) Patterns and processes in a landscape under stress: the study area, in *Landscape Ecology of a Stressed Environment* (eds C.C. Vos and P. Opdam), Chapman & Hall, London, pp. 1–28.

Wassen, M.J. (1990) *Water flow as a major landscape ecological factor in fen development*. Thesis, University Utrecht.

Wassen, M.J., Barendregt, A., Bootsma, M.C. and Schot, P.P. (1989) Groundwater chemistry and vegetation of gradients from rich fen to poor fen in the Naardermeer (The Netherlands). *Vegetatio*, **79**, 117–32.

Wassen, M.J., Barendregt, A. and De Smidt, J.T. (1988) Groundwater flow as conditioning factor in fen ecosystems in the Kortenhoef area, the Netherlands, in *Proceedings of the 8th International Symposium on Problems in Landscape Ecological Research* (eds M. Ruzicka, T. Hrnciarova and L. Miklos), Bratislava, CSSR, pp. 241–51.

Wassen, M.J., Barendregt, A., Schot, P.P. and Beltman, B. (1990) Dependency of local mesotrophic fens on a regional groundwater flow system in a poldered river plain in the Netherlands. *Landscape Ecology*, **5**: 21–38.

Waughman, G.J. (1980) Chemical aspects of the ecology of some South German peatlands. *Journal of Ecology*, **68**, 1025–46.

Waughman, G.J. and Bellamy, D.J. (1980) Nitrogen fixation and the nitrogen balance in peatland ecosystems. *Ecology*, **61**, 1185–98.

Wheeler, B.D. (1980a) Plant communities of rich-fen systems in England and Wales I. Introduction, tall sedge and reed communities. *Journal of Ecology*, **68**, 365–95.

Wheeler, B.D. (1980b) Plant communities of rich-fen systems in England and Wales II. Communities of calcareous mires. *Journal of Ecology*, **68**, 405–20.

Wheeler, B.D. (1980c) Plant communities of rich-fen systems in England and Wales III. Fen meadow, fen grassland and fen woodland communities, and contact communities. *Journal of Ecology*, **68**, 761–88.

Wilcox, A., Shedlock, R.J. and Hendrickson W.H. (1986) Hydrology, water chemistry and ecological relations in the raised mound of Cowles Bog. *Journal of Ecology*, **74**, 1103–17.

Wilson, K.A. and Fitter, A.H. (1984) The role of phosphorus in vegetational differentiation in a small mire. *Journal of Ecology*, **72**, 463–73.

Witmer, M.C.H. (1989) *Integral water management at regional level - an environmental study of the Gooi and the Vechtstreek*. Thesis, University Utrecht.

Part Two
Spatial Relations by Air Flows

As long as air flows move across scales that largely exceed the spatial pattern of landscapes and land units, the input from atmospheric sources will be equally distributed over large surfaces. Yet their effects can be locally very different, depending on the nutrient cycling at the local and regional scales.

This section focuses on the impact of N-deposition on ecosystems. Of course, other compounds (like heavy metals) have to be considered as possible factors to influence natural systems. Obviously, the impact of N-deposition receives so much attention in landscape ecology, partly because in terrestrial ecosystems N is often a key factor in determining the composition and structure of the vegetation, and partly because in a country with such a high density of cattle, N is an important environmental problem. Most research is directed at assessing how local processes are affected by N-deposition for various ecosystems (Berendse *et al.*, Chapter 5). Results of such studies lead to conclusions about critical levels of deposition allowing the sustainable conservation of particular types of ecosystems. Also, management regimes aiming at removing the adverse surplus of nutrients may be developed on the basis of these studies. Research has been focused on forest and heathland ecosystems on sandy soils, which are regarded as being poorly buffered against acid deposition. Measures to solve problems in natural systems or timber forest are, and should be, directed at lowering the emission at the source rather than at local attempts to remove or immobilize excessive N-contents. Berendse *et al.* conclude that usually management regimes aiming at removing the top soil with all vegetation, as is often practised in heathlands, are not effective under the current atmospheric deposition of nitrogen.

Measures aiming at reducing the emission of nitrogen usually do not consider spatial aspects. Wherever transport of nutrients by air is on a smaller spatial scale, the amount of N-deposition may vary between landscape units. Berendse *et al.* review research on the spatial effects of NH_3-emission from agroindustries like chicken breeding farms. Such research leads to planning rules for minimal distances between N-sources and the border of natural areas.

Melman and Van Strien (Chapter 6) consider N-input by air on a micro level: the input on the ditch banks bordering fertilized meadows. A detailed analysis, based on a model, shows that even at this small scale, spatial separation of agricultural units and units with nature conservation functions is possible. In such agricultural landscapes, ditch banks contain almost all remaining plant species that are regarded as valuable from a nature conservation point of view. This example demonstrates the fine spatial scale at which units with different functions are combined in landscapes where there is an urgent lack of space, as there is in The Netherlands.

Atmospheric nitrogen deposition and its impact on terrestrial ecosystems

5

Frank Berendse, Rien Aerts and Roland Bobbink

5.1 INTRODUCTION

Until recently, the impact of air pollutants on ecosystems was considered to result mainly from acidification (e.g. Hultberg, 1985). Some air pollutants, however, also cause eutrophication of formerly nutrient-poor ecosystems. In Western Europe, especially in The Netherlands, eutrophication of terrestrial ecosystems results largely from atmospheric ammonia deposition volatilized from intensive agricultural systems (dairy farming and intensive animal husbandry) (Asman, 1987; Buijsman *et al.*, 1987; Vos and Zonneveld, 1993, this volume).

Inputs of nitrogen into terrestrial ecosystems may follow different pathways: atmospheric deposition, nitrogen fixation, groundwater seepage, inundation and import by grazers, but in many natural ecosystems atmospheric deposition is the predominant nitrogen input. The accumulation of nitrogen in ecosystems is probably one of the main driving processes that determine the rate of succession and the dynamics of species composition (Crocker and Major, 1955; Olsen, 1958; Berendse, 1990). A consequence of the present high annual inputs is that succession in nutrient-poor ecosystems in The Netherlands proceeds extremely rapidly. This has led to many undesired changes in the

Landscape Ecology of a Stressed Environment
Edited by Claire C. Vos and Paul Opdam
Published in 1993 by Chapman and Hall, London. ISBN 0 412 44820 3

species composition of the ecosystems involved.

First, the spatial variation in the input of nitrogen through atmospheric deposition is discussed. Thereafter, we analyse the consequences of these inputs for the nitrogen balance of the ecosystem and the nitrogen supply to the plant. Finally, the effects of the increased nitrogen supply on the dynamics of the species composition are dealt with.

5.2 SPATIAL VARIATION IN THE DEPOSITION OF NITROGEN

Quantifying the input of nitrogen through atmospheric deposition is a prerequisite for understanding its adverse effects at the ecosystem level. Amounts of bulk precipitation are relatively unaffected by the characteristics of the surfaces on which they are deposited. However, amounts of dry deposition resulting from impaction, phoresis, and diffusion ('dry deposition s.s.') are strongly determined by the roughness length of the deposition surface (e.g. Fowler, 1980; Matzner, 1989). The roughness length is an important attribute of the vegetation in plant–atmosphere interactions and it is a measure of the momentum absorbing power of the surface. Vegetation is the most widespread deposition surface and its roughness length is one of the most important determinants of the deposition rate of air pollutants.

Forests have been shown to possess surface structures that strongly promote dry deposition of nitrogen (and sulphur) compounds. In the inland regions of The Netherlands the total nitrogen deposition in forests is high (mean 85 kg; range 50–170 kg N/ha/year), whereas only in the coastal regions is the nitrogen input to the forest canopy relatively low (c. 40–55 kg N/ha/year) (e.g. Van Breemen *et al.*, 1982, 1988; Ivens, 1990). In all Dutch forest sites studied more than 80% of total nitrogen input is due to deposition of ammonia or ammonium. Furthermore, the yearly nitrogen deposition is approximately 20–30% higher in coniferous stands than in deciduous stands, probably because of the high filtering efficiency of the conifer trees, especially in winter (Ivens, 1990). It is very likely that other canopy characteristics (e.g. stand density, age and height distribution of the trees) will influence the dry deposition of nitrogen. However, these relations are not yet quantified.

Forest edges were shown to receive relatively high nitrogen inputs. In a study in coniferous forest stands, Ivens (1990) demonstrated a significant increase of 30–50% of nitrogen deposition in the forest edge zone (50–100 m), if the forest stand borders an area with low vegetation. He suggested that this effect is caused by both horizontal inflow of polluted air and by enhanced air turbulence at the forest edge. In The Netherlands extensive forest areas are rather rare, so many edges exist in the Dutch landscape and higher nitrogen deposition can be expected caused by the relatively greater portion of the area occupied by edges.

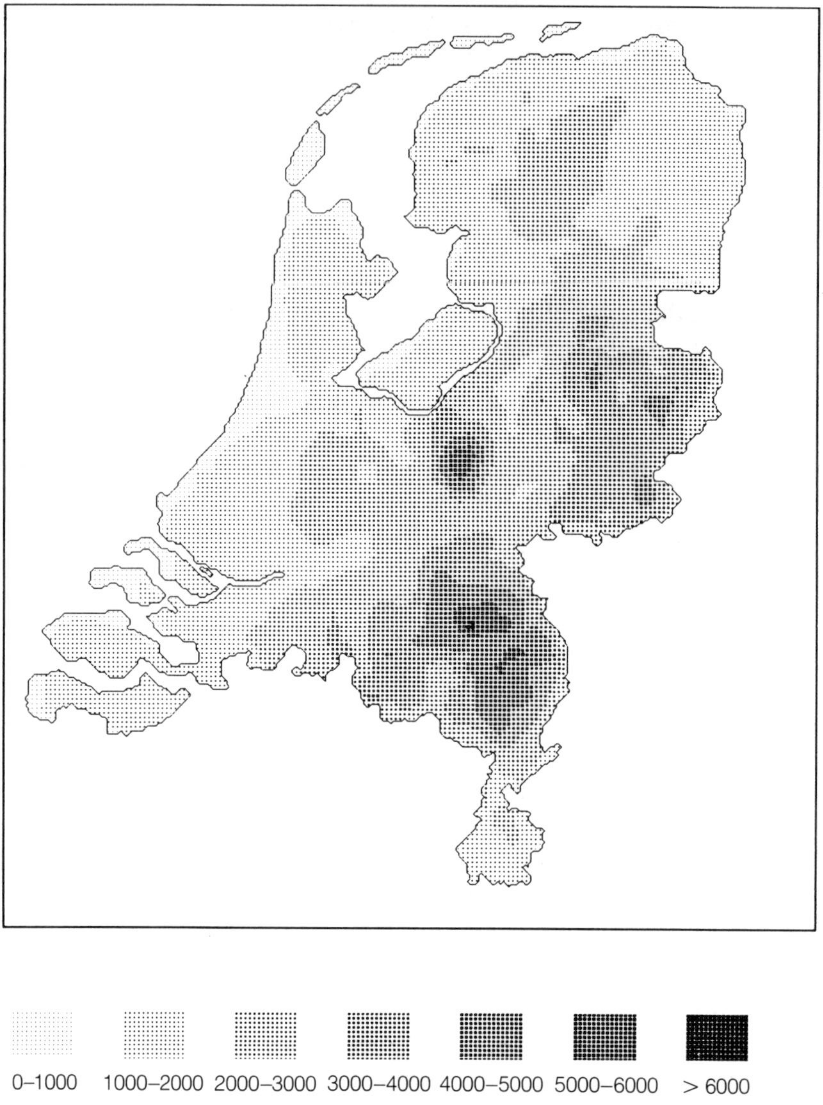

Figure 5.1 Total (wet and dry) deposition flux of NH_3 and NH_4^+ (mol/ha/year) in different parts of The Netherlands (from Erisman, 1991).

The distance from local ammonia sources has an important impact on the amount of nitrogen deposition. The ammonia deposition in The Netherlands is especially high in the eastern and southeastern parts that have large concentrations of intensive animal husbandry (Van Aalst and Erisman, 1990; Figure 5.1). Berendse *et al.* (1988) showed that in the immediate neighbourhood of a poultry farm (75-400 m) the nitrogen deposition to a *Pinus* forest increased significantly (from 42 to

68 kg N/ha/year at 400 and 75 m from the farm, respectively). Nitrate concentrations in the soil increased sharply towards the farm due to increased oxidation of the ammonium supplied through deposition, which led to increased soil acidification. The soil pH decreased from 5.7 to 4.9 going nearer to the farm (Figure 5.2). An increased input of ammonia/ammonium to the forest soil leads to soil acidification, if (1) the environmental conditions allow nitrification and (2) at least part of the nitrate produced is not absorbed by plant roots, but leaches to the groundwater or accumulates in the soil (Berendse *et al.*, 1988; Van

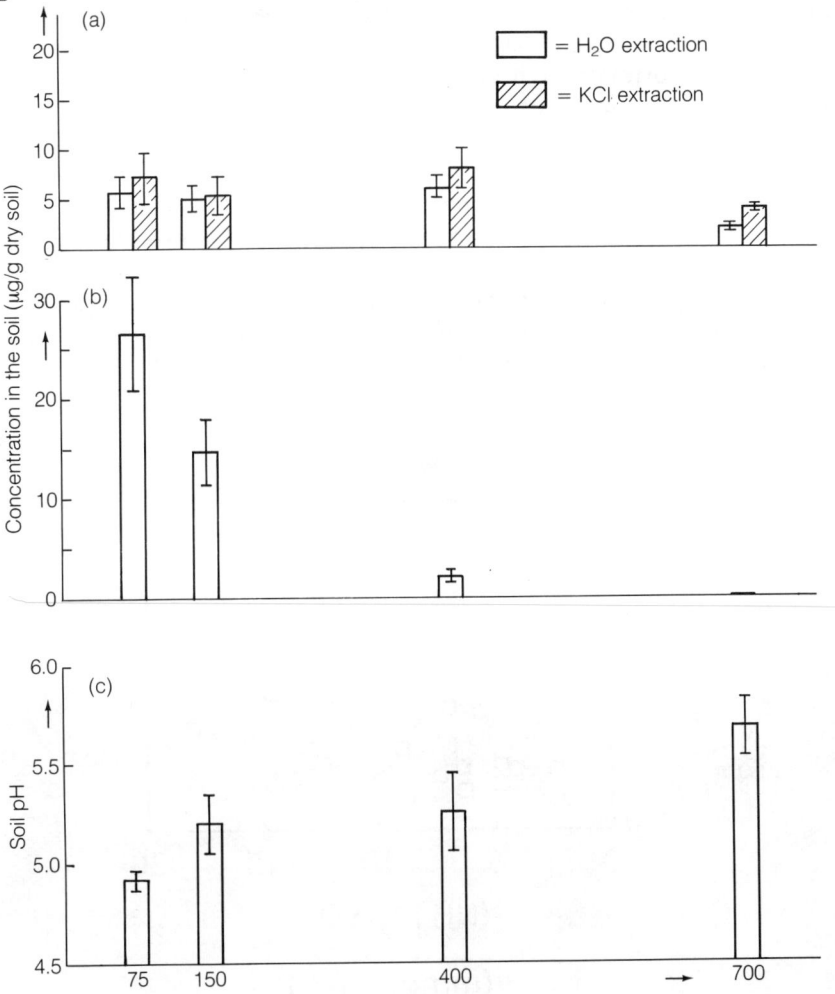

Figure 5.2 The average concentrations of (a) ammonium and (b) nitrate and (c) the pH(H_2O) in the upper 10 cm of the soil in a *Pinus sylvestris* stand at different distances from a poultry farm (after Berendse *et al.*, 1988).

Breemen *et al.*, 1988).

In contrast to forests, grasslands and heathlands were thought to have a rather smooth canopy structure and, therefore, to capture much smaller amounts of nitrogen. Measurements inside low vegetation canopies are rare, but methods have become available to quantify nitrogen deposition in grassland and heathland without disturbance of the vertical structure of the stand (Heil and Van Dam, 1986; Bobbink *et al.*, 1990a). Deposition measurements in The Netherlands clearly demonstrated that grassland canopies are also significant sinks for nitrogen pollutants, especially of ammonia or ammonium. Furthermore, it has been found that leaf area per unit ground surface (Leaf Area Index LAI) in grassland vegetation is significantly correlated with dry deposition rate: with increasing LAI of the vegetation total nitrogen input increases strongly (Heil, 1988; Heil *et al.*, 1988). Annual measurements show that nitrogen deposition is between 35 and 55 kg N/ha/year in grasslands (maximum LAI 3–4; cut once a year) in the central part of The Netherlands (Bobbink *et al.*, 1990b). These authors, using artificial canopies, also established that 30–90% of the dry deposited nitrogen compounds is directly assimilated by the green plant parts of the canopy. Even in chalk grassland vegetation with a low architecture atmospheric nitrogen input is quantitatively important (32–46 kg/ha/year) (Van Dam, 1990).

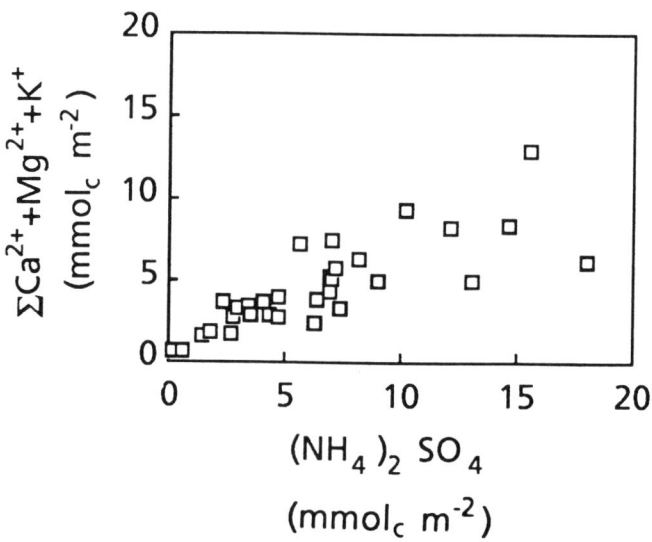

Figure 5.3 Relation between two-weekly canopy losses of potassium, calcium and magnesium and the deposition of ammonium sulphate in *Calluna*-dominated heathland in The Netherlands (after Bobbink *et al.*, 1991).

Bobbink *et al.* (1990a, 1991) studied canopy exchange and atmospheric deposition of nitrogen in dry inland heathland, dominated by *Calluna vulgaris*, in the eastern part of The Netherlands (The Veluwe). Their measurements in artificial and real *Calluna* canopies show that in the real *Calluna* canopy a significant part of the deposited ammonia or ammonium was directly assimilated by the green *Calluna* shoots, especially in wet periods. The concentrations of potassium, calcium and magnesium in throughfall after passage through the *Calluna* canopy increased significantly compared with bulk precipitation, especially in periods with high ammonium sulphate deposition (Figure 5.3). The amounts of these cations lost from the canopy were in good agreement with the observed ammonium uptake by the *Calluna* shoots. Furthermore, they demonstrated that dry deposition is also an important component of the atmospheric nitrogen input in these low *Calluna* stands. Bulk precipitation amounted to only about 35–40% of total atmospheric input and total atmospheric nitrogen deposition was 30–45 kg N/ha/year in the investigated heathland sites.

5.3 ACCUMULATION AND MINERALIZATION OF NITROGEN

The net change in the total amount of nitrogen in the ecosystem equals the difference between inputs and losses. Inputs and losses into and from terrestrial ecosystems have been measured at different locations in The Netherlands. Here we present as examples the nitrogen balance of some grassland, heathland and forest ecosystems (Table 5.1).

In the wet meadow, chalk grassland and heathland ecosystem the major export of nitrogen takes place through hay or sod removal (Bobbink and Willems, 1991; Berendse, 1990), whereas in the two forest ecosystems only losses through nitrate leaching and denitrification are significant (Tietema and Verstraten, 1991; Van Breemen *et al.*, 1988). The wet meadow and the chalk grassland were mown, respectively, twice and once a year. The wet meadow had been heavily fertilized up to 1977. Primary production declined from 12 ton/ha in 1977 to 7 ton/ha in the period 1987–1990. More than 10 years after fertilization had been stopped, the nitrogen balance still appears to be negative, which possibly will lead to a further decline in primary production in the near future.

In the dry heathland there is an accumulation of nitrogen. The studied heathland site was turf-stripped precisely 30 years before the measurements were taken. The total amount of nitrogen in soils and plants was measured in 17 heathland ecosystems dominated by either *Calluna vulgaris* or *Erica tetralix* with an age that varied between 1 and 30 years after turf-stripping (Berendse, 1990). The average nitrogen accumulation rate in these ecosystems was calculated to be 31.4 kg N/ha/year. This value is in the range of nitrogen inputs through

Table 5.1 Nitrogen balance (kg N/ha/year) of a wet meadow on clay-on-peat in the central part of The Netherlands; a dry chalk grassland in the southern part; a dry *Calluna*-dominated heathland in the central part; an oak–beech forest on moist boulder clay; and an oak forest on dry sand, both in the eastern part of The Netherlands.

	Wet meadow	Chalk grass land	Dry heath land	Wet forest[c]	Dry forest[e]
Inputs					
Atmospheric deposition	60[a]	40[a]	30–45[b]	53	56
Fixation of gaseous nitrogen	<1	1[a]	-	-	1
Outputs					
Exploitation	136	25[d]	[1003]	-	-
Denitrification	16	1	-	63	nm
Leaching	3	1[a]	-	10	28
Ammonia volatilization	-	-	-	-	-
Accumulation rate	-95	14	30–45	-20	28

[a]Van Dam (1990); [b]Bobbink *et al.* (1990a); [c]Tietema and Verstraten (1991); [d]Bobbink and Willems (1991); [e]Van Breemen *et al.* (1988). Other data were calculated on the basis of unpublished results (Berendse and Oomes).
Figure in brackets gives nitrogen removal through turf stripping, which occurs once in 30 years.
nm: not measured.

atmospheric deposition that have been measured in these heathlands. It seems that almost all nitrogen that enters these ecosystems accumulates, whereas losses from the ecosystem appear to be small. If it is assumed that by turf-stripping all above ground biomass and the litter and humus layer are removed, at this one moment 1003 kg N/ha is exported. An equilibrium between input and output will be established, if turf-stripping occurs once in 22–33 years, depending on the annual deposition of nitrogen compounds.

The denitrification rate in the wet forest is so high that the total amount of nitrogen in this ecosystem should be expected to decline. However, the denitrification at this site may have been overestimated. Denitrification rates were not measured at the dry site. Groffman and Tiedjee (1989) measured denitrification rates at nine different forest sites and concluded that annual denitrification rates on well-drained sandy soils did not exceed 1 kg N/ha/year.

It can be concluded that in all ecosystems under consideration atmospheric deposition is by far the most important input of nitrogen (Van Breeman *et al.*, 1988; Bobbink *et al.*, 1990a; Van Dam, 1990; Tietema and Verstraten, 1991). Nitrogen fixation by legumes or by blue–green algae on decomposing stems (Rosswall and Granhall, 1980) appears to be negligible. Nitrogen inputs through inundation, groundwater seepage or migration of grazers do not occur at these sites. The export of nitrogen in heathlands and grasslands occurs mainly through exploitation, but in forests losses through leaching and denitrification are important components of the nitrogen balance of the ecosystem.

The nitrogen that accumulates within the ecosystem is partitioned between the plant and the soil subsystem. In most cases animals can be neglected as a nitrogen sink. The accumulation of organic nitrogen in the soil has been observed during succession in a wide variety of different ecosystems: after glacier recession in Alaska (Crocker and Major, 1955), in sand dunes (Olsen, 1958) and even in old-fields after abandonment (Tilman, 1988).

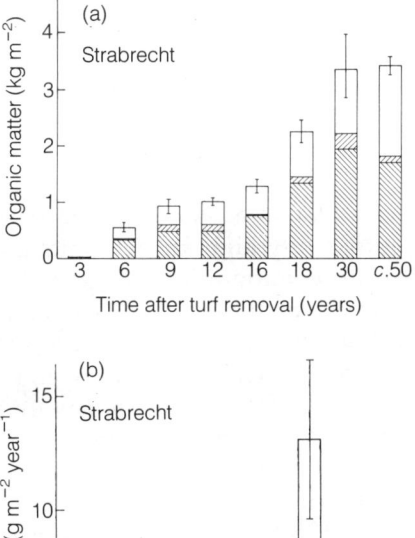

Figure 5.4 (a) The amounts of organic matter in the L layer (upper part of column), the FH layer (lower part of column) and the amount of living roots in the FH layer (middle part of column) in the plots where the vegetation and the humus and litter layer were removed in different years. Bars represent standard errors of the mean. (b) The annual mineralization of nitrogen. Bars represent standard deviations, which were estimated by the square root of the sum of the variances of the mineralization rates during the six periods. (After Berendse, 1990.)

The accumulation of organic soil nitrogen and its consequences for the nitrogen supply to plants through mineralization has been studied in detail in series of heathland plots for which the year in which turf-stripping had taken place was known. Both the amount of organic matter in the litter (L) and the humus (FH) layer and the annual nitrogen mineralization were measured (Berendse, 1990). The plots in the series at Strabrechtse heide (Figure 5.4) were completely dominated by *Calluna vulgaris*, except the ca. 50-year-old plot where *Calluna* had been replaced by the perennial grass *Molinia caerulea*. During secondary succession the amount of organic matter in the FH- and the L-layer gradually increased by the production of above-ground litter and the mortality of roots. After the first period of about 10 years the nitrogen mineralization rate increased with an increasing amount of soil organic matter. The mineralization rate increased six fold during the first 30 years, whereas during the later stages dominated by *Molinia* it even increased 12- to 13-fold (Figure 5.4). The increase in nitrogen release led to a fivefold increase in shoot production in this series of plots.

It appears that in many cases during the succession of both *Erica*- and *Calluna*-dominated heathlands a phase will be reached at which the nitrogen supply is sufficiently high for perennial grasses to be able to replace the dwarf-shrub communities (cf. next section). In the past, dwarf shrubs dominated Dutch heathlands for a long time because of the continuous export of nitrogen by turf stripping and burning. Moreover, nitrogen inputs were probably much lower, which might be expected to have been correlated with lower rates of humus accumulation. This caused mineralization rates to remain sufficiently low to prevent the rapid expansion of *Molinia* populations that is occurring nowadays. The increased input of nitrogen by atmospheric deposition probably affects the species composition of Dutch heathlands primarily through its positive effect on the rate of soil organic matter accumulation.

The amounts of organic matter and organic nitrogen in the soil have an important impact on the annual nitrogen mineralization, but also the chemical properties of the organic materials in the soil and the environmental conditions will influence strongly the rate of nitrogen mineralization per unit of soil nitrogen. Earlier it was shown that after invasion of wet *Erica tetralix*-dominated heathlands by *Molinia caerulea*, the nitrogen mineralization per unit soil nitrogen increased (Berendse, 1990). This positive effect was caused by the greater nitrogen and carbon losses from *Molinia* plants and by the higher decomposition rate of *Molinia* litter (Berendse *et al.*, 1989). Vice versa, after invasion of plant species which produce smaller quantities of litter, or litter that is decomposed at a low rate, nitrogen turn-over within the ecosystem could be expected to slow down.

5.4 EFFECTS ON THE COMPETITIVE BALANCE BETWEEN PLANT POPULATIONS AND THE DYNAMICS OF SPECIES COMPOSITION

In the preceding section the effects on the nitrogen balance and the nitrogen cycle in these systems were considered. This section shows how the increased availability of nitrogen affects the competitive balance and thereby the species composition of terrestrial ecosystems. These processes will be illustrated for heathlands, grasslands on calcareous soils, grasslands on heavy clay and forests.

Dutch wet and dry heathlands were formerly dominated by the evergreen dwarf shrubs *Erica tetralix* and *Calluna vulgaris* Hull, respectively. During the last few decades both species have largely been replaced by the grasses *Molinia caerulea* and *Deschampsia flexuosa* (L.) Trin, which has resulted in a dramatic decrease in the species diversity of these heathlands, because many (rare) plant species which were characteristic of the *Erica-* or *Calluna*-dominated heathlands have also disappeared. This involved, for example, *Drosera intermedia* L., *Gentiana pneumonanthe* L., *Narthecium ossifragum* (L.) Huds., *Cuscuta epithymum* (L.) L., *Scorzonera humilis* L., *Lycopodium clavatum* L. and *Lycopodium annotinum* L.

The effect of increased nitrogen availability on competition in Dutch heathlands was studied in a series of experiments in which nutrient availability was experimentally increased (Berendse and Aerts, 1984; Heil and Bruggink, 1987; Aerts and Berendse, 1988; Aerts *et al.*, 1990). One example of such a study (Aerts *et al.*, 1990) on the nutrient-poor dry heathland 'Deelense Veld' involved the competitive interactions between *Erica* and *Molinia* and between *Calluna* and *Molinia* and was carried out according to the replacement principle (De Wit, 1960). Nutrients (NPK) were supplied during three consecutive growing seasons. In 1984 the annual mineralization rate in unfertilized plots was around 5 kg N/ha/year. The range of nitrogen supply through fertilization corresponded to the range of nitrogen mineralization rates measured in Dutch heathlands during secondary succession (Figure 5.4b). The aboveground biomass in the plots was harvested at the end of the third growing season. We estimated the competitive ability of competing species by measuring the ratio between the per-plant yield of species x in mixture with species y and the per-plant yield of species x in monoculture (RY_{xy}). If $RY_{xy} = 1$, the composition of the mixture does not change. If RY_{xy} exceeds 1, species x gains on species y; if RY_{xy} is smaller than 1 the reverse is true. In the low nutrient treatments (≤ 100 kg N/ha/year) *Erica* gained on *Molinia* (Table 5.2), whereas in the highest nutrient treatment (200 kg N/ha/year) the situation was reversed. *Calluna* gained on *Molinia* in all nutrient treatments (Table 5.2). The results show that at a N-supply of more than 100 kg N/ha/year *Erica* is outcompeted by *Molinia*. This is in close agreement with earlier

Table 5.2 The competitive ability of *Erica tetralix* and *Calluna vulgaris* with respect to *Molinia caerulea* at different nitrogen supplies. The competitive ability is expressed by the relative yield (RY_{xy}; per-plant yield of species x in mixture with species y divided by per-plant yield of species x in mono- culture) of *Erica* with respect to *Molinia* (RY_{em}) and of *Calluna* with respect to *Molinia* (RY_{cm})

Nitrogen supply (kg N/ha/year)	0	50	100	200
RY_{em}	1.43*	1.52*	1.41*	0.65*
RY_{cm}	1.96*	1.91*	1.96*	1.88*

* indicates significant deviation from unity ($P<0.05$; $n=5$), (after Aerts *et al.*, 1990).

experiments (Sheikh, 1969; Berendse and Aerts, 1984; Aerts and Berendse, 1988). *Calluna*, however, is competitively superior to *Molinia* in this experiment even at a N-supply of 200 kg N/ha/year. This is not in agreement with the observed large-scale replacement of *Calluna* by *Molinia* in Dutch heathlands. It has been shown that this latter phenomenon is the combined result of increased levels of N-availability and disturbance events leading to opening of the *Calluna* canopy (e.g. frost damage; mowing; heather beetle attacks, cf. Brunsting and Heil, 1985). This mechanism differs from the one involved in the replacement of *Erica* by *Molinia*, because that process is dependent only on high levels of nutrient availability.

During the last decade an increase of *Brachypodium pinnatum* (L.) Beauv. has been observed in chalk grassland in The Netherlands. The increase of *Brachypodium* is negatively correlated with the number of plant species and is a serious threat to the high species diversity of this ecosystem. Especially, several short-lived species and perennial forbs with a low stature decline as a result of the increase of *Brachypodium*. In a fertilization experiment in the field, Bobbink *et al.*, (1988) demonstrated that nitrogen addition (70 kg N/ha/year as fertilizer) caused an increased productivity of *Brachypodium*, whereas the productivity of other grasses and forbs in these plots decreased (Figure 5.5). This resulted in a severe reduction of species diversity in these originally species-rich ecosystems. The vertical canopy structure of the competing species was found to be an important determinant of their competitive ability in this chalk grassland ecosystem. Mowing in August in chalk grasslands that were formerly mown in October led to a reduction of aboveground productivity and an increase in plant species diversity (Figure 5.6).

Elberse *et al.*, (1983) conducted a long term-study on the effects of mineral nutrient supply on the productivity and species diversity of hayfields and pastures on heavy river-clay soils. They found that NPK

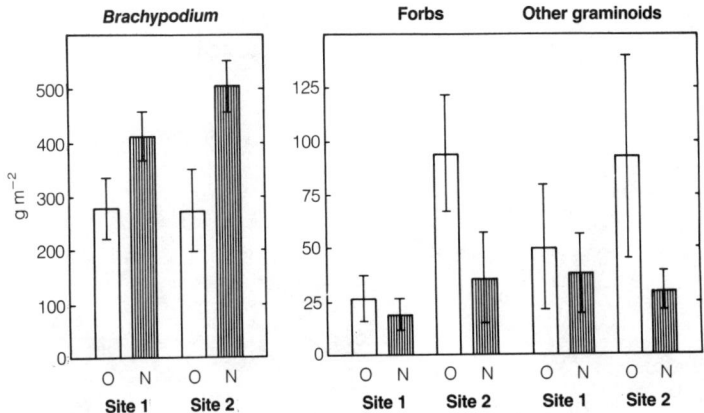

Figure 5.5 Phytomass of *Brachypodium pinnatum* (L.) Beauv., other graminoids and forbs (g dry weight/m^2) in August 1986 with (N) or without (O) nitrogen fertilization on two chalk grassland sites in South Limburg after three years of nitrogen fertilization (after Bobbink *et al.*, 1988). Bars represent standard deviations.

fertilization caused an increase in productivity and a decrease in species diversity in these grasslands. In general, the numerous low-productive species (grasses and forbs) were replaced by very few highly productive grasses with a tall, robust stature.

Most studies on the effects of atmospheric nitrogen deposition on forests are restricted to the effects on trees. However, there is increasing evidence that high loads of atmospheric nitrogen lead to considerable changes in the species composition of the herb layer in forests. The herb layer in dry *Pinus sylvestris* or *Quercus robur* forests often shows a shift towards increasing dominance of nitrophilous species, like *Dryopteris dilatata* (Hoffm.) A. Gray, *Corydalis claviculata* (L.) DC., *Galeopsis tetrahit* L. and *Rubus* spp. (Van Breemen and Van Dijk, 1988). Similar changes in the species composition of the herb layer of forests due to increased deposition of atmospheric nitrogen have been observed in Central Europe (Ellenberg, 1985) and in southern Sweden (Tyler, 1987; Falkengren-Grerup, 1989).

From these studies it can be concluded that an increased nitrogen supply leads to (a) an increase in primary production; (b) a decline in plant species diversity; and (c) increased competition for light, which favours species with a tall and robust stature.

5.5 CONCLUSIONS

Increased inputs of nitrogen through atmospheric deposition have led to many undesired shifts in the species composition of plant communi-

Figure 5.6 Photographs of plots in chalk grassland; (a) with usual management (cutting and hay removal in October); (b) after four years of experimental management (cutting and hay removal in August). Average aboveground productivity (±SE) was 350 (± 19) g/m^2/year in the plots with usual management and 235 (± 9) g/m^2/year in the plots that had received experimental management.

ties and to a strong decline of plant species diversity. The effects of the increased nitrogen inputs occur primarily through their accelerating effect on the rate of organic matter accumulation in the soil and hence on annual nitrogen mineralization. This process may lead to an increase in primary production and under certain conditions to soil acidification due to increased rates of ammonium oxidation.

Management of these ecosystems should be directed primarily towards manipulating their nutrient balance. First priority should be given to reduction of atmospheric nitrogen inputs, but in many cases this goal cannot be achieved within a short period. As long as this goal has not been achieved, preliminary measures could involve an increase in the output by harvesting part of the nitrogen stored in the ecosystem. Examples of management systems that lead to an increased export of nitrogen from the ecosystem are turf-stripping, burning and mowing including hay removal (cf. Table 5.1). Grazing may also lead to an increased nitrogen output by storage of nitrogen in animal tissues and ammonia volatilization from faeces and urine. The amounts of nitrogen lost through grazers, however, are in many cases quantitatively not significant and are always lower than the input through atmospheric deposition.

It is also possible to manipulate the structure of the vegetation in such a way that competition for light is reduced and the dominating species are deprived of their favourable canopy structure. These aims can be achieved by grazing or by mowing (Berendse, 1985; Bobbink and Willems, 1988; Bakker, 1989). Experimental evidence shows that in this way the original species diversity can be maintained or even restored (Bobbink and Willems, 1988, 1991, Bakker, 1989).

In old heathlands, which have high rates of nitrogen mineralization (Berendse, 1990), this type of management is insufficient (Bakker *et al.*, 1983). In these ecosystems it is necessary to remove the topsoil, together with the standing vegetation ('sod removal'). When a seed-bank of the original heathland species is still present, sod removal will lead to restoration of the original vegetation. However, the chances for successful heathland restoration depend on the nitrogen input through atmospheric deposition. At higher nitrogen inputs soil organic matter accumulation takes place more rapidly leading to a more rapid increase in nitrogen mineralization and *Calluna* or *Erica* being replaced by grasses after a shorter time period. A simulation model was developed for the nitrogen cycle in wet heathlands related to the competition between the two dominant plant species (*Erica* and *Molinia*) in these communities. This model was used to calculate the expected degree of dominance of *Erica* during a 50-year period after sod removal. It could be concluded from these preliminary calculations that nitrogen deposition rates should be reduced to at least c. 20 kg N/ha/year. The average nitrogen deposition in The Netherlands in 1986 was 41 kg N/ha/year, but in the eastern and southern parts of the country the deposition of ammonia is much higher than average values (Figure 5.1). It is clear that far-reaching measures should be taken to guarantee environmental conditions that will allow us to enjoy heathlands and many other plant communities characteristic for nutrient-poor environments in the near future.

REFERENCES

Aerts, R. and Berendse, F. (1988) The effect of increased nutrient availability on vegetation dynamics in wet heathlands. *Vegetatio*, **76**, 63–9.

Aerts, R., Berendse, F., De Caluwe, H. and Schmitz, M. (1990) Competition in heathland along an experimental gradient of nutrient availability. *Oikos*, **57**, 310–18.

Asman, W.A.H. (1987) *Atmospheric behaviour of ammonia and ammonium*. PhD thesis, Agricultural University of Wageningen.

Bakker, J.P. (1989) Nature management by grazing and cutting. On the ecological significance of grazing and cutting regimes to restore former species-rich grassland communities in the Netherlands. Kluwer Academic, Dordrecht.

Bakker, J.P., De Bie, S., Dallinga, J.H. *et al.*, (1983) Sheep-grazing as a management tool for heathland conservation and regeneration in The Netherlands. *Journal of Applied Ecology*, **20**, 541–60.

Berendse, F. (1985) The effect of grazing on the outcome of competition between plant populations with different nutrient requirements. *Oikos*, **44**, 35–9.

Berendse, F. (1990) Organic matter accumulation and nitrogen mineralization during secondary succession in heathland ecosystems. *Journal of Ecology*, **78**, 413–27.

Berendse, F. and Aerts, R. (1984) Competition between *Erica tetralix* L. and *Molinia caerulea* (L.) Moench as affected by the availability of nutrients. *Acta Oecologica/Oecologia Plantarum*, **5**, 3–14.

Berendse, F., Laurijsen, C. and Okkerman, P. (1988) The acidifying effect of ammonia volatilized from farm-manure on forest soils. *Ecological Bulletins*, **39**, 136–8.

Berendse, F., Bobbink, R. and Rouwenhorst, G. (1989) A comparative study on nutrient cycling in wet heathland ecosystems. II. Litter decomposition and nutrient mineralization. *Oecologia*, **78**, 338–48.

Bobbink, R. and Willems, J.H. (1988) Effects of management and nutrient availability on vegetation structure of chalk grassland, in *Diversity and Pattern in Plant Communities* (eds H.J. During, M.J.A. Werger and J.H. Willems), SPB Academic Publishing, The Hague, pp. 183–93.

Bobbink, R., and Willems, J.H. (1991) Impact of different cutting regimes on the performance of *Brachypodium pinnatum* in Dutch chalk grassland. *Biological Conservation*, **56**, 1–21.

Bobbink, R., Bik, L. and Willems, J.H. (1988) Effects of nitrogen fertilization on vegetation structure and dominance of *Brachypodium pinnatum* (L.) Beauv. in chalk grassland. *Acta Botanica Neerlandica*, **37**, 231–42.

Bobbink, R., Heil, G.W. and Raessen, M.B.A.G. (1990a) Atmospheric deposition and canopy exchange in heathland ecosystems. Report project 119 Dutch Priority Programme on Acidification, Department of Plant Ecology and Evolutionary Biology, University of Utrecht, 80 pp.

Bobbink, R., Heil, G.W. and Scheffers, M. (1990b) Atmospheric deposition of nitrogen in highway verges. Department of Plant Ecology and Evolutionary Biology, University of Utrecht, 64 pp. (in Dutch).

Bobbink, R., Heil, G.W. and Raessen, M.B.A.G. (1991) Atmospheric deposition and canopy exchange processes in heathland ecosystems. Environmental Pollution (in press).

Brunsting, A.M.H. and Heil, G.W. (1985) The role of nutrients in the interaction between a herbivorous beetle and some competing plant species in heathlands. *Oikos*, **44**, 23–6.

Buijsman, E., Maas, J.M. and Asman, W.A.H. (1987) Anthropogenic NH_3 emissions in Europe. *Atmospheric Environment*, **21**, 1009–22.

Crocker, R.L. and Major, J. (1955), Soil development in relation to vegetation and surface age at Glacier Bay, Alaska. *Journal of Ecology*, **43**, 427–48.

De Wit, C.T. (1960) On competition *Agricultural Research Report*, **66**, 1–82.

Elberse, W.Th, Van den Bergh, J.P. and Dirven, J.G.P. (1983) Effects of use and mineral supply on the botanical composition and yield of old grassland on heavy-clay soil. *Netherlands Journal of Agricultural Sciences*, **31**, 63–88.

Ellenberg, Jr, H. (1985) Veränderungen der Flora Mittteleuropas unter dem Einfluss von Düngung und Immisionen. *Schweizerische Zeitschrift fur Forstwissenschaft*, **136**, 19–39.

Erisman, J.W. (1991) Acid deposition in the Netherlands. National Institute of Public Health and Environmental Protection, Report 723001002, pp. 72.

Falkengren-Grerup, U. (1989) Soil acidification and its impact on ground vegetation. *Ambio*, **18**, 179–83.

Fowler, D. (1980) Removal of sulphur and nitrogen compounds from the atmosphere in rain by dry deposition. *Proceedings International Conference on the Ecological Impact of Acid Precipitation*, Norway 1980, SNSF Project. pp. 22–32.

Groffman, P.M. and Tiedjee, J.M. (1989) Denitrification in north temperate forest soils: spatial and temporal patterns at the landscape and seasonal scales. *Soil Biology and Biochemistry*, **21**, 613–20.

Heil, G.W. (1988) LAI of grasslands and their roughness length in *Vegetation Structure in Relation to Carbon and Nutrient Economy* (eds J.T.A. Verhoeven, G.W. Heil and M.J.A. Werger), SPB Academic Publishing, The Hague, pp.149–55.

Heil, G.W. and Van Dam, D. (1986) Vegetation structures and their roughness lengths with respect to atmospheric deposition. *Proceedings of the 7th World Clean Air Congress*, Sydney, Vol. 5, 16–21.

Heil, G.W. and Bruggink, M. (1987) Competition for nutrients between *Calluna vulgaris* (L.) Hull and *Molinia caerulea* (L.) Moench. *Oecologia*, **73**, 105–8.

Heil, G.W., Werger, M.J.A., De Mol, W., Van Dam, D. and Heijne B. (1988) Capture of atmospheric ammonium by grassland canopies. *Science*, **239**, 764–5.

Hultberg, H. (1985) Budgets of base cations, chloride, nitrogen and sulphur in the acid Lake Gårdsjon catchment. *Ecological Bulletin*, **37**, 133–57.

Ivens, W. (1990) Atmospheric deposition onto forests: an analysis of the deposition variability by means of throughfall measurements. PhD thesis, University of Utrecht.

Matzner, E. (1989) Acidic precipitation: Case Study Solling, in *Acidic Precipitation*, Vol. I: *Case Studies* (eds D.C. Adriano and M. Havas), Springer-Verlag, New York, pp.39–83.

Olsen,J.S. (1958) Rates of succession and soil changes on southern Lake Michigan sand dunes. *Botanical Gazette*, **119**, 125–70.

Rosswall, T. and Granhall, U. (1980) Nitrogen cycling in a subarctic ombrotrophic mire, in *Ecology of a subarctic mire* (ed. M. Sonesson) *Ecological Bulletin*, **30**, 209–34.

Sheikh, K.H. (1969) The effects of competition and nutrition on the interrelations of some wet-heath plants. *Journal of Ecology*, **57**, 87–99.

Tietema, A. and Verstraten, J.M. (1991) The nitrogen cycle of an oak–beech forest ecosystem in the Netherlands at increased nitrogen deposition: II the nitrogen and proton budget. *Biogeochemistry* (in press)

Tilman, D. (1988) *Plant Strategy and the Dynamics and Structure of Plant Communities*. Princeton University Press, Princeton, NJ.

Tyler, G. (1987) Probable effects of soil acidification and nitrogen deposition on the floristic composition of oak (*Quercus robur* L.) forest. *Flora*, **179**, 165–70.

Van Aalst, R.M. and Erisman, J.W. (1990) Atmospheric input. Thematic report on atmospheric input. Dutch Priority Programme on Acidification, 200-08.

Van Breemen, N., Burrough, P.A., Velthorst, E.J. *et al.* (1982) Soil acidification from atmospheric ammonium sulphate in forest canopy throughfall. *Nature*, **299**, 548–50.

Van Breemen, N. and Van Dijk, H.F.G. (1988) Ecosystem effects of atmospheric deposition of nitrogen in The Netherlands. *Environmental Pollution*, **54**, 249–74.

Van Breemen, N., Visser, W.F.J. and Pape, Th. (1988) Biogeochemistry of an oak-woodland ecosystem in the Netherlands affected by acid atmospheric deposition. *Agricultural Research Reports* 930. Pudoc, Wageningen.

Van Dam, D. (1990) *Atmospheric deposition and nutrient cycling in chalk grassland*. PhD thesis, University of Utrecht.

Vos, C.C. and Zonneveld, J.I.S. (1993) Patterns and processes in a landscape under stress: the study area, in: *Landscape Ecology of a Stressed Environment* (eds C.C. Vos and P. Opdam), Chapman & Hall, London, pp. 1–28.

Ditch banks as a conservation focus in intensively exploited peat farmland

6

Dick (Th.) C.P. Melman and Arco J. Van Strien

6.1 INTRODUCTION

Agricultural developments have led to a decrease in diversity of plant species in western Europe (Anonymus, 1985; Wolff-Straub, 1985a). This also holds for the peat district of the western Netherlands which is among the most intensively exploited areas of western Europe. Nowadays the peat grasslands, originating from the reclamation of wild fenlands, are used for dairy farming (Bijlsma, 1982). At present, the average livestock intensity is about two dairy cows per ha, nitrogen input (by application of fertilizers and animal manure) is 300–400 kg/ha/year and milk production is 12 000 kg/ha/year (Van Burg *et al.*, 1980; Clausman and Melman, 1991). The intensification has caused an impoverishment of the diversity of plant species on the dairy farms. Even plant species that until recently were common in the peat district are rapidly declining, such as *Trifolium pratense, Rumex acetosa* and *Caltha palustris* (De Boer, 1982; Clausman and Groen, 1987; Van Strien, 1991).

In The Netherlands an important strategy for conserving nature values in grassland areas is the establishment of nature reserves, which implies a segregation of agriculture and nature conservation (Baersel-

Landscape Ecology of a Stressed Environment
Edited by Claire C. Vos and Paul Opdam
Published in 1993 by Chapman and Hall, London. ISBN 0 412 44820 3

man and Vera, 1989). However, for financial and practical reasons nature reserves can never cover the whole area where nature-values are present. In The Netherlands about 700 000 ha of the grassland area used for agricultural purposes is marked as valuable with respect to nature conservation (Dijkstra, 1991). About 10 000 ha of this area has already been made into nature reserves and this area is increasing by about 1400 ha/year (Ministerie van Landbouw, Natuurbeheer en Visserij, 1990). The final aim is to arrive at a total area of 100 000 ha of grassland reserves (Ministerie van Landbouw, Natuurbeheer en Visserij, 1990). From this it may be concluded that the spatial effectivity of the strategy of segregation will be rather slow and limited. To overcome this, a strategy of integration of agriculture and nature conservation has been suggested, aimed at finding opportunities for wild plant and animal species to survive on intensive farms (Van der Weijden *et al.*, 1984; Ebel and Hentschel, 1987; Anonymus, 1989). This strategy of integration is partly covered by a governmental programme (the so-called 'Policy Document on Agriculture and Nature Conservation'; De Boer and Reyrink, 1987), in which farmers in selected regions have the opportunity to enter into so-called 'management agreements'. These agreements cover an area of 20 000 ha, increasing by 3000–4000 ha/year (Ministerie van Landbouw, Natuurbeheer en Visserij, 1992). The programme is targeted at a maximum of 100 000 ha. Besides management agreements there are several other initiatives to integrate agriculture and nature conservation, of which the voluntary meadow-bird protection programme is the most conspicuous (Clausman and Melman, 1991). Apart from the greater spatial reach, the strategy of integration may also have social benefits for the farmers, since less land is withdrawn from the agricultural sector.

Our studies were done to support the integration approach and were focused on the western peat district (Melman, 1991; Van Strien, 1991).

At the present high production levels of grasslands, a high plant species diversity cannot be obtained unless the nitrogen input is extremely reduced (Grime, 1979; Van Strien, 1991). This reduction is hardly feasible in current agricultural practice. Other prospects, however, may exist for endangered plant species. Field margins in particular offer opportunities for conserving many plant species as has been determined in certain arable regions in Germany (Ruthsatz and Haber, 1981; Wolff-Straub, 1985b). In the peat district the field margins include ditch banks, since the long and narrow lots are separated by shallow ditches which contain water throughout the year. Together, these banks offer an extensive potential refuge for plant species with a total length of thousands of km (ca 250 000), with a density of 400–800 m/ha. Many banks still support a species-rich vegetation on the 1 m-wide zone between the grassland and the ditch (Melman *et al.*, 1988).

The aim of this chapter is to assess whether it is possible to maintain or to restore species diversity on such narrow zones between intensively exploited grasslands on the one side and eutrophic ditches on the other. The spatial relations between banks and both bordering landscape elements form the basis of this assessment. The exploitation of grasslands may directly affect the banks, for example, by fertilizer application on the grassland or by cutting regimes that include bank vegetations. The management of ditches may also influence the banks, since waterplants and mud must be removed once a year, to prevent the eutrophic ditches becoming clogged, and the organic material is dumped on the banks. In addition, grassland management and ditch water might indirectly affect the vegetation of the banks by ground water flow.

To answer the questions related to the conservation of ditch bank vegetation (a) two effect studies were performed to assess the overall effects of management factors concerning grasslands and ditches, without studying their mechanisms thoroughly and (b) a model study of the nitrogen balance of the ditch bank was carried out to examine the effects of input of fertilizer, manure and sludge in more detail.

6.2 EFFECT STUDIES ON DITCH BANKS

6.2.1 METHODS

A *descriptive study* and an *experimental* study were performed to identify the factors determining the botanical composition of the ditch banks.

The *descriptive study* focused on about 320 ditch banks. The banks were selected so that it was possible to assess the individual effects of a number of interrelated factors; in addition only banks were selected that were known to have been under more or less the same management regime for the previous 5–10 years. Ditch bank vegetation sample plots of 50 m × 0.7 m were used; each plot running 0.7 m up the bank from the ditch water's edge. Data were collected on those environmental factors that might affect ditch bank vegetation: nitrogen supply on the grasslands, type of nitrogen application, type of grassland use, slope aspect, slope gradient, soil type, pH and P and K content of bank top soil, ditch water table, ditch cleaning method, ditch cleaning frequency and ditch dredging. Information about environmental factors was obtained from the farmers involved and from direct observations. Analysis of variance (ANOVA) was used to assess the effects of each factor, adjusted for the effects of the other factors (Van Strien, 1991). Effects of herbicides on ditch bank vegetation were not studied, since herbicides were applied on banks only locally and occasionally.

In the *experimental study* the effects of nutrient application, mowing, grazing and ditch cleaning were studied. In addition the importance of the physical shape of the bank was investigated. This field study covered eight locations, each with a number of experimental plots, representing a total ditch length of about 8 km, half of which was reprofiled. Four different profiles were compared: steep-profiled, terraced 1 m wide, terraced 3 m wide and undisturbed banks as a reference. Several management regimes were studied on different bank profiles. The management of the banks ranged from presumed nature-oriented (no fertilizer, late mowing, no cattle-trampling, no deposition of ditch sludge) up to presumed nature-unfriendly (the opposite conditions). The factors mentioned above were varied in such a way that their effects could be assessed separately. Apart from the ditch banks themselves, the fields at the experimental sites were managed according to standard modern agricultural practice with a nitrogen input of approx. 300–400 kg N/ha/year. Vegetation development was monitored for three seasons. Vegetation sample plots used were 25 m × ca 0.7 m; each plot ran up the bank from the ditch water's edge as far as the bank vegetation as such was recognized, which ranged from 0.5–1.2 m (up to 3 m on terraced 3 m banks).

Vegetation parameters

In both studies, the effects of management factors and bank shape were studied at both the sample plot and the species level. Two evaluation parameters were used to express significance for nature conservation on the sample plot level: (a) number of species, irrespective of the importance of the composing species, and (b) a nature-value index based on the rarity and decline of plant species which was developed by Clausman and Van Wijngaarden (1984). Data regarding the regional, national and worldwide rarity and the rate of decline on these spatial scale levels were combined into this index. Concerning the peat district the species index ranged from 10 for very common and increasing species, such as *Elymus repens,* to about 60 for a rare and decreasing species, such as *Cirsium dissectum.* The nature-value index of sample plots was calculated by summing the nature-values of all species present, taking into account their cover percentages. For ditch banks these sample indices ranged from 30 (species-poor, dominated by for instance *Glyceria maxima*) to about 60 (species-rich, including for instance *Carex panicea*) (see for more details also Van Strien, 1991).

At the species level, the analysis covered presence and abundance.

6.2.2 RESULTS

Below, we discuss some of the results of these studies, in particular the effects of nitrogen supply, ditch management and bank profile (Van

Strien *et al.*, 1989; Melman, 1991; Van Strien, 1991).

(a) Nitrogen supply

The negative relationship between species diversity in grassland vegetation and nitrogen supply has been assessed many times (e.g. Ennik,
1965; Heddle, 1967; Rorison, 1971; Grime 1979; Elberse *et al.*, 1983) and
is generally accepted. The *descriptive* study confirmed this for ditch
bank vegetation: a higher amount of nitrogen applied on the fields was
accompanied by a significantly lower number of species and a lower
nature-value on the banks (Figure 6.1). Species such as *Anthoxanthum
odoratum* and *Lychnis flos-cuculi* were found significantly more frequently
under conditions of low nitrogen supply; *Elymus repens* and *Phalaris
arundinacea* were two of the species favoured by a high nitrogen supply.

These results were confirmed in the experimental study. Three years
after the outset of the experiment, the vegetation of the plots kept free
from fertilizer and manure input differed from those not excluded from
the nitrogen supply regimes on the fields. On the N-fertilized plots the
cover of grasses increased, among which were *Agrostis stolonifera*, *Poa
trivialis* and *Phalaris arundinacea*; as a consequence the number of species decreased. Protection from fertilizer input favoured such species
as *Lychnis flos-cuculi*, *Carex disticha*, *Carex nigra*, *Anthoxanthum odoratum*,
Juncus effusus and *Holcus lanatus*.

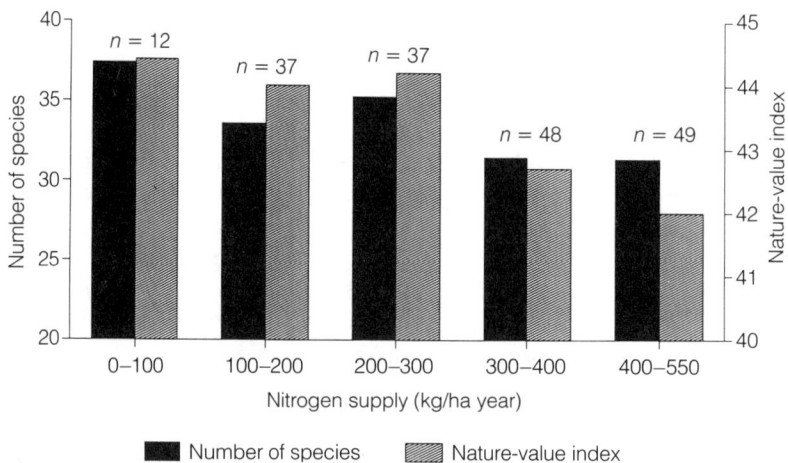

Figure 6.1 Mean number of species and mean nature-value index of banks
in relation to nitrogen supply on adjacent fields (with fertilizer and manure
input summed), corrected for other environmental factors by means of
ANOVA. ANOVA *F*-value = 15.2, *P* < 0.01 for species number; *F*-value = 14.8,
P < 0.01 for nature-value index.

(b) Ditch management

In order to ensure a proper discharge of water from the peat polders, regional water boards require regular ditch maintenance by farmers (Van der Linden, 1982). Almost all ditches are mechanically cleaned once a year, often in autumn. To prevent the ditches from becoming clogged, plants in and near the ditches are removed. Usually the organic matter is transferred to the banks, together with mud from the ditch bottom. Several aspects of ditch management were studied: (1) the deposition of ditch sludge, (2) the cleaning method, i.e. the machinery used, and (3) the cleaning frequency.

(i) *Deposition of sludge*

In the *experimental* study, deposition of sludge on the ditch bank after cleaning appeared to affect the bank vegetation. When the sludge was dumped on to the banks, the cover of *Ranunculus repens*, *Agrostis stolonifera*, *Cirsium arvense*, *Rorippa amphibia* and several other species increased, whereas the cover of *Juncus spp.*, *Ranunculus flammula* and *Lychnis flos-cuculi* decreased in comparison with untreated banks. On the whole, the vegetation of the treated banks indicated a higher trophic state and betrayed more disturbance. This resulted in a lower nature-value index, undoubtedly the result of the nutrients from the ditch sludge and the mechanical stress caused by the sludge.

(ii) *Cleaning method*

In the *descriptive* study, all current mechanical methods of ditch cleaning were compared. Cleaning by hand or by mowing-basket are often presumed to deposit less sludge on the banks and to affect the vegetation less severely than cleaning by ditch scoop or bottom auger. However, it was remarkable that, after correction for other factors, no differences in floristic parameters could be detected for the different cleaning methods used (ANOVA F-value = 0.2, $P > 0.05$ for species number; F-value = 1.0, $P > 0.05$ for nature-value index). The absence of a cleaning method effect might be due to the large variation in the way the machinery is used in practice. Both the amount of sludge as well as its composition may vary considerably within each method.

(iii) *Cleaning frequency*

The *descriptive* study showed that the frequency of ditch maintenance is important. Banks of ditches that were cleaned every year supported fewer plant species than banks that were cleaned once every two or three years (Figure 6.2). Also the nature-value index was lowest on the

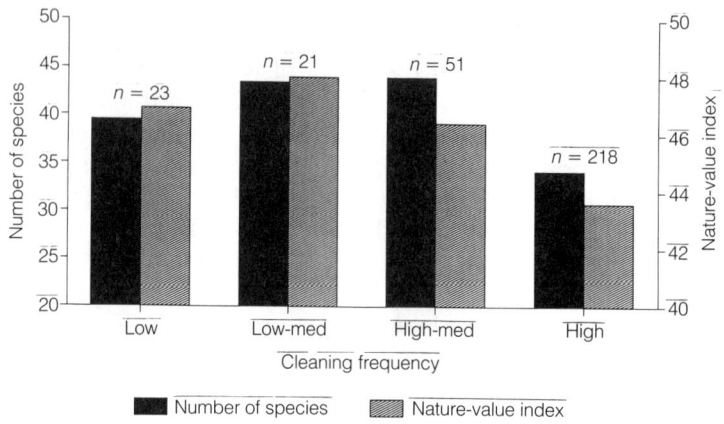

Figure 6.2 Mean number of species and mean nature-value index of banks in relation to cleaning frequency, corrected for other environmental factors by means of ANOVA. ANOVA F-value = 21.0, $P < 0.01$ for species number; F-value = 14.1, $P < 0.01$ for nature-value index. Low: cleaning less than once every 3 years. l-med: cleaning once every 3 years. h-med: cleaning once every 2 years. High: cleaning once a year.

banks of ditches cleaned annually.

Many species occurred more often and more abundantly along ditches that were cleaned less frequently than once a year, among which species of moderate eutrophic conditions such as *Caltha palustris* and *Lychnis flos-cuculi* and species which contribute to the terrestrialization of the ditches, such as *Acorus calamus*, *Cicuta virosa* and *Iris pseudacorus*. The number of species was also somewhat lower on banks along ditches that were cleaned less often than once every 2–3 years (Figure 6.2). This suggests that when succession is allowed to proceed undisturbed, species richness will eventually decline, possibly by the dominance of tall marsh plants (Van Strien, 1991).

(c) Bank profile

The influence of the bank profile was globally the same in all experimental sites (Melman, 1991) and is illustrated here with reference to the results obtained at two sites. At the first site three profile types were compared: steep profiled, terraced 1 m and undisturbed. At the second site were compared: terraced 1 m wide, terraced 3 m wide and undisturbed (Figure 6.3). Within the sites the management of the banks was identical.

On both sites the number of species on the 1 m and 3 m wide terraced banks was greater than on the steep profiled and undisturbed banks (Figure 6.4), partly due to the greater area appropriate for bank vegetation but mainly due to a greater species density. These differences

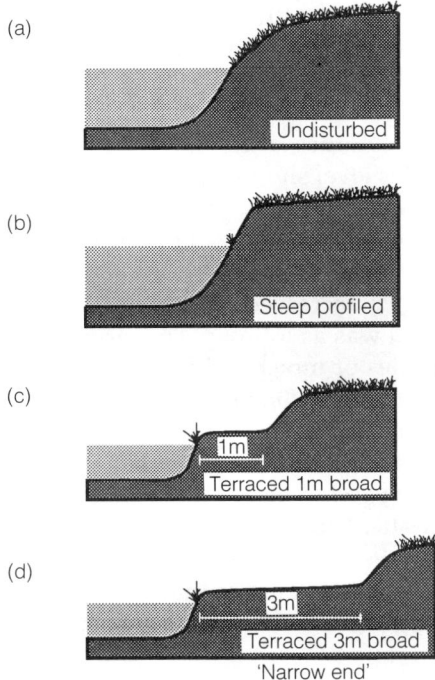

Figure 6.3 Bank profiles under study. (a) undisturbed; (b) steep profiled bank; (c) terraced 1 m wide; (d) terraced 3 m wide.

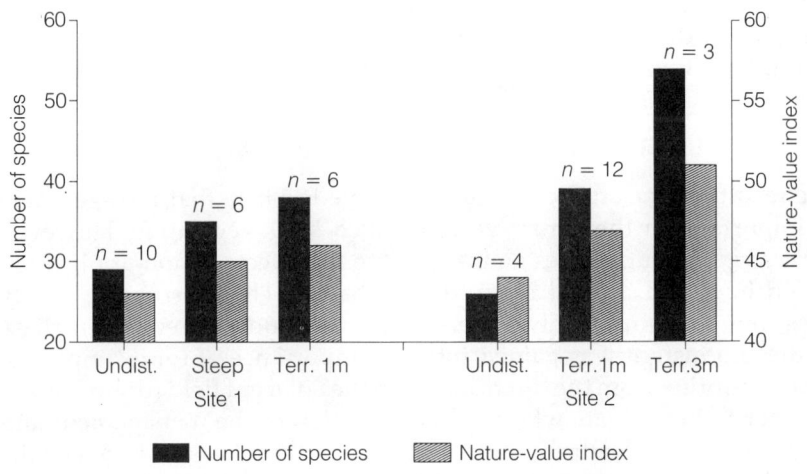

Figure 6.4 Mean number of species and mean nature-value index of bank vegetation in relation to profile type (ANOVA-tests, $P < 0.01$). Left: site 1 (undisturbed, steep, terraced 1m); right: site 2 (undisturbed, terraced 1m, terraced 3 m). The management of the sites was identical.

manifested themselves from the first season onwards and became more marked in the following two seasons: in 1988, on site 2 the average number of species at the narrow ends was 54 compared to 30 on the undisturbed banks (Figure 6.4). Measured against the nature-value index, a similar picture emerges, although somewhat less pronounced.

Analysis at the species level showed the process of succession on the reprofiled, denuded banks. In the first season after reprofiling, on the 1 m and 3 m wide terraced banks pioneer species such as *Bidens cernua*, *Polygonum hydropiper* and *Ranunculus sceleratus* occurred frequently; however, their abundance decreased rapidly in subsequent years. After three years the situation was as follows. The terraced bank profile (1m and 3m) favoured species of moist and wet environments, e.g. *Alisma plantago-aquatica, Berula erecta, Butomus umbellatus, Lychnis flos-cuculi, Peucedanum palustre* and *Triglochin palustris*. On the steep profiled banks, competitive species were favoured, e.g. *Agrostis stolonifera* and *Glyceria fluitans* on the lower part and *Cirsium arvense* on the upper part. On the undisturbed banks hay-field species like *Anthoxanthum odoratum, Rumex acetosa* and *Festuca rubra* were more abundant, reflecting drier conditions.

From these data it can be concluded that 1 m and 3 m wide terraced banks offer a promising perspective for nature conservation.

(d) Other factors

In addition to the factors mentioned above, several other factors proved to affect floristic values, for example, slope aspect, pH of the topsoil and type of grassland use. No effects of P and K of bank soil were detected, presumably due to the fact that the contents of P and K were rather high and did not limit the growth of bank vegetation (Van Strien, 1991).

6.3. MODEL STUDY OF THE NITROGEN BALANCE OF BANKS

The results of the studies discussed above indicate that nitrogen may be important for the nature-value of ditch-bank vegetation. However, the *descriptive* study gives no insight into the effects of non-application of fertilizer and animal manure near banks. The *experimental* study does, but this study only covers a short period. Moreover, in these studies no insight was gained into the importance of for example the nitrogen-influx from the ditch and from the adjacent field (drain water). In order to investigate whether manipulation of the management and shape of the banks substantially influence the uptake of nitrogen by the ditch-bank vegetation in the long term, a model simulation was performed of the nitrogen balance (Melman *et al.*, 1990). The manipulated factors of the banks concerned nitrogen input (fertilizer, manure and the deposition of the ditch sludge) and the physical shape (steep-sloped

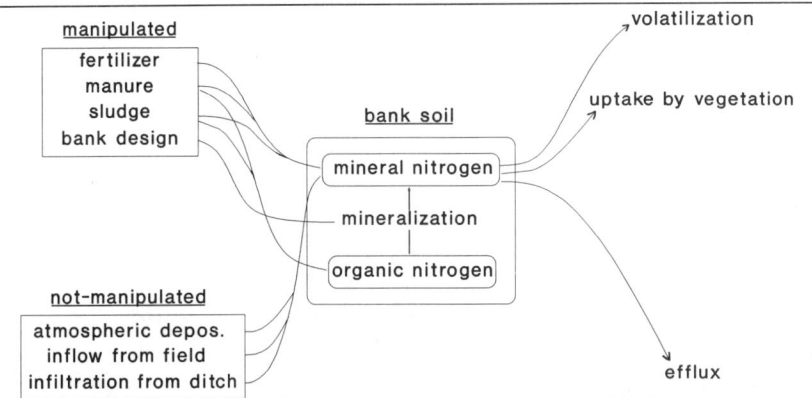

Figure 6.5 Nitrogen relations in ditch banks. Nitrogen input subdivided into manipulated and non-manipulated sources.

and terraced (1 m wide) (Figure 6.3)). The amount of nitrogen taken up by the ditch bank vegetation influenced by the manipulated factors is compared with that of other, non-manipulated sources, such as decomposition processes *of the peaty soil*, input via ditch-water, drain water, run-off and atmospheric deposition, together referred to as 'basic-supply' (Figure 6.5).

The model study used a grassland plot at Reeuwijk (near Gouda), situated on eutrophic lowland peat and exploited as an alternate pasture. The nitrogen use on the field was moderate by Dutch standards, ca. 350 kg/ha/year (160 kg in the form of fertilizer and 190 kg in the form of manure). It was assessed that in normal practice about half of this quantity was applied to the banks. Apart from this fertilizing regime two extra regimes were simulated: unfertilized and heavily fertilized (320 kg N/ha/year in the form of fertilizer and 380 kg N/ha/year in the form of manure). The deposition of the ditch sludge on the banks was about 13 ton/ha/year (fresh-weight).

The different import and export flows as represented in Figure 6.5 are linked with hydrological, biological and chemical processes (Harmsen and Schreven, 1955; Brady, 1984). For the simulation of these processes two models were used. The hydrological processes were simulated with WATBAL (a simple model for the WATer BALance in the saturated/unsaturated zone; Berghuijs-van Dijk, 1985). This model, in which the root-zone and the zone below are distinguished, calculates per point of time moistness, water-table and waterfluxes in the soil. It is a so-called one-dimensional, local model; a pseudo two-dimensional character is acquired by using the calculated fluxes which are leaving the system via drainage (Kemmers, 1986). The biological and chemical processes in which nitrogen is involved were simulated with ANIMO (Agricultural NItrogen MOdel; Berghuijs-van Dijk *et al.*, 1985). This dynamic, one-dimensional model simulates the turnover and transport

Table 6.1 Supply and uptake of mineral nitrogen (kg/ha/year) due to non-manipulated factors (= basic supply)

	Steep-profiled	Terraced 1 m
Supply		
Mineralization *of the soil*	129	65
Infiltration	8	11
Inflow	40	40
Atmospheric deposition	50	50
Total	227	166
Uptake		
Fraction of supply	0.48	0.45
Absolute	109	74

processes per point of time and per soil layer (Kemmers, 1986). The model accounts for soil-type, weather conditions, type-of-use (including its history), nutrient-input and hydrological situation (simulated by WATBAL). The results of the modelling were validated for the hydrological aspect by means of data of the groundwater level, and for the uptake of nitrogen by means of data of aerial plant production.

Assuming average weather conditions, the influence of the manage-

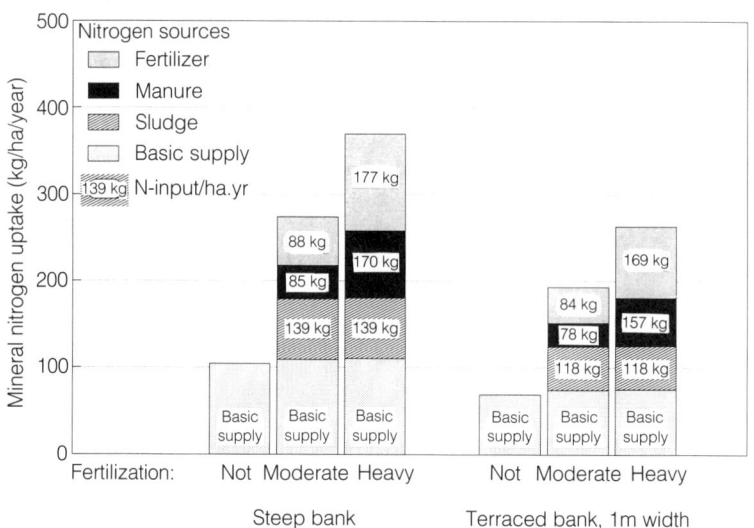

Figure 6.6 Mineral nitrogen uptake (kg N/ha/year) by the ditch bank vegetation, calculated for three fertilizing regimes: not, moderately and heavily fertilized. Left: steep profiled bank; right: terraced 1 m wide bank. The figures within the bars represent input quantities of nitrogen (kg N/ha/year), originating from fertilizer, manure and sludge. The basic supply originates from non-manipulated sources (see Figure 6.5).

ment regimes and profile type of the banks was computed for a period of 21 years. After this time-interval the model-simulation showed a stable situation, which is considered to be durable. This prediction is discussed below.

The non-manipulated, basic supply ranges from 170 to 220 kg N/ha/year (Table 6.1). About half this quantity (74–109 kg N/ha/year) is taken up by the vegetation, and is hardly affected by the fertilizing regime on the adjacent field (see Figure 6.6). The most important component of the basic supply is mineralization *of the soil* (65–129 kg/ha/year). The quantity of nitrogen reaching the banks via inflow (run-off and leaching) processes is found to be limited (approx. 40 kg/ha/year), one reason being that peaty meadows usually have a slightly concave shape, preventing the rainwater (with dissolved nutrients) from flowing over to the banks. Furthermore, nitrogen input from the ditch into the bank is low (ca 10 kg N/ha/year); the model-simulation shows that in peat there is hardly any water inflow from the ditch to the field.

The input via the manipulated nitrogen sources onto the steep banks amounts to ca 310 kg N/ha/year with a moderate fertilizing regime and up to 480 kg N/ha/year with a heavy fertilizing regime. This results in an uptake by the vegetation ranging from 160 kg N/ha/year with a moderate regime up to 260 kg N/ha/year with a heavy fertilizing regime (Figure 6.6). This is 50–55% of the total quantity applied to the banks.

From Figure 6.6 it can also be concluded that on steep banks a reduction of nitrogen uptake of 170–260 kg N/ha/year (depending on the fertilizing regime on the adjacent field) can be reached by avoiding direct input. For terraced banks the reduction amounts to 190 kg N/ha/year. This reduction might contribute significantly to preserving and developing species diversity and the nature-value index of the ditch bank vegetation (*cf.* Figure 6.1). Concerning the compatibility with prevailing agricultural practice, it should be emphasized that such nature-oriented ditch-bank management does not imply any reduction of fertilizer inputs to the fields themselves; it is only the margins that need to be spared.

Regarding the physical shape of the ditch bank the simulations show that on terraced banks the higher water table leads to a durable lower level of nitrogen supply to the plants than is the case for steep-banked ditches. This applies to both the basic (non-manipulated) and the manipulated supply (Table 6.1, Figure 6.6); the moist conditions prevailing on the terraced banks slow down the mineralization of the peaty soil and promote denitrification of the nitrogen entering the habitat (Harmsen and Schreven, 1955; Brady, 1984). This results in an approx. 33% lowering of the nitrogen uptake by the vegetation, corresponding to ca 35 and 120 kg N/ha/year in non-fertilized and heavily fertilized systems, respectively, which is a substantial quantity (Figure 6.6).

These findings indicate that, for identical management regimes in adjacent fields, terraced ditch banks may durably support vegetation of less eutrophic conditions (and thus more valued) than steep-banked ditches. Although the study focused on 1 m wide terraces, it is clear that the same holds for 3 m wide terraces.

The results of the model study agree with those of the empirical studies. Notwithstanding this it should be pointed out that model-simulation as such has no definite conclusive value. It is based on many suppositions, for the larger part originating from literature and some from the authors' own judgements (e.g. the ecophysiological properties of the ditch bank vegetation, Melman *et al.*, 1990). Additional research is needed to verify the results of the simulation.

6.4 COMPATIBILITY OF DITCH-BANK MANAGEMENT WITH CURRENT DAIRY FARMING PRACTICE

Conservation measures that are technically and financially compatible with current dairy farming practice have the greatest prospects. Therefore, we considered the fitting in of some measures in agricultural practice.

6.4.1 NITROGEN SUPPLY

From an agricultural point of view there is no great advantage in the application of fertilizer and manure on ditch banks, since they have little agricultural value. Therefore, economic and conservational inter-

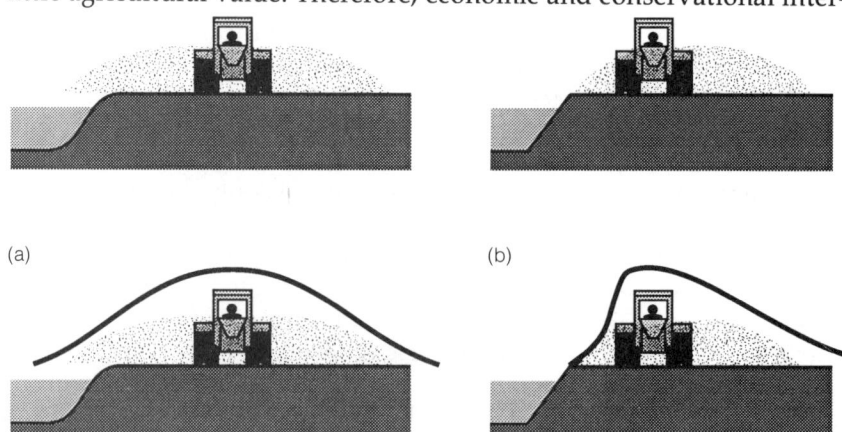

Figure 6.7 Distribution pattern of (a) commonly used fertilizer distributor and (b) fertilizer distributor with deflector provision for field margins.

ests run parallel for ditch banks: losses to the banks should thus be kept to a minimum.

It appears that the design of current fertilizer distributors is by no means optimal when it comes to operation along field margins (Melman and Van der Linden, 1988). Many distributors are designed to ensure an even spread over the field itself. However, along the margins this is counterproductive: fertilizer is spoiled in ditches and ditch-banks (Figure 6.7a).

This situation can be improved in either of two ways. The simplest approach is to keep a 1–2 m extra distance from ditch banks (compared to the direction of use), which is financially more or less neutral. A second approach is to use a special distributor designed for use along field margins, with which a sharp cut-off is achieved (Figure 6.7b). This may even lead to a modest financial advantage. Improvement in the design of the distributors to achieve this, deserves more attention.

6.4.2 DITCH CLEANING MANAGEMENT

It is shown that species diversity on ditch banks cannot simply be promoted by using a particular type of machine for ditch cleaning; it is also necessary that cleaning be carried out in an appropriate manner. To achieve this, minimalization is needed of (1) damage of the bank profile and sod and (2) eutrophication of the bank caused by sludge. Financially, this working method is likely to score neutral.

Furthermore, it may be financially attractive for farmers to reduce the frequency of cleaning from once a year to once every two or three years. This is only feasible if the water discharge is not obstructed too much. Additional research has shown that no serious problems are to be expected on this point (Twisk *et al.*, 1991).

6.4.3 BANK PROFILE

At first sight, terracing of ditch banks might not appear to be an attractive option, since it means a reduction in the area available for exploitation. Also, agricultural activities will be hindered, especially if the banks lengthwise to the field are terraced. This is offset by a number of benefits. First, there is, compared to the nature-oriented management of normal banks, considerable 'extra' advantage to nature-values, without the need to adapt fertilizer and manure regimes on adjacent fields. Second, the area lost to terracing is only small, implying only minor consequences for fodder production. Moreover, if only the banks on the narrow ends of the fields are terraced, the hindrance to agricultural activities is reduced to a minimum (Figure 6.8).

Using the narrow ends, 3 m wide terraces are preferred, assuming them to provide a more firm ecological base. These narrow ends may benefit the overall wetland infrastructure, such as the functioning as a species reservoir for the ditch-bank vegetation of the surrounding farmland (Figure 6.8).

In practice, construction of terraced banks might be implemented as a compensatory measure for further agricultural intensification, leav-

Figure 6.8 (a) and (b) Positioning of 3 m wide terraced ditch banks at the narrow ends of the fields minimizes the agricultural hindrance and yields a promising ecological perspective.

ing less opportunity for nature conservation on the fields. This is the case when, for example, the water table is lowered.

6.5 GENERAL DISCUSSION AND CONCLUDING REMARKS

To assess the prospects of species-rich banks, we should know which factors affect the bank vegetation, the ways in which they do so and whether or not the negative influences can be minimized. These aspects are briefly discussed here in relation to several of the factors related to nitrogen supply; other factors, which are discussed elsewhere (Melman, 1991; Van Strien, 1991), do not affect the main conclusions. First, we shall consider the factors associated with ditch and ditch management, and then the effects of grassland and bank management.

Because the ditch water hardly reaches the root zone of the bank vegetation, the input of nutrients from the ditch water into the bank is only small. This implies that even eutrophic ditches will hardly affect the bank vegetation.

Far more important are the nutrients from the ditch sludge brought on to the banks by the ditch-cleaning practice. Deposition of the ditch sludge causes eutrophication and suffocation of the vegetation, resulting in a lowering of the nature-value of the bank vegetation. The effects do not depend so much on the type of machinery used, but rather on the way it is applied. In addition, the cleaning frequency is important: once every two or three years is more beneficial to the bank vegetation than once a year.

Although ditch cleaning affects the bank vegetation considerably, these effects can be lowered by a more careful execution of the cleaning performance and by reducing its frequency to once every 2 or 3 years. Both actions seemed to be suitable in agricultural practice.

The amounts of nutrients that reach the bank indirectly from the grassland (run-off, groundwater flow) are limited. A heavier fertilizer regime hardly raises the indirect input on the bank zones. Actually, most of the nutrients from the grassland side reach the banks directly by fertilizing and manuring. These inputs affect the nature-value of the ditch bank vegetation negatively.

Nitrogen input of the banks can be lowered by excluding them while applying fertilizer and manure to the adjacent fields. With respect to fertilizer this can be financially neutral or even slightly positive to the farmers. The availability of nitrogen can be further reduced by changing the shape of the banks (terracing). In addition, terracing favours species of moist conditions.

Thus, the nitrogen inputs from non-manipulated, indirect sources are relatively small; the other sources are important, but can be reduced without severe problems for agricultural practice. Despite the fact that there is much nitrogen in the banks' vicinity, the amount of nitrogen that actually reaches the root zones can be kept relatively low. This suggests that it is feasible to maintain or restore species diversity in the narrow zones between eutrophic ditches and heavily fertilized grasslands.

Although the results reflect the situation on peat grassland, some features (ditch cleaning, bank profile) may also hold for other soil-types. Sparing the banks from fertilizer may also be effective, but it has to be assessed per soil type to what extent the nutrient household of the bank relates with the bordering field and ditch. Apart from this question of generalization it may be stated that there are possibilities to combine on a small spatial scale intensive grassland exploitation with succesful nature-oriented management of the bank vegetation. If ditch banks are managed with greater care by the farming community, they can become an important and extensive part of the ecological infrastructure in the midst of an intensively used agricultural area.

REFERENCES

Anonymus (1985), Auswirkungen extra: moderner Landbewirtschaftung auf die Schutzgüter der Umweltpolitik, in Rat von Sachverständigen für Umweltfragen. *Umweltprobleme der Landwirtschaft*, Verlag W. Kohlhammer GMBH Stuttgart, Mainz. pp. 161–79.

Anonymus (1989) *Agricultural and Environmental Policies. Opportunities for Integration.* Organisation for Economic Co-operation and Development (OECD) publications 2, Paris, France.

Baerselman, F. and Vera F.W.M. (1989) *Natuurontwikkeling; een verkennende studie.* Achtergrondreeks Natuurbeleidsplan 6, Ministerie van Landbouw, Natuurbeheer en Visserij, SDU, Den Haag.

Berghuijs - Van Dijk, J.T. (1985) *WATBAL, a Simple Water Balance Model for a Unsaturated/Saturated Soil Profile.* Nota 1670. ICW, Wageningen.

Berghuijs - Van Dijk, J.T., Rijtema, P.E. and Roest, C.W.J. (1985) *ANIMO, Agricultural Nitrogen Model.* Nota 1671. ICW, Wageningen.

Bijlsma, S. (1982) Geology of the Holocene in the Western Part of The Netherlands, in *Proceedings of a Symposium on Peat lands Below Sea Level* (eds H. de Bakker and M.W. van den Bergh), ILRI, Wageningen, pp. 11–30.

Brady, N.C. (1984) *The Nature and Properties of Soils.* Collier MacMillan, London.

Clausman, P.H.M.A. and Groen, C.L.G (1987) *Veranderingen in het vegetatiedek van de Alblasserwaard en de Vijfheerenlanden tussen 1977 en 1984.* Provincie Zuid-Holland, Dienst Ruimte en Groen. The Hague.

Clausman, P.H.M.A. and Melman, Th.C.P. (1991) Instruments for combining intensive dairy farming and nature conservation in The Netherlands. *Landscape and Urban Planning*, **20**, 205–10.

Clausman, P.H.M.A. and Van Wijngaarden, W. (1984) *Verspreiding en ecologie van wilde planten in Zuid-Holland.* Deel A. Waarderingsparameters. Provincie Zuid-Holland, Dienst Provinciale Planning, Den Haag.

De Boer, Th.A. (1982) The use of peat soils for grassland, in *Proceedings of the Symposium on Peat Lands Below Sea Level* (eds H. de Bakker and M.W. van den Bergh), ILRI, Wageningen, pp. 214–21.

De Boer, T.F. and Reyrink, L.A.F. (1987). The Netherlands, II: Policy, in *Environmental Management in Agriculture; European Perspectives* (ed J.R. Park), Proc. 14–17 July, Bristol UK, Belhaven, London.

Dijkstra, H. (1991). *Natuur- en landschapsbeheer door landbouwbedrijven.* Eindverslag COAL-onderzoek. COAL-publ. nr.60. Ministerie van

Landbouw, Natuurbeheer en Visserij, Den Haag.

Ebel, F. and Hentschel, A. (1987) Neue Wege des Naturschutzes in Nordrhein-Westfalen im Vergleich mit Naturschutzprogrammen anderer Bundesländer. *Berichte über Landwirtschaft*, **65**, 412–34.

Elberse, W.Th., Van den Bergh, J.P. and Dirven, J.G.P. (1983) Effects of use and mineral supply on the botanical composition on heavy-clay soil. *Netherlands Journal of Agricultural Science* **31**, 63–88.

Ennik, G.C. (1965) The influence of management and nitrogen application on the botanical composition of grassland. *Netherlands Journal of Agricultural Science*, **13**, 222–37.

Grime, J.P. (1979) *Plant Strategies and Vegetation Processes*. Wiley, London.

Harmsen, G.W. and Van Schreven, D.A. (1955) Mineralization of organic nitrogen in soils. *Advances in Agronomy*, **7**, 299–398.

Heddle, R.G. (1967) Long-term effects of fertilizers on herbage production. I. Yields and botanical composition. *Journal of Agricultural Science, Cambridge*, **69**, 425–31.

Kemmers, R.H. (1986) Perspectives in modelling of processes in the rootzone of spontaneous vegetation at wet and damp sites in relation to regional water management, in *Water Management in Relation to Nature, Forestry and Landscape Management*. Verslagen en Mededelingen Nr. 34 (ed. J.C. Hooghart), CHO, The Hague, pp. 91–116.

Melman, Th.C.P. (1991) *Slootkanten in het veenweidegebied; mogelijkheden voor behoud en ontwikkeling van natuur in agrarisch grasland*. Thesis, Centre for Environmental Studies, Leiden University.

Melman, Th.C.P. and Van der Linden, J. (1988) Kunstmeststrooien en natuurgericht slootkantbeheer. Over de betekenis van het opnemen van voorwaarden over perceelsrandbemesting in beheersovereenkomsten als praktische en natuurgerichte maatregel. *Landinrichting*, **28**, 37–43.

Melman, Th.C.P., Clausman, P.H.M.A. and Van Strien, A.J. (1988) Ditch banks in the western Netherlands as connectivity structure, in *Connectivity in Landscape Ecology* (ed K. Schreiber), Münstersche Geographische Arbeiten, **29**, 157–61

Melman, Th.C.P, Van Oers L.C.F.M. and Kemmers R.H. (1990) De stikstofbalans van slootkanten. *Landschap*, **7**, 183–201.

Ministerie van Landbouw, Natuurbeheer en Visserij (1989) *Jaarverslag 1989 Bureau Beheer Landbouwgronden*, Utrecht.

Ministerie van Landbouw, Natuurbeheer en Visserij (1990) *Natuurbeleidsplan*, SDU, Den Haag.

Ministerie van Landbouw, Natuurbeheer en Visserij (1992) *Evaluatieverslag beheersregelingen* Directie Beheer Landbouwgronden, Utrecht.

Rorison, I.H. (1971) The use of nutrients in the control of the floristic composition of grassland in *The Scientific Management of Animal and Plant Communities for Conservation* (eds E. Duffey and A.S. Watt), Blackwell, Oxford, pp. 65–77.

Ruthstaz, B. and Haber, W. (1981) The significance of small-scale landscape elements in rural areas as refuges for endangered plant species, in *Perspectives in Landscape Ecology* (eds S.P. Tjallingii and A.A. de Veer), Pudoc, Wageningen, pp. 117–24.

Twisk, W., Van Brussel, N.A. and Ter Keurs, W.J. (1991) *Minder vaak slootschonen: beter voor boer en natuur?* Rapport Milieubiologie, Rijksuniversiteit Leiden.

Van Burg, P.F.J., 't Hart, M.L. and Thomas, H. (1980) Nitrogen and Grassland. Past and present situation in the Netherlands, in *The Role of Nitrogen in*

Intensive Grassland Production (eds W.H. Prins and G.H. Arnolds), Pudoc, Wageningen, pp. 15–33.

Van der Linden, H. (1982) History of the reclamation of the Western Fenlands and of the organizations to keep them drained, in *Proceedings Symposium on Peat lands Below Sea Level* (eds. H. de Bakker and M.W. van den Bergh), ILRI, Wageningen, pp. 42–73.

Van der Weijden, W.J., Van der Wal, H., De Graaf, H.J. *et al.* (1984) *Towards an Integrated Agriculture*. Netherlands Scientific Council for Government Policy. W4. Staatsuitgeverij, The Hague.

Van Strien, A.J., Van der Linden, J., Melman, Th.C.P. and Noordervliet, M.A.W. (1989) Factors affecting the vegetation of ditch banks in peat areas in the Western Netherlands. *Journal of Applied Ecology*, **26**, 989–1004.

Van Strien, A.J. (1991) *Maintenance of plant species diversity on dairy farms*. Thesis, Environmental Biology, Leiden University.

Wolff-Straub, R. (1985a) Gefährdung und Schutz der Flora in der Bundesrepublik Deutschland, insbesondere in Nordrhein-Westfalen. *Publikatie Natuurhistorisch Genootschap Limburg*, **35**, 3/4, 50–5.

Wolff-Straub, R. (1985b) *Schutzprogram für Ackerwildkräuter*. Schriftenreihe des Ministers für Umwelt, Raumordnung und Landwirtschaft des Landes Nordrhein-Westfalen.

Part Three

Spatial Relations by Moving Organisms

It is the landscape pattern that determines flows of organisms through the landscape rather than that flows influence the spatial pattern of land units. In agricultural and other artificial landscape types, the spatial distribution of habitat is primarily determined by man. Chapters 7 and 8 discuss how size, shape, configuration and connectedness regulate the population processes, i.e. the persistence of populations, at the landscape level.

MacArthur and Wilson's concept of the dynamic equilibrium of species diversity of island communities was influential in the development of strategies for nature conservation. Although the application of the theory had been criticized vigorously, the general notion of the role of area and isolation was assimilated in recent conservation and landscape plans. For example, in the recently published Dutch nature conservation policy plan, the national ecological network was introduced as an answer to the adverse effects of landscape fragmentation on biodiversity.

However, this general notion of the importance of spatial relations in populations and communities is not sufficient to specify spatial problems at the species level, to differentiate among spatial scenarios or to predict the effects of landscape changes like the construction of a highway. Species differ in their response to landscape details, in the scale at which population processes take place and in their innate strategies for survival. This is why metapopulation theory was introduced as a theory which could account for population responses to landscape changes. Earlier, pioneering work by Den Boer and co-workers had stressed the role of heterogeneity in population dynamics, but their work was on a fine level of scale and not spatially explicit.

This section reviews the empirical and the modelling work initiated to elaborate the theory on metapopulations and find support for it in real world landscapes. Studies on the distribution pattern across landscapes were used to detect spatial effects of fragmentation and to describe the relations between distribution and landscape structure in statistical models. As the theory advanced, process studies were also initiated, mainly to measure metapopulation processes, but also to study the movements of individuals through the landscape. The latter type of studies are essential to produce data on dispersal distances under varying landscape conditions, but the level of detail used in these studies complicates generalizations to the level of populations across landscapes.

The studies reviewed by Opdam *et al.* (Chapter 7) mainly in agricultural landscapes, revealed that some fragmented populations behaved like metapopulations, whereas others did not. The latter were assumed to have either a high dispersal rate across the landscape or a very small dispersal flow. The role of large areas of habitat as population nuclei stabilizing locally the dynamics of metapopulations was demonstrated. In spite of their flight capacities, forest birds were clearly shown to be affected by the spatial structure of the landscape.

Empirical studies are restricted in the choice of species, variation in landscape structure and period of time. Hence, in applied studies, results of empirical studies often have to be extrapolated to other spatial scales, to other landscape patterns and other species, and also prediction of the effects of landscape changes is often required. Here, mechanistic models of fragmented populations are indispensible. They are based on the processes measured, but predictions are made in terms of persistence as a function of spatial landscape parameters. Hence, mechanistic models link spatial processes to pattern. Also, they can be used as tools in impact assessment and comparing landscape scenario's (Verboom *et al*. Chapter 8), especially when linked to geographical data bases.

At present, under the influence of questions by nature conservationists and landscape planners, Dutch research on fragmented populations is extending to other types of habitat, particularly heathland and marshland ecosystems. Also, the interest in the genetical implications of landscape fragmentation is rapidly growing, although the emphasis is still largely on studies in completely isolated small populations. Little attention is paid to possible effects of landscape fragmentation on species interactions, like predator–prey relations or relations between competitors in a community of an isolated patch, one of which is temporarily absent. Both aspects of the ecology of fragmented landscape are neglected here.

Population responses to landscape fragmentation

7

Paul Opdam, Rob van Apeldoorn, Alex Schotman and Jan Kalkhoven

7.1 FRAGMENTATION IN THE DUTCH LANDSCAPE

Fragmentation of habitat can be defined as a process: the destruction of habitat leaving the remaining fragments scattered throughout the newly created landscape. For species restricted to the original type of habitat, fragmentation means a disintegration into small, spatially disjunct patches, separated by land which is unsuitable to reproduce or find food or shelter. Fragmentation may also be perceived as a pattern, the result of this process of disintegration. It may then be described functionally as a spatially distributed set of habitat patches, characterized by patch area and shape, by patch configuration and by the resistance of the intermediate land to movements of individuals of a particular species. Obviously the perception of this pattern, and hence the response to the fragmentation process of a population, will vary widely among species.

The Dutch landscape had already been changed by man long before the Middle Ages. In describing the fragmentation of natural and semi-natural habitat, we will focus on the last 100 years or so, assuming this period to be most relevant to the actual distribution of species. The following trends are most conspicuous:

Landscape Ecology of a Stressed Environment
Edited by Claire C. Vos and Paul Opdam
Published in 1993 by Chapman and Hall, London. ISBN 0 412 44820 3

1. An increase of the area of timber woodland (Figure 7.1), mainly due to afforestation of heathland and peatmoors;
2. A decrease in the density of small woody elements in the agricultural landscape, due to removal of hedgerows and woodlots in the process of enlarging the size of parcels (Table 7.1) (Vos and Zonneveld, 1993, this volume, Figure 1.5);

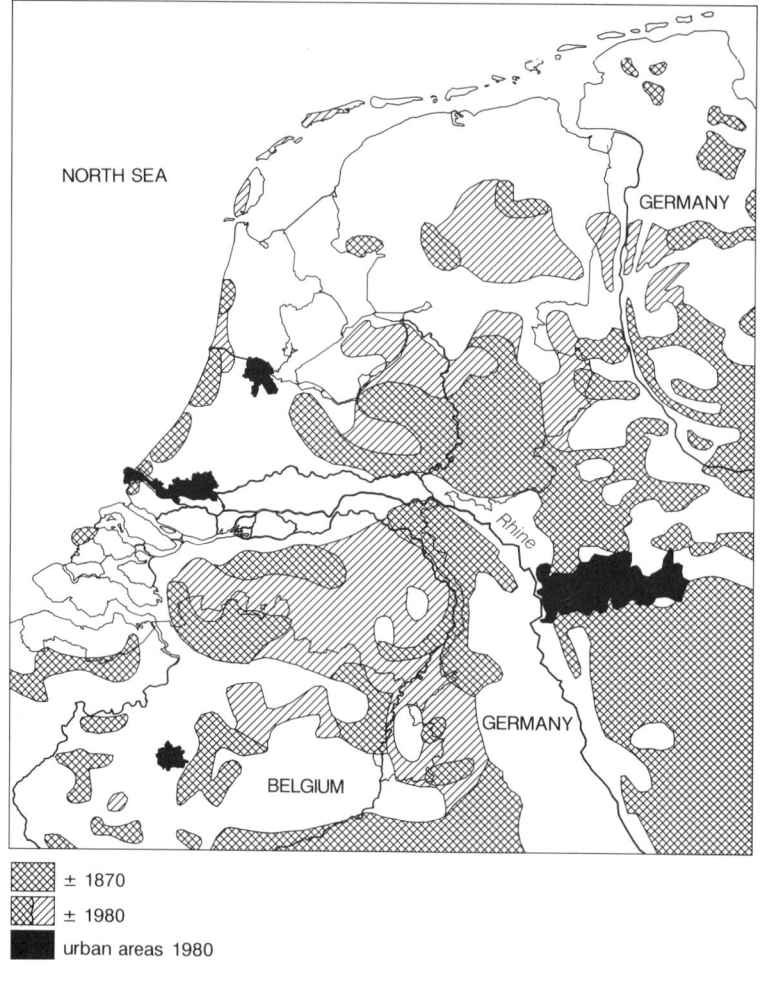

Figure 7.1 Wooded landscape around 1870 and 1980 in The Netherlands and surroundings. Areas with more than 2 km^2 forest per 50 km^2 are roughly indicated, based on old and recent topographical maps. The woods in the middle and the south-east were connected with old wooded landscapes in Germany and Belgium, already more then a century ago. New woods are mostly coniferous plantations on former heathland.

3. A decrease in the density of small landscape elements in the agricultural landscape, like cattle ponds, marshes and seminatural grassland, down to 10% or less (Weinreich and Musters, 1989);
4. A dramatic decrease in the area of peatmoors and heathland due to reclamation and afforestation (Vos and Zonneveld, 1993, this volume, Figure 1.3). Between 1800 and 1980, the area of heathland declined from 800 000 ha to 42 000 ha (Weinreich and Musters, 1989);
5. a decrease in the density of small marshland elements between large freshwater marshes (Weinreich and Musters, 1989).

Thus, fragmentation occurred in most non-agricultural habitat types and on both local and regional scales. The exception is forest on the regional scale: there are more larger tracts of forest than a century ago, but in between these tracts the network of small woody elements has disintegrated.

Intuitively, one expects such a landscape change to have affected spatial relations between landscape units. This may have resulted in a decrease of species frequencies or abundance, or even in (local) extinction. This chapter, first develops a theoretical basis to assess the effects of changing the spatial distribution of habitat. Second, we will review our studies in the agricultural landscape with woodlots and hedgerows, and then discuss possible generalizations and extrapolations to other habitat types and spatial scales.

This chapter is restricted to the effects of the configuration and size of (seminatural) habitat patches, habitat quality and species characteristics on the functioning of spatially structured populations, and neglects other implications of fragmentation, such as boundary effects (Wiens, 1990). Furthermore, we disregard implications for the community level (Bengtsson, 1991; Taylor, 1991) as well as for the genetics of fragmented populations (Gilpin, 1991; Olivieri *et al.*, 1990).

Table 7.1 Linear woody vegetation along field parcels in 1900 and 1980. Mean amount of woody elements as a percentage of the total length of parcel edges per 5 × 5 km^2 (Barends, 1987)

Landscape type	Percentage in different years	
	1900	*1980*
Sandy soil landscape	25.6	13.6
Fluvial clay landscape	26.8	10.5
Lowland peat landscape	6.3	3.1
Marine clay landscape	8.4	4.9
Coastal dune landscape	12.3	7.0

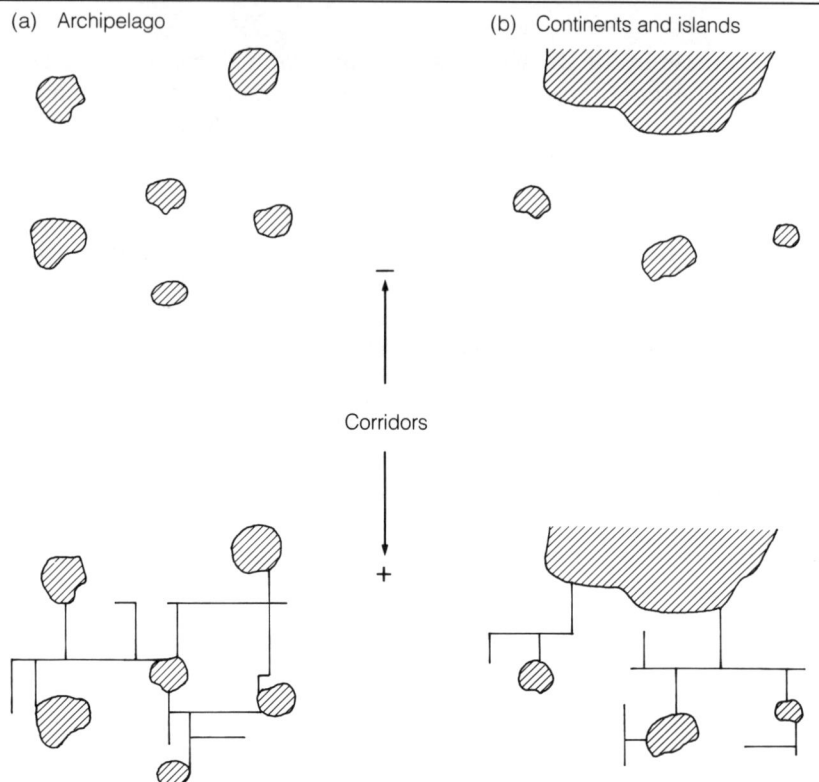

(a)   Archipelago

(b)   Continents and islands

Corridors

Figure 7.2 Four basic types of fragmented pattern: the archipelago landscape (a) and the continent–islands landscape (b), with and without connecting linear elements. These linear elements may also function as marginal habitat patches.

7.2 ECOLOGY OF SPATIALLY STRUCTURED POPULATIONS

7.2.1 THEORETICAL MODELS

Ecological theory has produced two spatial population concepts, the metapopulation concept (Levins, 1970; Boorman and Levitt, 1973) and the source–sink concept (Wiens and Rotenberry, 1981; Pulliam, 1988). Both can be regarded as a follow-up at the population level of the MacArthur and Wilson (1967) dynamic biogeography theory, although Andrewartha and Birch (1954) had already stressed the occurrence of dynamic turnover of local populations. Originally, Levins' theoretical model referred to a system composed of an infinite number of identical patches with asynchronous local dynamics, dispersal between patches and density-dependent local dynamics (Hanski, 1991). Hence, individuals and subpopulations are assumed to be identical (Verboom *et al.*,

1993, this volume). Within this system, subpopulations become extinct and are re-established. Of course, a population in the real world is composed of adults and juveniles, of males and females, and the number of patches in a landscape is finite. Moreover, the spatial configuration and the dimensions of the patch are variable, as was pointed out by Boorman and Levitt (1973) in their modelling of group selection in small boundary populations along a large population with zero extinction probability. Such configurations converge towards source–sink configurations.

We will extend Levins' theoretical concept to relate the functioning of spatially structured populations to the spatial structure of the landscape. We will consider two extremes of habitat distribution (Figure 7.2): an archipelago landscape where patches are of the same order of size, versus a continent–islands landscape where at least one patch is much larger than the others (Opdam, 1990). The first type resembles a Levins' type situation. The second type resembles a Boorman and Levitt type of landscape, and may be more properly considered in the light of a source–sink relationship. In each of these types, connecting elements may be introduced. These elements can function as secondary habitat as well, giving rise to source–sink relationships. Source–sink relationships also come into play when differences in habitat quality are introduced into Levins' theoretical model (Harrison, 1991).

7.2.2 EFFECTS OF LANDSCAPE STRUCTURE

With an increasing loss and disintegration of habitat, the area of the remaining fragments diminishes, and the interpatch distance grows. The effect of this change in habitat distribution may be twofold: local populations become smaller and interpatch dispersal gradually decreases down to (almost) zero. Both effects lower the probability of an average habitat patch being occupied by a local population. Populations may become extinct purely by coincidence of demographic processes, or by a combination of environmental disturbance and stochastic demographic processes (Goodman, 1987; Leigh, 1981; Verboom *et al*, 1993, this volume). As long as dispersal is frequent, local extinctions may be prevented by immigrants, and if they do occur, they will not last for long. The period of vacancy will expand the more a growing average patch distance lowers the probability of interpatch dispersal. At this stage of fragmentation, patches may remain empty for some period and become reoccupied after some time, whereas other patches become empty in their turn, and so on. Then, the fragmented population shows the spatial dynamics characteristic for metapopulations (Levins, 1970; Merriam, 1988; Opdam, 1987, 1990). With ongoing fragmentation, a further increase in local extinction rate and a decline in the chance to recolonize empty patches may cause a further decrease in the proportion of occupied patches (Verboom *et al.*, 1993, this volume), to a level where the stochastic processes in sub-populations are likely to cause metapopulation extinction.

If the fragmentation process becomes stabilized, the metapopulation may reach an equilibrium in the average number of occupied patches. This regional equilibrium contrasts with a lack of equilibrium on the local scale (May and Southwood, 1990). The time to reach equilibrium, the response time of the metapopulation to a landscape change, will depend on the average rates of local extinction and recolonization (Opdam, 1990, 1991).

Characteristically, metapopulations show a continuous change in the distribution over patches (Figure 7.3b). This turnover will be observed in a series of repeated censuses. The two extremes in the fragmentation gradient of Figure 7.2 might fail to show this winking pattern of presence–absence. A slight degree of fragmentation (Figure 7.3a) may still permit a strong dispersal flow: as a result, vacancies will be brief, if occurring at all. However, at the other end of the fragmentation gradient (Figure 7.3c), successful dispersal between habitat sites may become so improbable that the coherence between the subpopulations desintegrates. Then, the metapopulation falls apart into a set of uncon-nected populations, which will go extinct one after the other and will not be re-established. Which of these possibilities is applicable in a certain landscape depends on how the spatial configuration of habitat patches relates to the species characteristics.

It has been pointed out that local extinction rate is governed by population size and, hence, by patch area, whereas recolonization is governed by the degree of isolation to other patches. Having a charac-teristic combination of these spatial features, each patch will have its

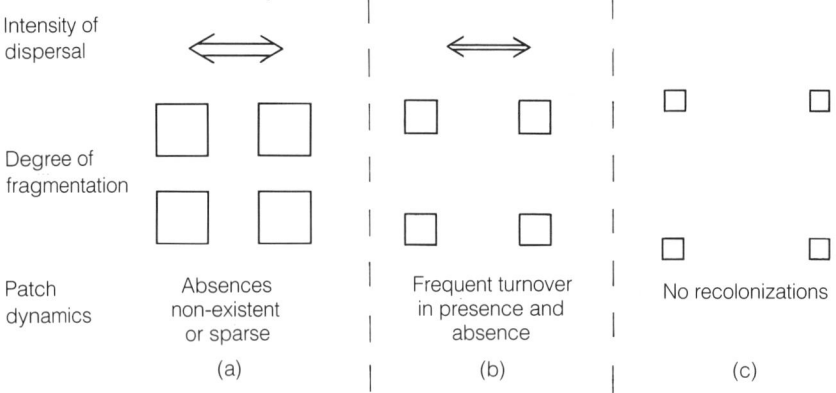

Figure 7.3 In a gradient of fragmentation, dispersal and patch dynamics gradually change.

characteristic probability of being occupied. A habitat area may be so large that this probability may be close to one. In such a 'continent' (Figure 7.2), the local population will serve as a large, stable source sending out dispersers to the surrounding satellite patches. Here, patch occupancy will be partially governed by the distance from this source, and we may expect a gradient in patch occupation.

A similar phenomenon may occur along the edge of a cluster of habitat patches, because edge patches may have lower recolonization probabilities than central patches due to greater isolation. Again, a gradient of a decreasing chance of being occupied may be observed towards the edge.

7.2.3 EFFECT OF HABITAT HETEROGENEITY ON LOCAL POPULATION PROCESSES.

Animals may find their home range or territory heterogeneous (Figure 7.4a). Different parts of it may serve different demands in various times of the year (e.g. breeding sites and overwintering sites) or may vary in quality, as expressed in the quantity or quality of vital resources. This is habitat heterogeneity on the scale of the individual home range (Figure 7.4a). On a higher scale level habitat heterogeneity of patches may be defined in terms of the variation of the average quality of home ranges or sites. (Figure 7.4b). Here, heterogeneity is considered on the landscape level (Figure 7.4c), the between-patch heterogeneity. At this spatial scale, within-patch heterogeneity is neglected. Heterogeneity is defined then as the variation in average habitat quality among patches. Patches are classified into marginal and optimal quality patches, according to the ratio between mortality and reproductive rate. In marginal habitat this ratio is, on average, smaller than unity, whereas in optimal patches, reproduction usually exceeds annual mortality.

Marginal habitat patches will be unoccupied more often than optimal sites, due to higher extinction rates and a smaller chance that immigrants will establish a new local population. Additionally, marginal habitat may also be actively deserted by species. Birds that failed to breed successfully in their first reproductive season tend to leave their territory and try to find a vacant site in better quality patches (Greenwood and Harvey, 1982).

In populations of species with territorial behaviour, a marginal habitat patch adjacent to an optimal site provides temporary shelter and food for survival to surplus individuals expelled from the optimal sites by territory-holders. Later on, these individuals fill-in open places in the optimal habitat that arise in the course of the non-reproductive period (Matthijsen, 1987a). Consequently, a maximal reproductive potential is maintained. However, this functioning of sinks may be inter-

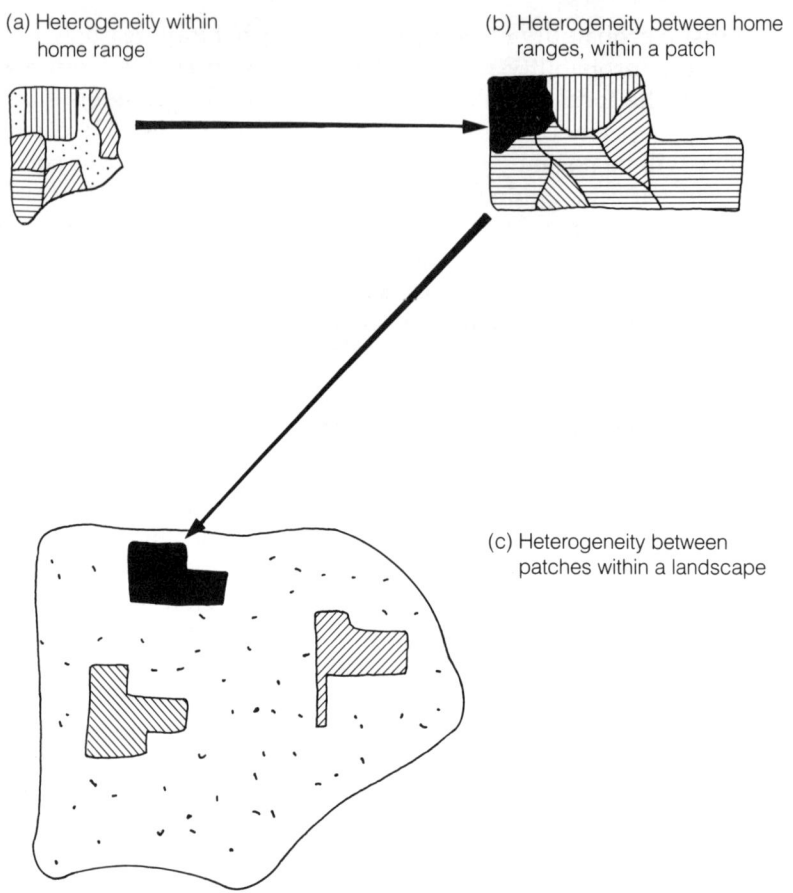

(a) Heterogeneity within home range

(b) Heterogeneity between home ranges, within a patch

(c) Heterogeneity between patches within a landscape

Figure 7.4 Heterogeneity in habitat quality on three spatial scales.

rupted by barriers. Due to isolation, individuals may stay longer in low quality sites, and at the same time optimal patches may remain empty. Throughout a landscape, this effect of isolation will result in a smaller overall reproductive output and, hence, in a higher probability of overall extinction.

Since, on average, mortality exceeds reproduction, local populations in marginal habitat patches will persist only by the support of local populations in optimal habitat patches (Pulliam, 1988). In a mixed archipelago of marginal and optimal patches, the latter will be crucial for metapopulation persistence. Sometimes, however, low quality patches can give rise to a temporary dispersal outflow of adults or subadults, as was observed in the bank vole *Clethrionomys glareolus* and the field vole, *Microtus agrestis*, on habitat islands as well as on real islands (Hansson, 1977; Pokki, 1981; Gliwicz, 1989). Therefore, low quality patches may still contribute to the persistence of the metapopulation. The relation between patches of different quality may be inhibited due to fragmentation. In combination with environmental stochasticity, this may lead to an increased extinction chance.

Varying atmospheric conditions may change the pattern of good and poor sites. For example, in dry summers, wet sites which are marginal under average conditions, may turn into optimal habitat, whereas local populations in optimal patches may fail to reproduce or even go extinct (Den Boer, 1986 gives more examples). Barriers may prevent individuals responding to such changes by shifting between habitat types. This may result in decreased mean population densities.

7.2.4 EFFECTS OF SPECIES CHARACTERISTICS

In using stochastic models of isolated populations, Goodman (1987) showed that the probability of extinction is primarily governed by the following factors:

1. the maximum number of reproductive females in the habitat area;
2. the intrinsic growth-rate of the population;
3. a density-dependent relation between growth-rate and population density;
4. the variation in growth-rate due to stochastic variation in environmental conditions.

With growth rates being constant over time, high growth rates resulted in long survival times of very small populations. However, in general, species with a high potential population growth are very susceptible to environmental variation, resulting in high standard deviations of average growth rate. Implementing this variation into the models had a dramatic effect on survival time. Density-dependent effects on growth rate also lowered the survival time of populations considerably, but less dramatically than environmental variation. Obviously, the extent to which these factors play a role varies greatly among species.

The recolonization process is composed of a dispersal phase and a phase of population establishment. The dispersal flow is a function of the number of dispersers leaving an average occupied patch and the resistance of the landscape matrix. Corridors and barriers may play a role in most taxonomic groups, though the evidence is quite scanty (e.g. Getz *et al.*, 1978; Mader, 1984; Johnson and Adkisson, 1985; Szacki, 1987; Van Dorp and Opdam 1987; Korn and Pitzke, 1988). Also, the social organization of a species may interact with the dispersal flow. In both mammals and birds, the dispersal flow is often sex-biased. Whether males or females are the dominant sex is regulated by the territorial system (Gaines and McClenaghan, 1980; Greenwood, 1980). In voles, isolated woodlots were found to be inhabited by males only (Gottfried, 1979).

Dispersal curves, describing the number of individuals to reach a certain distance, often resemble a negative exponential function (e.g. Mook, 1971; Karban, 1981; Den Boer, 1983; Matthijsen and Schmidt, 1987; Johnson, 1988). In small mammals, some dispersal curves show a small peak at longer dispersal ranges (Crawley, 1969). These long-distance movements might be important particularly in relation to inter-patch dispersal in fragmented landscapes.

In the initial phase of establishment, populations will still be below a density level where the growth rate is restrained. Growth rate, and its variation due to external factors like variable weather conditions, are important factors determining whether a species is a successful colonizer or not. The overall result could be that invertebrates, after having reached an empty patch, have a good chance to grow away from low population densities still liable to demographic stochasticity, but only in years with favourable environmental conditions. Birds and mammals will be less susceptible to variation in weather conditions, but due to their lower growth rate the attempt to recolonize has a considerable chance to fail due to stochastic demographic events (Verboom *et al.* unpublished).

At least in mammals, social attraction by olfactory signals may enhance settlement (Smith and Peacock, 1989). For the ground squirrel *Spermophilus columbianus*, this phenomenon was claimed by Weddell (1991). Squirrels did reach vacant patches, but emigrants tend to settle near other squirrels in occupied patches. Such conspecific attraction may lower the chance of recolonization, but at the same time the extinction probability of existing subpopulations may also be lowered. A simulation based on the Levins model by Ray *et al.* (1991) showed that conspecific attraction resulted in a decrease in the proportion of occupied patches.

7.3 OBSERVATIONS OF FRAGMENTED POPULATIONS

7.3.1 SPATIAL DYNAMICS IN AN ARCHIPELAGO LANDSCAPE

Small and medium-sized woods are scattered throughout many parts of the western European agricultural landscape. They may or may not be linked to wooded banks or hedgerows. For species restricted to woody vegetation for most of their life, they constitute an archipelago-like pattern of habitat fragments, sometimes connected by potential corridors. We investigated the distribution and functioning of fragmented populations of various forest-dwelling species, which vary with respect to the maximum size of local populations and to their mobility during dispersal. Here, we summarize and compare results obtained for songbirds (the nuthatch, *Sitta europaea*, in particular), the red squirrel, *Sciurus vulgaris*, the bank vole, and millipedes and wood ants.

(a) Forest birds

Songbirds of mature deciduous and mixed forest were investigated most thoroughly. Spatial analyses of distribution patterns (Opdam *et al.*, 1985; Van Dorp and Opdam, 1987) were followed by repeated censuses to assess local extinction and recolonization rates (A. Schotman, unpublished results; Van Noorden, 1986; Verboom *et al.*, 1991b) and studies with a simulation model of a nuthatch metapopulation (Verboom *et al.*, 1993, this volume). Under the assumption that fragmented populations of these songbirds function as metapopulations, we would expect the probability of occurrence to increase with patch area as well as with the density of hedgerows and woodlots in the surrounding landscape. Using species number as the integrated parameter of probability of occurrence for all 16 species of mature forest, regression analyses produced various alternative models including area as the primary factor explaining most of the variation in species number. In addition, the models included one or two parameters for spatial configuration, either the area of wood or the density of wooded banks in the close surroundings, or a combination of these parameters. Van Dorp and Opdam also measured habitat variation, but due to their efforts to keep habitat differences in their sample of 234 woodlots as small as possible, obviously this factor played a minor role in their regression models.

A more detailed examination of the underlying processes for nuthatch populations by Verboom *et al.* (1991b) supported the empirical evidence for the effects of forest fragmentation on forest birds. Their analysis of local extinction frequency showed a negative correlation with the estimated maximum number of territories and, additionally, with the proportion of optimal habitat territories. Recolonization chance was related to a connectivity index which combined both distance to and size of woodlots within a 2-km range (cf. dispersal curve in Matthijsen and Schmidt, 1987). Notably, local extinction showed no relationship with this connectivity index, suggesting that local immigration did not exert a measurable effect on the chance to go extinct.

Thus, both distribution patterns and processes in local populations are in accordance with predictions from metapopulation theory. Accordingly, a considerable proportion of habitat patches was unoccupied. For the nuthatch, 40–50% of the suitable woodlots remained

Table 7.2 Patch occupation and patch dynamics in a metapopulation of the nuthatch in an archipelago of small woodlots in an agricultural landscape in Twente (in the eastern part of The Netherlands) (Schotman, unpublished results),

Parameters	Years of observation		
	1988	1989	1990
Number of woodlots examined	68	73	73
Number of empty woodlots	29	32	32
Number of local extinctions	-	9	7
Number of recolonizations	-	6	7

without nuthatch recordings in a single year (Table 7.2). In the study by Verboom *et al.* (1991b), patch quality was estimated on the basis of wood structure characteristics and detailed habitat descriptions by Matthijsen (1987a). Marginal patches showed a higher risk of being unoccupied than optimal patches. On the basis of their results, Verboom *et al.* suggested that the shifting from low quality to high quality territories, which is a frequent phenomenon in nuthatch populations in extensive forest (Matthijsen, 1987b), is inhibited in fragmented landscapes.

(b) Mammals

Like the nuthatch, the red squirrel occurs in very small populations in the average woodlot that is typical for the kind of agricultural landscape studied. Little is known of the distances covered during dispersal, though circumstantial evidence suggests that they resemble those of the nuthatch (Gurnell, 1987).

A single-year census yielded results that were quite similar to those obtained for nuthatches (Verboom and Van Apeldoorn, 1990). Again, about half the habitat patches were apparently unoccupied. The squirrel's probability of occurrence was correlated to patch area, to the density of hedgerows in the immediate surroundings (up to 600 m), and to either the distance to the nearest woodlot larger than 30 ha or to a measure of the coverage of wood in the surrounding landscape (up to 600 m). Also, a habitat quality parameter was found to play a role, in this case the amount of coniferous woods within a patch (the optimal habitat type for red squirrels).

Bank voles are smaller in size and they form local populations 10–50 times larger than those of red squirrels or nuthatches, but at the same time their range of dispersal movements is supposed to be much shorter. Van Apeldoorn *et al.* (1992) summarized evidence indicating that movements up to 500 m are not exceptional, but that little is known how these movements are inhibited by the landscape matrix. The fragmented population of bank voles in our study area showed little spatial dynamics in patch occupancy. Six out of 51 woodlots were found to be unoccupied in early spring or became empty in autumn, but (re)colonizations were observed to occur within a few months. The authors concluded that local extinctions are more likely in woodlots smaller than 0.5 ha (which compares to 2 ha optimal nuthatch habitat).

Demographic parameters of local populations were influenced by the spatial configuration of woodlots, but only in combination with patch quality. A model with four habitat factors explained about half of the variation in male and female abundance. Additionally, the area of the nearest woodlot positively influenced local abundance of sexes, whereas the distance to a wood of at least 25 ha (a supposed permanent source of immigrants) negatively affected various local abundance parameters. Indications for a smaller effect of distance on males are consistent with the generally higher male mobility and larger home ranges. Thus, the fragmented populations of this small rodent did show some spatial dynamics in distribution patterns. However, the strongest effects were found on a local demographic level and hence on local extinction chance.

(c) Invertebrates

Ground-dwelling invertebrates seem to cover only small distances during dispersal, although (with the exception of carabid beetles) little is known about their dispersal behaviour in agricultural environments (Welch, 1990). A regression analysis of presence/absence patterns of various groups of ground-dwelling invertebrates in a small sample (n=22) yielded equivocal results (Mabelis, 1990). After accounting for habitat variation, a positive impact of surrounding woodlots was suggested only for carabid beetles and for millipedes, but not for centipedes, isopods and ants. A problem with invertebrates is the technique of sampling a vast number of patches. A transect of pitfall-traps, which is normally applied, produces information on presence, rather than on absence, because large parts of the patch area cannot be sampled. Actually, some groups showed a negative relation between number of species present and woodlot area (Mabelis, 1990), which may be attributed to a sampling artefact rather than to an ecological phenomenon.

A comparison of the distribution of species with different dispersal capacities gives more insight into possible fragmentation effects. For example, the stenotopic forest carabid *Cychrus caraboides*, which does not fly, was found in only 4 out of 22 woodlots, whereas the stenotopic forest carabid *Pterostichus oblongopunctatus*, which occasionally disperses by air, was found in every woodlot. The ant species *Myrmica ruginodis*, dispersing by winged queens, was found in all of 32 habitat patches irrespective of size and degree of isolation. Other ant species with poor colonization capacities were less abundant. *Lasius fuliginosus* occurred in about half the woodlots, and *Formica polyctena* in only six of the largest woodlots, and in three small ones connected to one of the larger woods by hedgerows. These differences in colonization behaviour were linked to the type of social organization (Mabelis, 1986).

Mabelis also stressed the interrelation with habitat quality: very small woodlots were much affected in habitat quality for ants by intrusion of fertilizers during neighbouring farming activities. He was able to show that in a similar landscape in Poland, with low intensity land use, wood ants tend to inhabit smaller woodlots as compared to average Dutch conditions (Mabelis, unpublished results).

7.3.2 DISTRIBUTION IN A CONTINENT–ISLANDS LANDSCAPE

In the preceding section, we considered a landscape with small wood fragments. However, where such agricultural landscapes are adjacent to more extensive tracts of forest, different patterns should be expected. Since these forests may function as stable sources of immigrants, the probability of occurrence in small patches will diminish with an increasing distance to the source, down to an equilibrium level beyond the range of influence of the source. Verboom *et al.* (1991a) considered such a landscape by means of a simulation model of a nuthatch

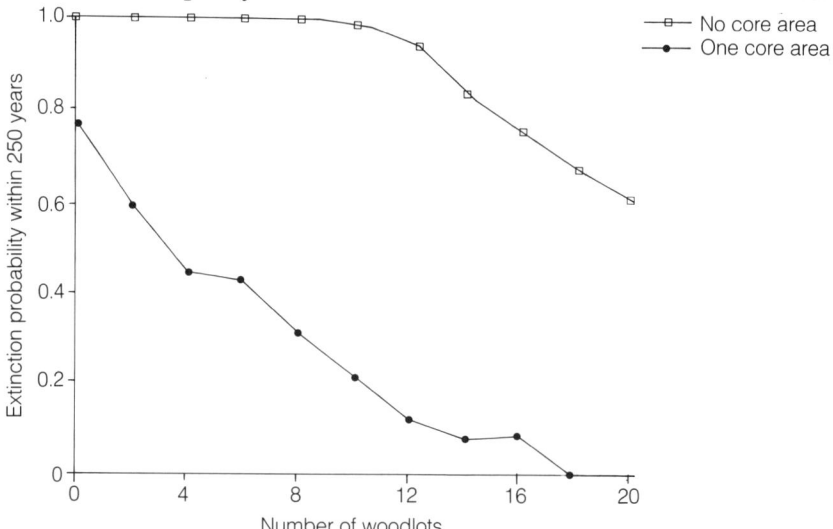

Figure 7.5 The probability of extinction within 250 years of a metapopulation of nuthatches, distributed over woodlots and core area, plotted against the number of woodlots in the landscape. The two lines compare a situation without a continental area (with the number of woodlots varying between 0 and 20) to one where a core area of max. 20 territories is added.

metapopulation and obtained obvious results (Figure 7.5). They also showed that the probability of occurrence decreased from 76% at 1 km distance to 64% at 18 km from the source.

Empirical support for such a distance effect comes from studies discussed in the preceding section. Red squirrel presence was more probable in the vicinity of forest tracts exceeding 30 ha in size (Verboom and Van Apeldoorn, 1990). Similar results were reported by Opdam *et al.* (1985) regarding bird species of mature forest. Bank vole abundance decreased with an increasing distance to woods larger than 25 ha (Van Apeldoorn *et al.*, 1992). An increase in distance from 25 to 3800 m was correlated to a decline in female abundance of 10–50%, and a fluctuation in male abundance of 10–25%.

7.3.3 UPGRADING METAPOPULATION PROCESSES TO A BIOGEOGRAPHICAL SCALE: THE IMPACT OF EXTENSIVE REAFFORESTATIONS

For a century, the network of woodlots and hedgerows in the historic agricultural landscapes is disintegrating. On the other hand large areas of heathland were afforested for timber production (Section 7.1). Most of this timber woodland was planted as coniferous stands, but scattered deciduous and mixed stands became suitable as nuthatch habitat in recent decades. At the same time, these forest tracts may serve as land bridges towards remote historic forest sites, which used to be unoccupied in former times (Harms and Opdam, 1990). Around 1975, the nuthatch was almost restricted to the forest landscapes of the central and eastern provinces. Since then, its distribution has expanded to the provinces that once were largely covered by open heathland landscapes.

The expansion of nuthatch distribution is unmistakeable from a comparison of the national breeding bird census schemes of 1973–1977 and 1979–1983 (SOVON, 1987). This trend is also apparent from regional censuses in the northeastern and southern provinces. In an area of 725 km^2 in the south, nuthatch density increased from 20–30 breeding

Table 7.3 Changes in rates of local extinction and (re)colonization in habitat patches of the nuthatch, in an area that became invaded by the species from 1972 onwards. Based on data supplied by A.J. van Dijk. Rates calculated according to Verboom *et al.* (1991b)

Period	Colonization rate	Extinction rate
1972–75	0.10	0.36
1976–78	0.08	0.08
1979–81	0.17	0.23
1982–84	0.18	0.06
1985–87	0.50	0.02

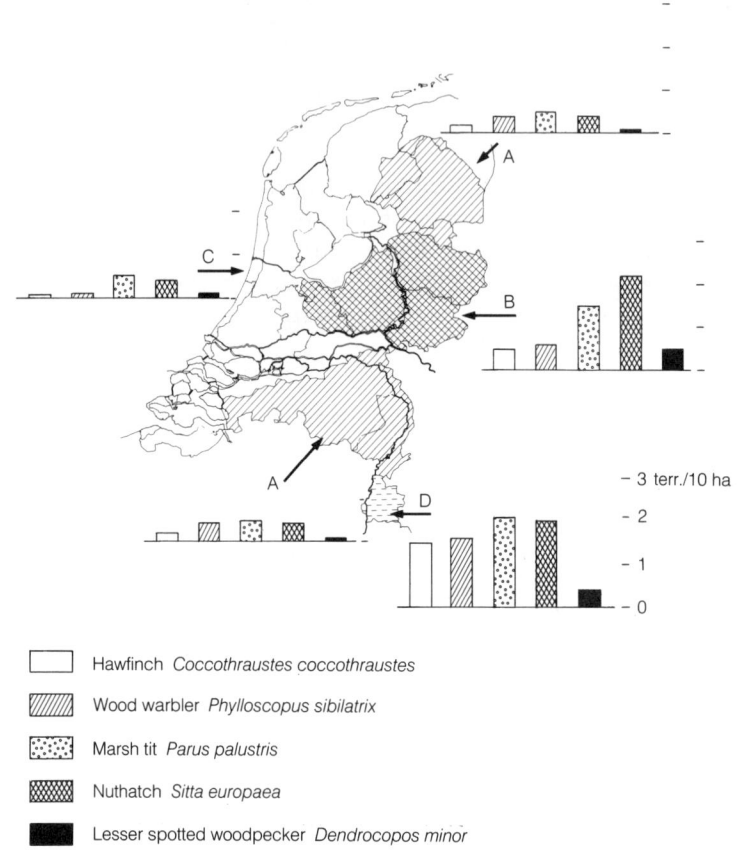

Figure 7.6 Densities of some birds of deciduous forest in 1973–1984, expressed in number of territories per 10 ha. The densities in the four regions A–D are corrected for habitat and landscape factors, by means of a regression model. The division into regions is partly based on the breeding bird districts of The Netherlands (Kwak and Reijrink, 1987).

pairs in 1965 to 200–225 in 1990 (Post and Ongenae, 1990), and in another area in the northeastern province of Drenthe (200 km^2) it increased from zero in 1972 to 47 in 1988 (A.J. van Dijk, unpublished data). More interestingly, most of the woods in the latter area must have been suitable for nuthatches since at least the early 1950s. Table 7.3 shows the trends in the observed local extinction and recolonization rates. It remains unclear to what extent this range expansion can be attributed to an increase in the dispersal stream caused by a growing population in the source areas, or to the changes in the local landscape (more habitat, lower resistance).

Along with the nuthatches, several other bird species restricted to mature woodland showed the same expansion in the same areas. Two of these species (marsh tit, *Parus palustris*, hawfinch *Coccothraustes coccothraustes*) were claimed to be susceptible to woodland fragmentation (Van Dorp and Opdam, 1987).

The range expansion of forest-dwelling species over newly developed habitat is not yet complete. This can be concluded from an analysis of geographical variation in breeding density by Schotman *et al.* (1992). Breeding bird census data from over 400 homogeneous stands of woodlands were treated with multiple regression techniques to account for the variation in forest type and forest structure among the plots, as well as for differences in the surrounding landscape. The residual variation in species' densities was interpreted with reference to the biogeographical scale (Figure 7.6). A comparison with Figure 7.1 shows that the highest densities of species restricted to mature forest were found in plots situated in regions that were already connected to German and Belgian forest landscapes a century ago. The time needed for complete transmission depends on the growth rate and dispersal distance of the expanding species (Hengeveld and Van den Bosch, 1991) as well as on the density of suitable habitat and the landscape permeability in the area of expansion. After two or three decades of range expansion, average densities in the newly colonized regions are still lower than expected from habitat and local landscape features alone.

7.4 DISCUSSION

7.4.1 LANDSCAPE CONFIGURATION AND METAPOPULATIONS

The empirical studies reviewed revealed parallels between woodland birds, bank voles, red squirrels and several invertebrate species, particularly with respect to the role of habitat patches in the surrounding landscape. The supposed local processes of extinction and recolonization were especially distinct in the nuthatch, resulting in the characteristic shifting distribution pattern of metapopulations. The role of larger core areas to stabilize metapopulation dynamics was illustrated by empirical results (forest birds, red squirrel, bank vole, wood ants) and supported by modelling results. These processes on a landscape scale are being translated to the larger scale of the species range, as was illustrated for the nuthatch and other forest-dwelling bird species.

Quantifying the role of variation in habitat quality among patches is being complicated by difficulties with transferring quality into demographical parameters. A system of a continuously changing mosaic of optimal and marginal habitats was extensively analysed and described by Den Boer (1986) using carabid beetles in heathland systems. However, the system he investigated was heterogeneous rather than fragmented: spatial variation in habitat quality with no apparent barriers between the patches. It has already been pointed out that adding such barriers to the system will reduce the 'spreading-of-risk effect' of dampening population fluctuations (Den Boer, 1986). When interpatch movement is inhibited, patches that have changed to optimal may remain underpopulated or empty, while former optimal patches still hold the majority of individuals reluctant to disperse, and having low chances of reproduction.

In our study area at the landscape scale, the degree of connectivity has decreased since the beginning of the 20th century. However, changes in species distribution were not documented. The study is merely a snapshot somewhere in the course of the process of landscape change. The results were consistent with theoretical predictions, and some more detailed process studies further supported our ideas. However, it had been assumed that the fragmented population had completely responded to the change in the landscape pattern, with the proportion of empty patches being constant over time. This assumption can be questioned, particularly in the case of forest invertebrates, which might have low rates of local extinction (e.g. carabid beetles, Den Boer, 1990). It is not unlikely that the observed frequencies of patch occupation are too optimistic and will decrease gradually over the next decades (cf. Wiens, 1990). It is even possible that this decrease continues to regional extinction, simply because the recolonization rate might be too small to be able to keep pace with the local extinction rate. In other words, it is questionable whether we are dealing with metapopulations of invertebrates here (cf. Figure 7.3). Nuthatches, were found to have high turnover rates (Tables 7.1 and 7.2) so it is more likely that their distribution pattern is in equilibrium with the landscape structure in our study area. Judging from the distribution pattern, the red squirrel could very well be in the same position. Bank voles also seemed to have a high recolonization chance (Van Apeldoorn unpublished data) with all observed patches occupied in September, at the end of the dispersal period. A small time scale of observation clearly revealed local extinctions. For this bank vole metapopulation we expect that a further increase in fragmentation would result in typical metapopulation patch dynamics, as indicated by data from Belgium (Bauchau and Le Boulengé, 1990).

We argue that whether or not a fragmented population shows characteristics typical for Levins' type metapopulations depends on the spatial scale of the investigated landscape (including the variation in patch sizes) and the temporal scale of observation. Of course, these factors must be considered in relation to the time scale of the underlying population processes which are strongly related to species-specific traits. We propose to extend existing definitions of a metapopulation (Levins, 1970; Opdam, 1990; Hanski and Gilpin, 1991) (in a demographical rather than a genetical sense) to any set of spatially defined local populations, which are demographically affected by the spatial arrangement of habitat patches and the resistance of the non-habitat of the landscape matrix. This effect may, but should not always, result in turnover in patch occupation. This definition excludes a set of completely independent local populations, with no significant exchange of individuals. It also excludes a collection of patch populations where the exchange of individuals is so intense that variation in spatial configuration does not affect the local demographic history. With a decrease in exchange rate, such a spatially structured population will turn into a collection of local populations, the more independent the populations in the patches will become. Drawing sharp boundaries between different kinds of spatially structured populations would be misleading (Hanski and Gilpin, 1991).

7.4.2 APPLICATION IN LANDSCAPE PLANNING AND CONSERVATION: EXTRAPOLATING THE RESULT

This chapter summarizes and extends the theory of spatially structured populations, and gives an overview of data that were collected in fragmented Dutch landscapes. For application, these data have to be extrapolated to other regions, other species and other ecosystems. Most empirical results were acquired in descriptive studies and consequently, causal relations can be questioned, and extrapolation beyond the observed range of landscape variation is not justified. Alternatively, mechanistic models are to be preferred as a tool for application (Verboom *et al.*, 1993, this volume). Other problems with extrapolating these results to other geographical areas are caused by differences in the history of fragmentation between landscapes and in the specific traits of species. In landscapes with a different history, fragmented populations may be in a different stage of responding to the change in spatial pattern. Also, species may differ as to their flexibility to adapt their behaviour, particularly their dispersal behaviour, to the fragmented spatial structure of their habitat. Such responses on an evolutionary time-scale have not been reported in literature, but are quite likely in those landscapes with a slow to moderate pace of fragmentation. Hengeveld (1990) argued that in the border zone of a species range, individuals tend to have lower reproductive rates and poorer dispersal capacities. Such spatial variation will limit quantitative applications of the results reviewed here.

However, applying the results to other systems and species is seriously restricted by two sorts of bias: the emphasis on forest ecotopes and the emphasis on birds and mammals. Most bird and mammal species are spatially organized on a territorial basis, show moderate fluctuations in response to variability in environmental factors, and are physically quite capable of crossing unsuitable tracts of vegetation up to at least a few kilometres. Intuitively, this seems in contrast to many ground-dwelling invertebrates, to reptiles and amphibians and to most plants. In these groups, which may fluctuate strongly, movements among habitat remnants over 1–5 km may no longer be possible, or very unlikely.

Even within taxonomic groups considerable variance among species has to be considered. For instance, in forest birds, species differences were quite substantial (Opdam et al., 1985; Van Dorp and Opdam, 1987), but are as yet unexplained. The study on voles by Van Apeldoorn et al. (1992) and Mabelis' work on ants with different dispersal strategies stressed the role of autecological characteristics in dispersal, both in a quantitative and qualitative sense, but we are far from understanding to what extent behavioural traits are attributed to the susceptibility to fragmentation. Very little is known about how individuals actually move through the landscape, whether they use linear elements as corridors, or simply follow a straight or random route, or how they react to boundaries (Opdam, 1991).

Despite such difficulties of interpretation, application is still possible. At least for birds and mammals, metapopulation theory seems a good scientific basis to develop general guidelines for application (Harms et al., 1993, this volume; Lankester et al., 1991; Schafer, 1991; Verboom et al., 1993, this volume) and for problem detection. Whether or not fragmentation has proceeded to a degree that metapopulations begin to run high extinction risks may be quickly assessed by the proportion of empty, but suitable patches (Verboom et al., 1993, this volume). Also, it is very clear that as many patches as possible in an archipelago system must be protected, instead of only those that were found to be occupied in a one-year census.

Other types of habitat than mature forest will be usually less constant over time. Succession rates and the frequency, intensity and area of disturbances will affect, on an evolutionary time-scale, the site-tenacity and the dispersal capacity of species (Mikkonen, 1983; Svensson, 1985; Turin and Den Boer, 1988; McPeek, 1989). If this assumption is right, species of early successional stages must be better adapted to fragmentation, on a particular spatial scale, than species of later stages. Likewise, species of marshland and especially coastal habitat will be better adapted to fragmentation than species restricted to heathland, peat-moor and forest (Opdam, 1991). However, no systematic empiral studies in systems other than forests have been published. Circumstantial evidence suggests that spatial effects in fragmented bird populations occur in marshland on a national scale (bittern *Botaurus stellaris*, bearded tit *Panurus biarmicus*, marsh harrier *Circus aeruginosus*, bluethroat *Cyanosylvia svecica*, unpublished data).

Extrapolation to other taxonomic groups seems to be questionable. Little is known about effects of habitat fragmentation on plants, though it seems quite obvious that many species of mature forests, chalk grasslands and grasslands on poor acid soils are quite susceptible (Møller and Rørdam, 1985; Dzwonko and Loster, 1988; Van Ruremonde and Kalkhoven, 1991). Dispersal seems to be the critical factor here, since many plant species have relatively long life-expectancies and low dispersal capacities. Intuitively, this may imply that on the landscape scale metapopulation theory is not quite relevant to plants.

The same may be true for small ground-dwelling invertebrates. Local populations of such species will be quite large, but extremely variable in size, whereas distances between patches in a landscape can not be covered within a lifetime. If so, corridor systems which are suitable for establishing reproductive populations (at least temporarily) will be indispensable in order to conserve these species in fragmented landscapes.

In conclusion, theory and empirical data have developed far enough to produce general planning guidelines, and to detect detrimental effects of fragmentation in metapopulations. Quantitative application in landscape scenarios is only possible for forest birds, if one accepts the possible flaws mentioned above.

REFERENCES

Andrewartha, H.G. and Birch, L.C. (1954) *The Distribution and Abundance of Animals*. University of Chicago Press, Chicago.

Barends, S. (1987) *Steekproefsgewijze inventarisatie van perceelsvormen in Nederland*. Rapport nr. 1927, Stichting voor Bodemkartering, Wageningen.

Bauchau, V. and Le Boulengé, E. (1990) Population biology of woodland rodents in a patchy landscape, in *Le Rongeur et l'espace* (eds. M. Le Berre and L. Le Guelte), Chabaud, Paris.

Bengtsson, J. (1991) Interspecific competition in metapopulations. *Biological Journal of the Linnean Society*, **42**, 219–37.

Boorman, S.A. and Levitt, P.R. (1973) Group selection on the boundary of a stable population. *Theoretical Population Biology*, **4**, 85–128.

Crawley, M.C. (1969) Movements and homeranges of *Clethrionomys glareolus* Schreber and *Apodemus sylvaticus* L. in north-east England. *Oikos*, **20**, 310–19.

Den Boer, P.J. (1983) Het verschijnsel dispersie. *Vakblad voor Biologen*, **63**, 310–14.

Den Boer, P.J. (1986) Environmental heterogeneity and the survival of natural populations, in *Proceedings 3rd European Congress of Entomology* (ed. H.H.W. Velthuis), Amsterdam, pp. 345–56.

Den Boer P.J. (1990) Density limits and survival of local populations in 64 carabid species with powers of dispersal. *Journal of Evolutionary Biology*, **3**, 19–48.

Dzwonko, Z. and Loster, S. (1989) Species richness of small woodlands on the western Carpathian foothills. *Vegetatio*, **76**, 15–27.

Gaines, M.S. and McClenaghan, L.R. (1980) Dispersal in small mammals. *Annual Review of Ecology and Systematics*, **11**, 163–96.

Getz, L.L., Cole, F.R. and Gates, D.L. (1978) Interstate roadsides as dispersal routes for *Microtus pennsylvanicus*. *Journal of Mammalogy*, **59**, 208–12.

Gilpin, M. (1991) The genetic effective size of a metapopulaton. *Biological Journal of the Linnean Society*, **42**, 165–75.

Gliwicz, J. (1989) Individuals and populations of the bank vole in optimal, suboptimal and insular habitats. *Journal of Animal Ecology*, **58**, 237–47.

Goodman, D. (1987) The demography of chance extinction, in *Viable Populations for Conservation* (ed. M.E. Soulé), Cambridge University Press, Cambridge, pp. 11–34.

Gottfried, B.M. (1979) Small mammal populations in woodlot islands. *American Midland Naturalist*, **102**, 105–12.

Greenwood, P.J. (1980) Mating systems, philopatry and dispersal in birds and mammals. *Animal Behaviour*, **28**, 1140–62.

Greenwood, P.J. and Harvey, P.H. (1982) The adaptive significance of variation in breeding area fidelity of the blackbird (*Turdus merula* L.). *Journal of Animal Ecology*, **45**, 887–91.

Gurnell, J. (1987) *The Natural History of Squirrels*, Helm Publishers, London.

Hanski, I. (1991) Single-species metapopulation dynamics: concepts, models and observations. *Biological Journal of the Linnean Society*, **42**, 17–38.

Hanski, I, and Gilpin, M. (1991) Metapopulation dynamics: brief history and conceptual domain. *Biological Journal of the Linnean Society*, **42**, 3–16.

Hansson, L. (1977) Spatial dynamics of field voles *Microtus agrestis* in heterogeneous landscapes. *Oikos*, **29**, 539–44.

Harms, W.B. and Opdam, P. (1990) Woods as habitat patches for birds: application in landscape planning in The Netherlands, in *Changing Landscapes: an Ecological Perspective* (eds. I.S. Zonneveld and R.T.T. Forman), Springer-Verlag, New York, pp. 73–97.

Harms, W.B., Knaapen, J.P. and Rademakers, I.G.M. (1993) Landscape planning for nature restoration: comparing regional scenarios, in *Landscape Ecology of a Stressed Environment* (eds. C.C. Vos and P. Opdam), Chapman & Hall, London, pp. 197–218.

Harrison, S. (1991) Local extinction in a metapopulation context: an empirical evaluation. *Biological Journal of the Linnean Society*, **42**, 73–88

Hengeveld, R. (1990) *Dynamic Biogeography*, Cambridge University Press, Cambridge.

Hengeveld, R. and Van den Bosch, F. (1991) The expansion velocity of the collared dove *Streptopelia decaocto* population in Europe. *Ardea* **79**, 67–72.

Johnson, W.C. (1988) Estimating dispersibility of *Acer, Fraxinus* and *Tilia* in fragmented landscapes from patterns of seedling establishment. *Landscape Ecology*, **1**, 175–88.

Johnson, W.C. and Adkisson, C.S. (1985) Dispersal of beech nuts by blue jays in fragmented landscapes. *American Midland Naturalist*, **113**, 319–24.

Karban, R. (1981) Flight and dispersal of periodical cicadas. *Oecologia*, **49**, 385–90.

Korn, H. and Pitzke, Chr. (1988) Stellen Strassen eine Ausbreitungs-Barriere für kleinsäuger dar? *Berichte Akademie für Naturschutz und Landschaftspflege*, **12**, 189–95.

Kwak, R.G.M. and Reijrink, L.A.F. (1984) Broedvogeldistricten van Nederland en de samenhang met het landschap. *Landschap*, **1**, 251–60.

Lankester, K., Van Apeldoorn, R.C., Meelis, E. and Verboom, J. (1991) Management perspectives for populations of the Eurasian badger *Meles meles* in a fragmented landscape. *Journal of Applied Ecology*, **28**, 561–73.

Leigh, E.G. (1981) The average lifetime of a population in a varying environment. *Journal of Theoretical Biology*, **90**, 213–39.

Levins, R. (1970) Extinctions, in *Some Mathematical Questions in Biology*, Vol. 2, Lectures on Mathematics in Life Sciences. American Mathematical Society, Providence, Rhode Island, pp. 77–107.

Mabelis, A.A. (1986) Why do queens fly?, in *Proceedings 3rd European Congress of Entomology* (ed. H. Velthuis), Amsterdam, pp. 461–64.

Mabelis, A. (1990) Natuurwaarden in cultuurlandschappen. *Landschap*, **7**, 253–68.

MacArthur, R.H. and Wilson, E.D. (1967) *The Theory of Island Biogeography*, Monographs in Population Biology Princeton, New York.

Mader, H.J. (1984) Animal habitat isolation by roads and agricultural fields. *Biological Conservation*, **29**, 81–96.

Matthijsen, E. (1987a) A longterm study on a population of the European nuthatch, *Sitta europaea caesia* Wolf. *Sitta*, **1**, 2–17.

Matthijsen, E. (1987b) Territory establishment of juvenile nuthatches after fledging. *Ardea*, **75**, 53–7.

Matthijsen, E. and Schmidt, K.H. (1987) Natal dispersal in the nuthatch. *Ornis Scandinavica*, **18**, 313–16.

May, R.M. and Southwood, T.R.E. (1990) Introduction, in *Living in a Patchy Environment* (eds. B. Shorrocks and I.R. Swingland), Oxford University Press, Oxford, pp. 1–22.

McPeek, M.A. (1989) Differential tendencies among Enallagma damselflies (*Odonata*) inhabiting different habitats. *Oikos*, **56**, 187–195.

Merriam, H.G. (1988) Landscape dynamics in farmland. *Trends in Ecology and Evolution*, **3**, 16–20.

Mikkonen, A.V. (1983) Breeding site tenacity of the Chaffinch *Fringilla coelebs* and the brambling *F. montifringilla* in Northern Finland. *Ornis Scandinavica*, **14**, 36–47.

Møller, T.R. and Rørdam, C.P. (1985) Species numbers of vascular plants in relation to area, isolation and age of ponds in Denmark. *Oikos*, **45**, 8–16.

Mook, J.H. (1971) Observations on the colonization of the new IJsselmeer-polders by animals, in *Dispersal and Dispersal Power of Carabid Beetles* (ed. P.J. den Boer), Miscellaneous papers Landbouwhogeschool Wageningen 8, 13–31.

Olivieri, I., Couvet, D. and Gouyon, P.H. (1990) The genetics of transient populations: research at the metapopulation level. *Trends in Ecology and Evolution*, **5**, 207–10

Opdam, P. (1987) De metapopulatie, model van een populatie in een versnipperd landschap. *Landschap*, **4**, 289–306.

Opdam, P. (1990) Understanding the ecology of populations in fragmented landscapes. Transactions of the 19th IUGB Congress (ed. S. Myrberget), NINA, Trondheim, pp. 373–380.

Opdam, P. (1991) Metapopulation theory and habitat fragmentation: a review of holarctic breeding bird studies. *Landscape Ecology*, **4**, 93–106.

Opdam, P., Rijsdijk, G. and Hustings, F. (1985) Bird communities in small woods in an agricultural landscape: effects of area and isolation. *Biological Conservation*, **34**, 333–52.

Pokki, J. (1981) Distribution, demography and dispersal of the field vole, *Microtus agrestis* (L), in the Tvärminne archipelago, Finland. *Acta Zoologica Fennica*, **164**, 1–48.

Post, F. and Ongenae, J.P. (1990) Over de Boomklever en andere zware jongens, in *Vogels in Midden-brabant* (eds. F. Post, A. Braam and R. Buskens), Werkgroep voor Vogel- en Natuurbescherming Midden-Brabant, Oisterwijk, pp. 34–40.

Pulliam, H.R. (1988) Sources, sinks, and population regulation. *American Naturalist*, **132**, 652–61.

Ray, C., Gilpin, M. and Smith, A.T. (1991) The effect of conspecific attraction on metapopulation dynamics. *Biological Journal of the Linnean Society*, **42**, 123–34.

Schotman, A.G.M., Opdam, P. and Ter Braak, C.J.F. (1992) Over de geografische variatie in bosvogeldichtheden in Nederland. *Landschap*, (in press).

Shafer, C.L. (1991) *Nature Reserves: Island Theory and Conservation Practice*, Smithsonian Institution Press, Washington D.C.

Smith, A.T. and Peacock, M.M. (1989) Conspecific attraction and the determination of metapopulation colonization rates. *Conservation Biology*, **4**, 320–3.

SOVON (1987) *Atlas van de Nederlandse vogels*. SOVON, Arnhem.

Svensson, B.W. (1985) Local extinction and re-immigration of whirligig beetles (*Coleoptera, Gyrinidae*). *Ecology*, **66**, 1837–48.

Szacki, J. (1987) Ecological corridors as a factor determining the structure and organization of a bank vole population. *Acta Theriologica*, **32**, 31–44.

Taylor, A.D. (1991) Studying metapopulation effects in predator–prey systems. *Biological Journal of the Linnean Society*, **42**, 305–23.

Turin, H. and Den Boer, P.J. (1988) Changes in the distribution of carabid beetles in The Netherlands since 1880. II. Isolation of habitats and long-term trends in the occurrence of carabid species with different powers of dispersal (*Coleoptera, Carabidae*). *Biological, Conservation*, **44**, 179–200.

Van Apeldoorn, R., Oostenbrink, W.T., Van Winden, A. and Van der Zee, S. (1992) Effects of habitat fragmentation on the bank vole, *Clethrionomys glareolus*, in agricultural landscape. *Oikos*, (in press.).

Van Dorp, D. and Opdam, P.F.M. (1987) Effects of patch size, isolation and regional abundance on forest bird communities. *Landscape Ecology*, **1**, 59–73.

Van Noorden, B. (1986) *Dynamiek en dichtheid van bosvogels in geïsoleerde loofbosfragmenten*. Report Research Institute for Nature Management 86/19, Leersum.

Van Ruremonde, R.H.A.C. and Kalkhoven, J.T.R. (1991) Effects of woodlot-isolation on the dispersion of fleshy-fruited species. *Journal of Vegetation Science*, **2**, 377–84.

Verboom, B. and Van Apeldoorn, R. (1990) Effects of habitat fragmentation on the red squirrel, *Sciurus vulgaris* L. *Landscape Ecology*, **4**, 171–6.

Verboom, J., Opdam, P. and Schotman, A. (1991a) Kerngebieden en kleinschalig landschap. Een benadering met een metapopulatiemodel. *Landschap*, **8**, 3–14.

Verboom, J., Schotman, A., Opdam, P. and Metz, J.A.J. (1991b) European nuthatch metapopulations in a fragmented agricultural landscape. *Oikos*, **61**, 149–56.

Verboom, J., Metz, J.A.J. and Meelis, E. (1993) Metapopulation models for impact assessment of fragmentation, in *Landscape Ecology of a Stressed Environment* (eds. C.C. Vos and P. Opdam), Chapman & Hall, London, pp. 172–92.

Vos, C.C. and Zonneveld, J.I.S. (1992) Patterns and processes in a landscape under stress: the study area, in *Landscape Ecology of a Stressed Environment*, (eds. C.C. Vos and P. Opdam), Chapman & Hall, London, pp. 1–29.

Weddell, B.J. (1991) Distribution and movements of Columbian ground squirrels (*Spermophilus columbianus* (Ord)): are habitat patches like islands? *Journal of, Biogeography*, **18**, 385–94.

Weinreich, J.A. and Musters, C.J.M. (1989) *Toestand van de natuur. Veranderingen in de Nederlandse natuur*. SDU, The Hague.

Welch, R.C. (1990) Dispersal of invertebrates in the agricultural environment, in *Species Dispersal in Agricultural Habitats* (eds. R.C.H. Bunce and D.C. Howard), Belhaven Press, London, pp. 203–18.

Wiens, J.A. (1990) Habitat fragmentation and wildlife populations: the importance of autecology, time and landscape structure. *Transactions of the 19th IUGB Congress* (ed. S. Myrberget), NINA, Trondheim, pp. 381–91.

Wiens, J.A. and Rotenberry, J.T. (1981) Censusing and the evaluation of avian habitat occupancy. *Studies in Avian Biology*, **6**, 522–32.

Metapopulation models for impact assessment of fragmentation

8

Jana Verboom, Johan A.J. Metz and Evert Meelis

8.1 INTRODUCTION

In Western Europe, fragmentation of natural habitat types is a considerable problem for nature conservation (Opdam *et al.*, 1993, this volume). Guidelines for landscape design and management are badly needed. To develop these guidelines, we need a better understanding of the quantitative relation between landscape characteristics (such as size and configuration of natural elements) and the prospects for survival of plant and animal populations restricted to those elements. Moreover, we need to assess the effects of mitigating measures (e.g. restoring linear landscape elements) or measures that add to fragmentation (e.g. construction of roads and railways). Metapopulation models are an indispensable tool for understanding the dynamics of fragmented populations and for impact assessment of (de)fragmentation.

A metapopulation is a set of local populations which interact via individuals moving among them (Levins, 1970). For many species in man-dominated landscapes, natural habitats occur in small, spatially separated fragments. Species that inhabit these fragments may function as metapopulations (Opdam, 1988).

Levins (1970) was the first to quantify the metapopulation idea. His metapopulation model describes the dynamics of an infinite number of identical patches, in which all occupied patches have a constant extinc-

Landscape Ecology of a Stressed Environment
Edited by Claire C. Vos and Paul Opdam
Published in 1993 by Chapman and Hall, London. ISBN 0 412 44820 3

tion rate, whereas all empty patches have a colonization rate that is proportional to the fraction of patches occupied and a colonization parameter. Subsequently, metapopulation models were derived that relax one or more of the assumptions. As in the original model, in some models patches can be classified as occupied by a local population (potentially subject to extinction and, at the same time, a source of dispersers) or unoccupied (with a probability of being colonized) (Goodman, 1987; Quinn and Hastings, 1987; Hanski, 1989; Harrison and Quinn, 1989; Ray et al., 1991; Verboom et al., 1991c). This binary classification of patches is relaxed by others, such that within-patch dynamics can be taken into account (Den Boer, 1981; Fahrig and Merriam, 1985; Urban and Shugart, 1986; Burkey, 1989; Hastings and Wolin, 1989; Verboom et al., 1991b; Gyllenberg and Hanski, 1992). We use the term 'winking patches metapopulation model' if the binary classification of occupied and empty patches is employed, and 'structured metapopulation model', if within-patch dynamics is taken into account.

This chapter discusses the merits of metapopulation models for impact assessment of (de)fragmentation measures. Some general remarks on models and modelling are given in section 8.2, and extinction, a crucial aspect of (meta)population models in section 8.3. Subsequently models of the winking patches type are discussed section 8.4 and structured metapopulation models in section 8.5. The main results are listed and their applicability in landscape design and management is discussed.

8.2 MODELLING STRATEGY

Before proceeding with the discussion of mathematical models, let us point out the difference between descriptive, regression type models and mechanistic models, which both have their merits in applied ecology. Regression techniques can be employed to detect relationships in empirical data sets (Opdam et al., 1993, this volume). However, regression models are purely descriptive, and therefore of limited use for impact assessment. Since regression equations do not necessarily represent causal effects, they do not allow extrapolation outside the range of parameter values of the specific situation the model was tuned on.

Mechanistic models are based on underlying mechanisms or processes. They can be used for impact assessment, by modifying the input parameter values and surveying the change in the relevant model output. However, there are some problems with the use of mechanistic models, as these are always a simplification of reality. Models differ in the generality, realism and precision of the results. These differences are associated with the purpose of the model, which can be to gain better insight in a process (cf. the Lotka–Volterra model for predator–prey interaction) or to produce exact, quantitative predictions (cf. the weather forecast). A model can never at the same time be general, realistic and precise. To survey this dilemma, a general discussion of (mechanistic) models is provided (section 8.2.1), and some tools for impact assessment are discussed (section 8.2.2).

8.2.1 STRATEGIC AND TACTICAL MODELS

May (1973) classified models into strategic and tactical ones. Strategic models are general, simple and parameter sparse. Though unrealistic for any specific situation, and hence unsuitable for exact predictions, they may lead to general insight. Tactical models, on the other hand, are specific, complex, detailed and have many parameters. If input processes are qualitatively well understood and input parameters are quantitatively well known (which is not often the case), the models are realistic and suitable for exact predictions. Tactical models, however, do not lead to general insight.

Metz (1990; Metz and de Roos, 1992) modified May's classification to obtain a better framework for providing a coherent and general picture of robust relations between mechanisms and phenomena, as opposed to the consideration of particular cases only. Connected through a general and encompassing class of strategic models, Metz distinguishes tactical models with a strategic goal, which are mathematically as simple as possible and constructed to uncover potential generalities, and tactical models with a practical goal, constructed for prediction or testing and usually incorporating lots of technically awkward detail.

In this field of tension between simple and complex models, compromise is inevitable. A model should have just enough realism and accuracy for its purpose, yet the results should be generalizable. As no model can ever embody the full truth, any specific problem can be tackled through a series of models, ranging from simple models, that provide a better route to understanding, to complex models, that yield more specific results. Furthermore, it is important to work in close connection to empirical research: it only makes sense to include those parameters into the model, of which we have or can obtain reasonable estimates.

8.2.2 SENSITIVITY ANALYSIS AND SCENARIO STUDIES AS TOOLS

Metapopulation models will usually not be able to provide exact, quantitative solutions for management problems (see discussion above). Instead, they can be used to identify crucial parameters, qualitatively simulate effects of landscape changes, direct field research and compare scenarios.

Sensitivity analysis is an important tool for assessing the relative importance of model parameters: a small change in some parameters may have a great effect on population viability, whereas the population may be relatively insensitive to changes in other parameters. The combined effect of changing two or more parameters at a time can be studied. Sensitivity analysis results can suggest what measures should be taken to preserve certain species; the measures that affect the crucial parameters are the ones that will have most effect. For example, for long-lived birds and mammals, adult survival is often identified as the key parameter for population viability (e.g. Lande, 1988; Lebreton and Clobert, 1991, see also section 8.5.3 and Figure 8.2 for an example). Furthermore, sensitivity analysis results can lead empirical research to focus on the parameter that most affects the population viability. Both the precision of the model and a general ecological understanding of the species under study will benefit most if knowledge on the crucial parameter is gathered.

A second important tool for impact assessment are scenario studies. Interference in the landscape, management measures and other advantageous or adverse human activities, can be simulated by translating them into changes in model parameters. Even when no data quantifying the impact of measures on the input parameters are available, best guesses and a safety range can be used. Although the exact quantitative model outcome is not necessarily correct, the qualitative results may be robust (insensitive to details in model specification): the best alternative as predicted by the model is likely to be the best one in real life, provided the model captures the essential qualitative behaviour of species and landscape under study. Harms *et al.* (1993, this volume) gives examples of scenario studies.

8.3. ON THE RISK OF EXTINCTION

Extinction is a crucial aspect of metapopulation studies. While local extinction is a key feature of metapopulation dynamics, regional (or metapopulation) extinction is what we want to prevent by management measures. Some causes of extinction are

1. deterioration of the environment, leading to an overall birth-deficit;
2. demographic stochasticity;
3. environmental stochasticity;
4. genetic effects.

The latter three are all related to small local carrying capacity, and therefore of particular interest to metapopulation studies. All can operate at a local or regional (metapopulation) spatial scale.

Environmental deterioration and genetic effects (inbreeding depression) can cause populations to decline. If average mortality outnumbers reproduction at all densities, this will lead to deterministic local extinction. At the regional scale, if extinctions outnumber colonizations systematically, this will result in deterministic metapopulation extinction. It may take a long time between the onset of adverse conditions and the death of the last individual (Quinn and Hastings, 1987).

Demographic stochasticity affects local populations particularly if numbers are small (say, below 20 adult females (Richter-Dyn and Goel, 1972)). Even in a favourable environment, chance events may lead to local extinction. At the metapopulation level, an equivalent of demographic stochasticity (immigration–extinction stochasticity sensu Hanski, 1991) may lead to chance extinction of small metapopulations, consisting of on the average less than, say, 20 local populations (Hanski, 1989).

Environmental stochasticity affects vital rates of all individuals in a population at the same time, the metapopulation level equivalent (regional stochasticity sensu Hanski, 1991) affects all metapopulation rates simultaneously. A sequence of adverse conditions (e.g. severe winters) can drive both local populations and metapopulations in an environment which is favourable on the average, to chance extinction. There are no safe numbers for environmental stochasticity (Goodman, 1987).

In summary, three types of metapopulation extinction can be distinguished. (1) If local extinction is deterministic, metapopulation extinction will be deterministic as well. (2) In case local extinction is due to chance events, (2a) colonization rate may be too small to compensate for local extinctions, leading to deterministic metapopulation extinction, or (2b) colonization rate may be sufficiently large, but the metapopulation may suffer chance extinction. In Figure 8.1, situations 2a and 2b correspond to values of the (colonization parameter)/extinction ratio to the left, respectively to the right of the critical values at which the curves intersect the x-axis.

In practice, it may be impossible to distinguish between the three types of metapopulation extinction. Chance events make it difficult to interpret field data, which are a single realization from a range of possible outcomes, and to make predictions. Because of the ever-present risk of chance extinction, most models discussed in this chapter are stochastic. In some situations, extinction occurs quickly, and would have occurred even in the absence of stochasticity (deterministic extinction). In other situations, extinction is due to chance events. Model results may be expressed in terms of the mean time to extinction, or, equivalently, the probability of persistence over, say, 100 years.

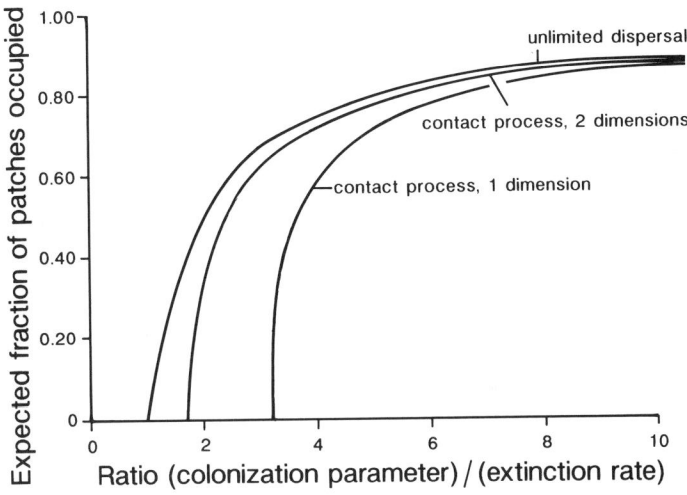

Figure 8.1 Expected fraction of patches occupied in a metapopulation with an infinite number of equal patches, against the ratio of colonization parameter and extinction rate. Three cases are distinguished: (a) unlimited dispersal (original Levins model), (b) only nearest neighbour interactions, patches on a square grid (contact process in two dimensions), and (c) only nearest neighbour interactions, patches on a line (contact process in one dimension). The curve in (a) is obtained by analytical techniques, (b) and (c) are obtained by simulation.

8.4 WINKING PATCHES METAPOPULATION MODELS

This section discusses models in which patches are classified as occupied by a population or empty. Four conditions should be satisfied for a winking patches metapopulation model to be a fair approximation of reality.

1. Patches of habitat are scattered in a matrix of non-habitat.
2. Extinctions and colonizations occur locally.
3. The time between actual colonization (the moment immigrants arrive in an empty patch) and reaching a local population equilibrium, is short relative to the mean occupation time of the patch.
4. The local dynamics are relatively independent of the regional dynamics; there is no rescue effect (cf. Brown and Kodric-Brown, 1977).

The first two conditions are satisfied for all species that are confined to natural types of habitat, and have small local populations and some interpatch dispersal. The last two conditions are usually satisfied if the local population growth rate is strongly density dependent, as in territorial animals, and if the dispersal frequency is of the same order of magnitude as the extinction rate of local populations. Population equilibrium in this chapter is used for a statistically stationary distribution of states that may vary in time (Tuljapurkar, 1989).

Below, we discuss the original Levins (1970) model with unlimited dispersal, the contact process with only nearest-neighbour dispersal (section 8.4.1), models with locally variable rates, reflecting differences in patch size, quality, and configuration (section 8.4.2), and models which take environmental fluctuation into account (section 8.4.3).

8.4.1 THE LEVINS MODEL AND THE CONTACT PROCESS

Levins (1970) introduced a metapopulation model, in which populations wink in and out of existence in an infinite number of equal patches. Any subpopulation has a fixed extinction rate, while any empty patch has a colonization rate that depends only on the fraction of occupied patches and a colonization parameter. The critical parameter of the model is the ratio (colonization parameter)/(extinction rate): the metapopulation persists forever if the ratio exceeds the critical value one, and otherwise becomes extinct.

In spite of its simplicity (the Levins model is a typical strategic model, sensu May), the model teaches us to think in terms of metapopulation dynamics, with instability at the local level and stability at the landscape level. Hence, it shows that a species may be (temporarily) absent from some suitable habitat patches. It also demonstrates the significance of dispersal in fragmented landscapes: species may become extinct in the presence of suitable habitat if the rate of local extinction exceeds the rate of colonization.

An interesting related model is the contact process, which only allows for colonization of nearest-neighbour patches, while all other assumptions remain unchanged. Patches may lie on a square grid (every patch has four nearest neighbours) or a line (every patch has two nearest neighbours). In these cases, the contact process in two and one dimension, the critical values of the ratio (colonization parameter)/(extinction rate) are 1.65 and 3.30 respectively (Brower *et al.*, 1978; Mollison, 1986). If this ratio permanently drops below the critical value, the metapopulation will go extinct.

The original model (with unlimited dispersal) and the contact process (with very restricted dispersal) may be seen as limiting cases of dispersal distance: individuals generally may be expected to have a dispersal range somewhere between the nearest neighbour patch and infinity. In Figure 8.1 the expected fraction of patches occupied is plotted against the ratio (colonization parameter)/(extinction rate) for the three cases: unlimited dispersal, contact process in two and in one dimension. Note that to the left of the threshold values, metapopulation extinction is certain (deterministic extinction). The Levins model predicts a higher occupation fraction and a lower extinction threshold than the other two models. Increasing overall abundance of a species can be achieved by two equivalent measures: by increasing the colonization parameter (e.g. construction of corridors) or by decreasing local extinction rate (e.g. increasing patch size or habitat quality). Note that for all three models the colonization parameter is defined as the rate of colonization of empty patches if only one patch is occupied and all others are empty.

8.4.2 EFFECTS OF PATCH SIZE, QUALITY, AND CONFIGURATION

The Levins model is based on the simplifying assumptions that all patches are equal, and their configuration is unimportant. This section discusses how the conclusions from the Levins model are altered if variations in patch size, quality and configuration are taken into account. Patches may differ in extinction rate (mimicking the effect of patch size and habitat quality), and the colonization probability of an empty patch may be a complex function of the number of occupied patches, interpatch distance, and the density of possible dispersal corridors in the surrounding landscape.

Surprisingly, not many people have explored this path of modelling systematically. Perhaps this is due to the almost infinite number of possible situations and algorithms on the one hand, and the lack of empirical data to implement the model on the other. Gathering the essential parameter values is time-consuming, as one needs time series of spatially explicit presence and absence data. For example, at the scale of the agricultural landscape of The Netherlands, local extinction rate in the nuthatch *Sitta europaea* was found to be a function of patch size and habitat quality. An overall colonization parameter could be calculated, each empty patch having a rate of being colonized equal to the colonization parameter times a measure of nuthatch density in the surrounding landscape (Verboom *et al.*, 1991c; Opdam *et al.*, 1993, this volume).

Heterogeneity in patch size and habitat quality has a stabilizing effect on metapopulations (Hanski, 1991): if the number of occupied patches drops, the species will tend to occur in the most favourable patches, causing the average extinction rate to decrease. Hence, the Levins model yields over-pessimistic results for metapopulations with heterogeneity of habitat patches.

Distance, configuration, and connectivity, and their effect on colonization rate have so far not been examined extensively in the context of winking patches models. If extinction rate increases with isolation (rescue effect), the Levins model may be over-optimistic (Hanski, 1991): if the number of occupied patches drops, the average extinction rate will increase. Conspecific attraction may have a similar effect (Ray *et al.*, 1991). Section 8.5.3 lists some of the results of structured metapopulation models. Opdam *et al.* (1993, this volume) discusses the empirical evidence for effects of patch size, habitat quality and configuration.

8.4.3 ENVIRONMENTAL FLUCTUATIONS AND SPREADING OF RISK

The assumption in the Levins model that local populations have constant extinction rates is unrealistic. In this section, we consider the effect of environmental fluctuations (both abiotic and biotic) on metapopulation dynamics. Attention is restricted to fluctuations in extinction rate. Colonization rate can be analyzed in the same way.

Extinction rates may fluctuate in time and space, due to varying weather conditions, diseases, or other factors. In the absence of fluctuations, an unfragmented population is always better off than a fragmented one. In a fragmented population, some local populations will be near carrying capacity (with a small growth rate) whereas others will be missing (with zero growth rate). Resources will not be optimally utilized, and dispersal may bring extra costs, in terms of energy or survival. This assumption does not hold for metapopulations in a fluctuating environment. If patches have independently fluctuating extinction rates, some degree of fragmentation may benefit the (meta)population, by means of spreading of risk: bad conditions occur only in a subset of patches at a time (Den Boer, 1981; Goodman, 1987).

Quinn and Hastings (1987) suggest that if a fixed area of habitat is divided into a number of reserve fragments, there is an optimum in reserve size and number. Their model, which assumes strongly but independently fluctuating extinction rates and zero colonization rate, predicts an optimum number of fragments equal to the square root of total carrying capacity, provided that local carrying capacity does not drop below 20 breeding females (Quinn and Hastings, 1988).

Contrary to the above assumption, environmental fluctuation in extinction rates will, at least partially, be correlated in space. For example, a severe winter will affect all patches in the same manner. However, due to differences between patches and stochastic effects, the effects on local populations may vary locally. Harrison and Quinn (1989) show that spatial correlation between local extinction rates decreases the mean time to extinction dramatically, especially in metapopulations that consist of many equal local populations. In highly correlated systems, an unfragmented population may again have better prospects for survival than a fragmented population. Hanski (1989) suggests, that in many species both exchange rate and degree of correlation in local dynamics change with distance. For example, the spread of an infectious disease through a metapopulation will be closely linked to dispersal rate and distance. He concludes that rare and endangered species may be the ones with low dispersal rates but high correlation of local dynamics.

8.5 STRUCTURED METAPOPULATION MODELS

A structured metapopulation model is in effect a population model with several populations and a dispersal algorithm. Structured models can be used to test whether the conditions for use of winking patches models (section 8.4) are met (Verboom *et al.*, 1991a). For impact assessment, structured metapopulation models can be employed if the use of winking patches models is impossible, or in addition to winking patches models.

To test whether the winking patches models accurately describe the metapopulation dynamics, two tests can be performed. The first test (cf. condition 3 in section 8.4) is for constancy of the extinction rate. Under constant extinction rate, the probability distribution of extinction times is exponential (and the log-survivor function is linear). A probability distribution of extinction times derived with a structured model may show deviations from an exponential distribution: shortly after colonization, the extinction rate may be higher due to small population size. If the period in which the extinction probability deviates from an exponential distribution is relatively short (which means that the equilibrium population size is reached rather quickly), condition 3 is satisfied.

The second test (cf. condition 4 in section 8.4) is for rescue effects. In winking patches models, populations with and without immigration have the same extinction rate. With a structured model, mean persistence times of a population with and without (realistic levels of) immigration can be compared to reveal the impact of overall abundance (through immigration rate) on local population dynamics.

If local populations grow to equilibrium shortly after immigration and are little affected by immigrants, much computer time can be saved by using a winking patches model. Moreover, such a model will yield results that better allow generalization. Verboom *et al.* (1991a) have shown for the European badger (*Meles meles*), that model results were essentially the same for the structured metapopulation model and the derived winking patches model. In such cases, a combination of winking patches and structured metapopulation models can be used for impact assessment of (de)fragmentation. Structured models can be used to estimate extinction and colonization rates, if these cannot be estimated from field data. Moreover, changes in these parameters, due to changes in the landscape, can be implemented through the change in the structured model parameters birth, death and dispersal rate.

Below, we discuss the two key processes of metapopulation dynamics, local extinction and (re)colonization (sections 8.5.1 and 8.5.2). We end with some examples of 'complete' structured metapopulation models, taking both local and metapopulation dynamics into account (section 8.5.3).

8.5.1. LOCAL EXTINCTION

Population viability depends on birth, death and dispersal rates and patch carrying capacity, and how these vary in time. Section 8.3 lists some of the causes of extinction. Here we list some results linking population parameters to local extinction probability.

1. Probability of extinction from chance events decreases with increasing carrying capacity, actual population size, and population growth rate at low local densities (Richter-Dyn and Goel, 1972). Therefore, newly founded and slowly growing populations have a higher extinction risk than established populations or populations with a large growth rate. Small populations are more endangered than large ones.

2. Populations experiencing large fluctuations in vital rates are more likely to become extinct (Goodman, 1987). Minimum viable population size is for example, much larger for invertebrates and small vertebrates than for large vertebrates. Mean persistence time is predicted to be approximately proportional to exp(carrying capacity) if environmental fluctuation has little effect on population dynamics (as may be the case in large, long-lived birds and mammals), and to increase somewhat less than linearly with carrying capacity if environmental fluctuations do have an effect (as in many small and/or short-lived organisms) (Leigh, 1981; Goodman, 1987; Quinn and Hastings, 1987).

3. Local extinction rate may be decreased by a frequent influx of immigrants into the population (Brown and Kodric-Brown, 1977). If this is the case, fragmentation may have an extra severe effect if it reduces dispersal flows between the patches.

4. Local extinction rate in large, long-lived species is most sensitive to the adult survival rate. For example, in the European badger, adult mortality was found to be the key parameter for local persistence (Figure 8.2, after Lankester *et al.*, 1991). In small, short-lived species,

extinction rate seems to be most sensitive to parameters associated with reproduction (Lebreton and Clobert, 1991).

8.5.2 COLONIZATION

The probability of (re)colonization depends on two processes: dispersal (immigration) and establishment. A propagule may be a fertilized female, or a group of at least one female and a male. The probability per time unit that a patch is reached depends on distance to nearby populations as well as on the presence of dispersal corridors and barriers.

To assess the colonization rate of an empty patch, information must be available on dispersal rate and distance, and on the behaviour of dispersing organisms in relation to landscape elements. Unfortunately, the available information is usually insufficient. Individuals may react to non-habitat by for example, turning back (non-dispersing) or moving fast, and with an increased mortality hazard, to the nearest patch of habitat within sight. Known dispersal curves are often the result of averaging dispersal movements through many different landscape types, and consequently can not simply be applied in a specific fragmented landscape.

Possible simplifications are:

1. dispersal is equally likely to all patches in the metapopulation (Levins model);
2. dispersal is only possible to the nearest neighbouring patches on a grid (contact process);
3. dispersal is only possible if corridors are present (Fahrig and Merriam, 1985).

Harms *et al.* attempted to quantify the relative accessibility of habitat patches for dispersing organisms (Harms and Opdam, 1989; Harms *et al.* 1993, this volume). They dissected the landscape into grid cells with a resistance value, using a GIS (Geographical Information System). Their model yields species-specific relative accessibility maps that, although based on a simple algorithm, are of great value for landscape designers.

Mean growth rate, variation in growth rate, and initial population size are the key parameters affecting the probability of establishment (Lawton and Brown, 1986). The general results for local extinctions listed above also hold for the probability of establishment. Single population models can be used to study the probability of establishment, starting with a propagule (e.g. Verboom *et al.*, 1991a).

Some attention should be given to the problem of two sexes: Do fertilized females disperse? Does the settlement of a female in a new patch suffice to start off a population (if males disperse more often than females and she is able to attract one with some certainty)? Or, alternatively, does a single singing male bird (frog, cricket) attract a female easily to an isolated patch? If both a male and a female have to arrive in a patch independently, the colonization rate is expected to be much smaller than if there is some attraction between the sexes. This aspect may possibly be more important than others (Lande, 1987; Verboom *et al.*, 1991a).

8.5.3 EXAMPLES OF STRUCTURED METAPOPULATION MODELS

A population model can be extended to a metapopulation model by allowing dispersal between a set of patches with local dynamics. The resulting models are often complicated (of the tactical model with practical goal type, sensu Metz) and do not allow generalization easily. Some general results and specific examples are given below.

Overall density may affect local density (Hastings, 1991; Hanski, 1991; Gyllenberg and Hanski, 1992), and a metapopulation may have to surpass a threshold of overall density to be viable. Above this threshold, due to rescue effect, mean extinction rate is smaller than mean colonization rate. Beneath the threshold, this relation is reversed, and the metapopulation is non-viable. As a result, some metapopulations may be more endangered than their size suggests. This result has also implications for reintroduction programmes: a minimum number of released individuals may be necessary for a non-zero probability of success.

The proportion of the landscape which is suitable habitat influences the prospects for survival. Lande (1987) derived an extinction threshold for territorial animals, depending on birth, survival and dispersal characteristics. This threshold value is increased if Allee-effects, edge effects, or environmental fluctuations are taken into account. The model was applied to the northern spotted owl (*Strix occidentalis caurina*), for which the habitat (old growth forest) has recently decreased from 60–70% to ca. 38% (Lande, 1988). Under Forest Service guidelines the future proportion would be 7–16%, which is far below below the extinction threshold estimated at 21% (Lande, 1988). Doak (1989) added explicit spatial structure to the model. He suggests three strategies to aid owl populations: (1) facilitation of dispersal, (2) increased clumping of territories (patch size) and (3) allow second-growth forest to mature.

Urban and Shugart (1986) have demonstrated the synergistic effect of restricted vagility and reduced natality for a typical neotropical migrant bird. Small, isolated patches were found to be unoccupied most of the time, large patches rarely suffered extinction, and small or inter-

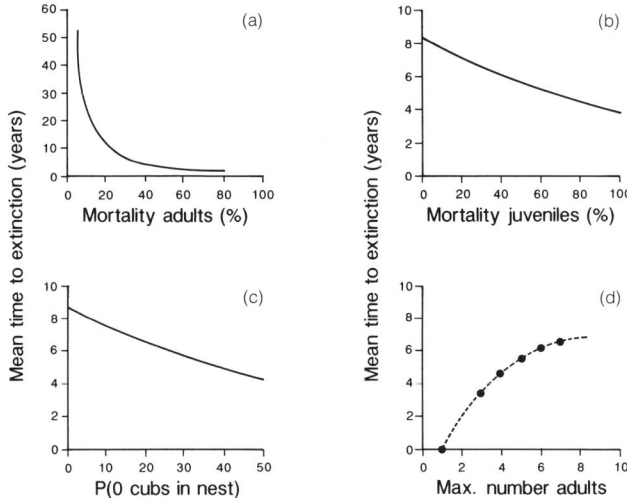

Figure 8.2 Sensitivity analysis results for a model of a local badger population (a badger clan). The mean time to extinction, a measure of population viability, is plotted against (a) adult mortality, (b) juvenile mortality, (c) the probability of zero cubs (equivalent to the mean number of cubs), and (d) the maximum number of adults (the carrying capacity) in a clan. Default values, as estimated for The Netherlands, are respectively 25% (per year), 50% (per year), 30% and 5. Note that the sensitivity to adult mortality dominates. Results after Lankester *et al.* (1991).

mediate patches adjacent to other patches showed dynamics of colonization and extinction.

Fahrig and Merriam (1985) have shown the importance of dispersal, connectivity (presence of corridors), and patch configuration for the white-footed mouse *Peromyscus leucopus*. Although good quality patches and corridors are important for metapopulation survival, bad quality patches and corridors may serve as a metapopulation sink, if mortality is high (Henein and Merriam, 1990).

Prospects for conservation of the European badger have been studied (Lankester *et al.*, 1991; Verboom *et al.*, 1991a), and some conclusions from this study are the necessity to decrease adult (traffic) mortality for local (Figure 8.2) and regional survival, and the importance of dispersal and number of patches (group territories) connected by dispersal. It is essential to conserve uninhabited badger setts, and allow for future recolonization (Lankester *et al.*, 1991).

The interaction between large core areas and small forest fragments was studied with a structured metapopulation model for the European nuthatch by Verboom *et al.* (1991b), who found that both the large core areas had a stabilizing role for the population in the fragments and vice

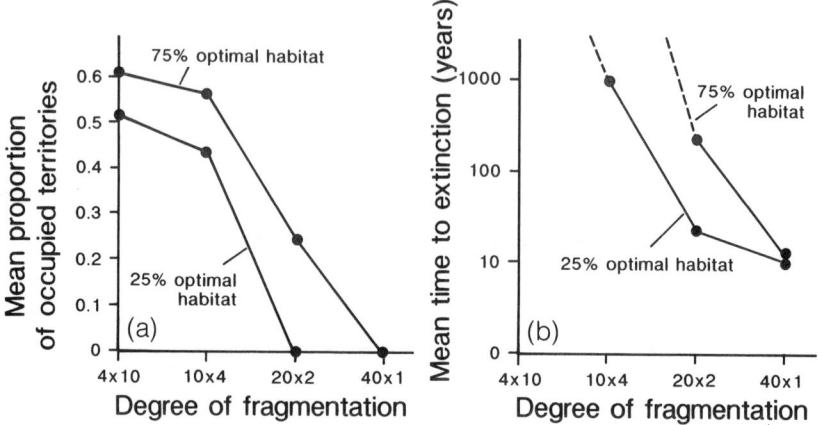

Figure 8.3 Metapopulation viability of the European nuthatch and the degree of fragmentation of an area of habitat with a carrying capacity of 40 pairs. Habitat configuration ranges from four patches with ten territories each to 40 patches with a single territory. Interpatch distances are kept equal to 3 km. Either 75% or 25% of the habitat is optimal, the rest is suboptimal (with higher mortality and lower reproductive success). (a) The mean proportion of territories occupied. (b) The mean time to extinction, starting with all territories occupied. The simulation model is described by Verboom *et al.* (1991b).

versa: the fragment populations buffered the fluctuations in the core areas and served as stepping stones between the core areas.

For illustration, Figure 8.3 shows the effects of fragmentation on territory occupation and mean time to metapopulation extinction as predicted by the nuthatch model. Figure 8.3 shows the results of the fragmentation of an area of habitat with a carrying capacity of 40 pairs (territories), starting with four patches containing ten territories each and ending with 40 patches containing one territory. The patches are regularly spaced out with an equal interpatch distance of 3km. Both simulations with 75% and 25% of optimal habitat were carried out. In suboptimal habitat, mortality is higher (especially for juveniles and in severe winters) and reproduction is lower. Note (Figure 8.3a) that the territory occupation decreases with increasing fragmentation. An extinction threshold exists, below which no equilibrium territory occupation exists and metapopulations decrease to extinction deterministically. In favourable landscape (75% optimal habitat), this threshold is traversed when fragmenting the habitat into patches smaller than two territories, whereas in unfavourable landscape (25% optimal habitat) metapopulations with less than four territories per path are non-viable. The mean time to extinction, starting from carrying capacity, is near infinite (over 1000 years) for some situations and decreases with increasing fragmentation, being lower for the unfavourable landscape (Figure 8.3b). Note that even the subcritical metapopulations (below the extinction threshold) have a mean time of one or several decennia before the last patch population becomes extinct.

8.6 DISCUSSION

Many species that are interesting from a conservation point of view, either charismatic and threatened species, or species characteristic for the landscape in which they occur, occur in metapopulations. They have declined as a result of habitat fragmentation, although some patches of apparently good habitat still remain. Characteristically, some patches are crowded while others remain unexploited because they cannot be reached. The species have small local populations, prone to local chance extinction, while the interpatch dispersal flow is too weak for all suitable patches to be reached and colonized quickly to counteract local extinctions. For such species, metapopulation models can be used.

Species that do not occur in typical metapopulations may also be affected by fragmentation, if some patches are crowded while others are underexploited (Burkey, 1989). Even if dispersal is frequent and dispersal distances are large, edge effects and an increased dispersal effort (in terms of energy or mortality) may threaten populations with fragmented habitat. Most of the results and methods for metapopulations presented here are expected to hold for all populations with fragmented habitat.

This chapter has concentrated on single species models. Of course, interactions between species are often important and for some species should be taken into account when studying the effects of fragmentation. By studying a variety of species with different life histories, generalities can be revealed: species characteristics can be linked to specific responses to (de)fragmentation. Thus, extrapolation to other species becomes possible.

Metapopulation models for impact assessment of (de)fragmentation yield certain strategic results, i.e. qualitative, model-independent rules of thumb. These rules can be expressed in terms of size, nature and configuration of landscape elements, for the species we want to conserve. Here, some of the general results are summarized.

1. The ratio of colonization parameter and extinction rate is the driving parameter for metapopulation dynamics (Figure 8.1). To ameliorate the prospects of a metapopulation, local extinction rates should be minimized and opportunities for colonization should be maximized.

2. Low local extinction rates can be achieved if patches are large and contain good quality habitat. In other words, local population growth rate should be optimized and local carrying capacity should be high.

3. Colonization rates will be high if the habitat patches have a configuration such that interpatch distances are small. Colonization can be stimulated by dispersal corridors, stepping stones, or devices that counteract the effect of barriers (like badger tunnels and cerviducts). Management action for local reintroduction (transportation of propagules by man) could be considered if isolation makes natural colonization impossible. However, this should be seen as a temporary solution, and action should be taken to increase natural colonization rates (Opdam and Verboom, 1991). Moreover, interpatch dispersal may benefit the population due to the rescue effect.

4. Large numbers of habitat patches should be available, and empty patches should also be preserved. Especially patches that due to their size and habitat quality serve as a stable source for dispersing propagules, may play an important stabilizing role in metapopulations and should be conserved.

5. The choice between several large or many small patches, for any species, can only be made after consideration of its particular life history characteristics, the role of interpatch dispersal, and the relative merits of spreading of risk and local density dependence effects. There are no general rules here.

6. For specific problems, tactical models can be derived, imbedded in general metapopulation theory, and based on specific empirical data. Sensitivity analysis and scenario studies can serve as tools for impact assessment.

ACKNOWLEDGEMENTS

The authors thank the editors and Wim van der Steen, Ilkka Hanski, and Donald DeAngelis for their constructive criticism.

REFERENCES

Brower, R.C., Furman, M.A. and Moshe, M. (1978) Critical exponents for the reggeon quantum spin model. *Physics Letters*, **76**(B), 213–19.

Brown, J.H. and Kodric-Brown, A. (1977) Turnover rates in insular biogeography: effect of immigration on extinction. *Ecology*, **58**, 445–9.

Burkey, T.V. (1989) Extinction in nature reserves: the effect of fragmentation and the importance of migration between reserve fragments. *Oikos*, **55**, 75–81.

Den Boer, P.J. (1981) On the survival of populations in a heterogeneous and variable environment. *Oecologia*, **50**, 39–53.

Doak, D. (1989) Spotted owls and old growth logging in the pacific northwest. *Conservation Biology*, **3**, 389–96.

Fahrig, L. and Merriam, G. (1985) Habitat patch connectivity and population survival. *Ecology*, **66**, 1762–8.

Goodman, D. (1987) Consideration of stochastic demography in the design and management of biological reserves. *Natural Resources Modelling*, **1**, 205–34.

Gyllenberg, M. and Hanski, I. (1992) Single-species metapopulation dynamics: a structured model. *Theoretical Population Biology* (in press).

Hanski, I. (1989) Metapopulation dynamics: does it help to have more of the same? *Trends in Ecology and Evolution*, **4**, 113–14.

Hanski, I. (1991) Single species metapopulation dynamics: concepts, models and observation. *Biological Journal of the Linnean Society*, **42**, 17–38.

Harms, W.B. and Opdam, P. (1989) Woods as habitat patches for birds: application in landscape planning in the Netherlands, in *Changing Landscapes: an Ecological Perspective* (eds. I.S. Zonneveld and R.T.T. Forman), Springer-Verlag, New York, pp. 73–97.

Harms, W.B., Knaapen, J.P. and Rademakers, J.G.M. (1993) Landscape planning for nature restoration: comparing regional scenarios in *Landscape Ecology of a Stressed Environment* (eds. C.C. Vos and P. Opdam), Chapman & Hall, London, pp. 197–218.

Harrison, S. and Quinn, J.F. (1989) Correlated environments and the persistence of metapopulation. *Oikos*, **56**, 293–8.

Hastings, A. (1991) Structured models of population dynamics. *Biological Journal of the Linnean Society*, **42**, 57–71.

Hastings A. and Wolin, C.L. (1989) Within-patch dynamics in a metapopulation. *Ecology*, **70**(5), 1261–6.

Henein, K. and Merriam, G. (1990) The elements of connectivity where corridor quality is variable. *Landscape Ecology*, **4**, 157–70.

Lande, R. (1987) Extinction thresholds in demographic models of territorial populations. *American Naturalist*, **130**, 624–35.

Lande, R. (1988) Demographic models of the northern spotted owl (*Strix occidentalis caurina*), *Oecologia*, **75**, 601–7.

Lankester, K., van Apeldoorn, R.C., Meelis, E. and Verboom J. (1991) Management perspectives for populations of the Eurasian badger *Meles meles* in a fragmented landscape. *Journal of Applied Ecology*, **28**, 561–73.

Lawton, J.H. and Brown, K.C. (1986) The population and community ecology of invading insects. *Philosophical Transactions of the Royal Society of London*, **B314**, 607–17.

Lebreton, J.-D. and Clobert, J. (1991) Bird population dynamics, management, and conservation: the role of mathematical modelling, in *Bird Population Studies* (eds. C.M. Perrins, J.-D. Lebreton, and G.J.M. Hirons), Oxford Ornithology Series, Oxford University Press, Oxford pp 105–25.

Leigh, E.G.J. (1981) The average lifetime of a population. *Journal of Theoretical Biology*, **90**, 213–39.

Levins, R. (1970) Extinction. In: *Some Mathematical Questions in Biology*, vol. II, pp. 75–108. American Mathematical Society, Rhode Island.

May, R.M. (1973) *Stability and Complexity in Model Ecosystems*, Princeton University Press, Princeton.

Metz, J.A.J. (1990) Chaos en populatiebiologie, in *Dynamische Systemen en Chaos: een Revolutie Vanuit de Wiskunde* (eds. H.W. Broer and F. Verhulst), Epsilon, Utrecht, pp. 320–44.

Metz, J.A.J. and de Roos, A.M. (1992) The role of physiologically structured population models within a general individual-based modelling perspective, in *Individual-Based Approaches in Ecology: Concepts and Models* (eds. D.L. DeAngelis and L.J. Gross), Chapman & Hall, New York (in press).

Mollison, D. (1986) Modelling biological invasions: chance, explanation, prediction. *Philosophical Transactions of the Royal Society of London*, **B314**, 675–93.

Opdam, P. (1988) Populations in fragmented landscape, in *Connectivity in Landscape Ecology* (ed. K.-F. Schreiber), Proceedings of the 2nd International Seminar of the International Association for Landscape Ecology. Münster: Münstersche Geographische Arbeiten 29, pp. 75–7.

Opdam, P., Van Apeldoorn, R., Schotman, A. and Kalkhoven, J. (1993) Population responses to landscape fragmentation, in *Landscape Ecology of a Stressed Environment* (eds. C.C. Vos and P. Opdam), Chapman & Hall, London, pp. 140–70.

Opdam, P. and Verboom, J. (1991) Is uitsterven een probleem, en herintroductie een oplossing? *De Levende Natuur*, **1991/5**, 149–55.

Quinn, J.F. and Hastings, A. (1987) Extinction in subdivided habitats. *Conservation Biology*, **1**, 198–208.

Quinn, J.F. and Hastings, A. (1988) Extinction in subdivided habitats: reply to Gilpin. *Conservation Biology*, **2**, 293–6.

Ray, C., Gilpin, M. and Smith, A.T. (1991) The effect of conspecific attraction on metapopulation dynamics. *Biological Journal of the Linnean Society*, **42**, 123–34.

Richter-Dyn, N. and Goel, N.S. (1972) On the extinction of a colonizing species. *Theoretical Population Biology*, **3**, 406–33.

Tuljapurkar, S. (1989) An uncertain life: demography in random environments. *Theoretical Population Biology*, **35**, 227–94.

Urban, D.L. and Shugart, H.H. (1986) Avian demography in mosaic landscapes: modelling paradigm and preliminary results, in *Wildlife 2000* (eds. J. Verner, M. Morrison, and C.J. Ralph), University of Wisconsin Press, Madison, pp. 273–9.

Verboom, J., Lankester, K. and Metz J.A.J. (1991a) Linking local and regional dynamics in stochastic metapopulation models. *Biological Journal of the Linnean Society*, **42**, 39–55.

Verboom, J., Opdam, P. and Schotman, A. (1991b) Kerngebieden en kleinschalig landschap, een benadering met een metapopulatiemodel. *Landschap*, **8/1**, 3–13.

Verboom, J., Schotman, A., Opdam, P. and Metz, J.A.J. (1991c) European nuthatch metapopulations in a fragmented agricultural landscape. *Oikos*, 149–56.

Part Four

Methods and Concepts of Landscape Planning

In The Netherlands, different kinds of land use compete for space, which is in short supply . Out of necessity, physical planning is highly developed. Activities that can not coincide have to be spatially separated in such a way that spatial relations do not adversely affect the functioning of separated land units. There are three planning levels: the national level, the regional level and the local level, corresponding to the levels of the national government, the provincial authorities and the local municipality. Land consolidation plans usually encompass several municipalities. Originally, these plans were developed for the improvement of the landscape from the point of view of agriculture. However, recently they are evolving toward multifunctional landscape planning, in which restoration of ecological fuctions may be a prominent issue. At all levels, spatially explicit plans are being developed, sometimes for a single type of land use (e.g. nature conservation), sometimes encompassing various types of land use. At the national level, the character of the plans is more a strategic outline indicating locations only roughly. The national nature policy plan, published recently, is a good example of a strategic plan. On the local level, plans are or should be detailed, indicating the location and function of land units.

In recent times it has become customary to include nature conservation goals in planning. Very recently, nature restoration is becoming a popular issue. Nowadays the usual planning strategy is to separate spatially functions that affect each other adversely, as is often the case with nature conservation and agriculture. The framework planning concept proposed by Van Buren and Kerkstra is a good example. Drinking water extraction and nature conservation can be spatially combined as long as the groundwater level is not affected. Where drinking water is in short supply and artificial measures are being taken to replenish groundwater loss, these functions can not be integrated simply. Bakker and Stuyfzand develop a basis for spatially separating drinking water and nature conservation functions in the coastal dunes, which are designated as nature areas on a national level.

Depending on the level of scale, landscape ecological knowledge may have different functions in landscape planning. On the national and regional level, the emphasis is on detecting problems resulting from the actual configuration of land units, identifying the causes and indicating spatially explicit possibilities for solving these problems. Often, qualitative assessments and planning rules are sufficient. On the local level, the evaluation of detailed plans with reference to the goals set for the conservation or restoration of natural systems is relatively more important. This requires detailed knowledge of quantitative relations and spatially explicit tools to apply that knowledge in comparing plan scenarios.

These different types of applications of landscape ecological knowledge can also be distinguished within the cyclic planning procedure proposed by Harms *et al*. Several planning cycles, design and evaluation, are gone through, separated by an evaluation step in which the best scenario is selected, going from rough to detailed, from qualitative to quantitative. Tjallingii presents a non-cyclic method of developing plans on the basis of landscape ecological analysis, systematically proceeding from a global analysis to a detailed plan that integrates various functions of an urban system.

A detailed analysis of spatial relations is indispensable for comprehensive planning on an ecological basis. Several of the previous chapters have shown the complexity of spatial relations in the landscape, due to the interaction of flows on various spatial scales and the species-specific responses to changes in the landscape structure. However, in planning procedures there is usually no time for laborious empirical studies, and planners have to be content with generalized knowledge derived from studies elsewhere. That is why it is so important to develop an extensive set of ecological planning rules and a collection of tools to apply, quantitatively or qualitatively, landscape ecological knowledge. Some of these tools have been discussed earlier in this book (e.g. Verboom *et al*, Barendregt *et al.*). In this section, most of the knowledge summarized in preceding chapters is extended to ecologically based planning methods and concepts.

These planning concepts, although based on landscape ecological knowledge, often nicely demonstrate the lack of knowledge when it comes to application. Such attempts to generalize existing knowledge within a spatially explicit concept are sometimes discarded by researchers, who stress the uncertainties and weaknesses. However, planning concepts in which the assumptions about ecological relations and effects are made explicit, should rather be a challenge to landscape researchers. In fact, maps on which concepts are visualized ask for translating knowledge of spatial processes into predictions on population or ecosystem functioning, given an explicit set of pattern characteristics. Hence, maps for landscape planning ask to link process to pattern, and that is what landscape ecology is all about.

Landscape planning for nature restoration: comparing regional scenarios

9

Bert Harms, Jan P. Knaapen and Jos G. Rademakers

9.1 INTRODUCTION

Recently, ecologists have become increasingly convinced that an emphasis on nature conservation only is not sufficient to maintain natural values. In spite of a strong tradition in nature conservation in The Netherlands, only 6% (about 200 000 ha) of the country's area is in designated nature reserves. There are more than 500 such reserves and most of them are smaller than 200 ha. The growing awareness of the adverse effects of fragmented landscapes in general and of the threat to small natural areas from the surroundings in particular, has led to a more aggressive initiative to claim additional space for nature restoration, in order to maintain phenomena of ecological value in the long run. This was manifest in the Nature Policy Plan (Ministerie van Landbouw. Natuurbeheer en Visserij, 1990) issued recently by the Dutch government, which paid due attention to nature restoration. For the first time in the history of Dutch national physical planning it was proposed to compensate for the deterioration of nature by allocating a considerable area (50 000 ha) for nature restoration.

Landscape Ecology of a Stressed Environment
Edited by Claire C. Vos and Paul Opdam
Published in 1993 by Chapman & Hall London. ISBN 0 412 44820 3

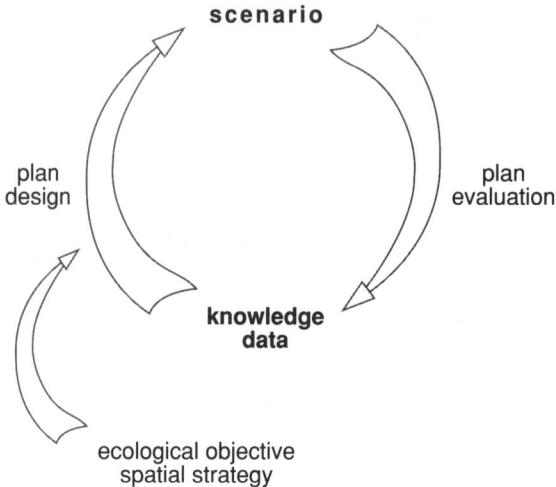

Figure 9.1 Cyclical planning procedure.

The shift in emphasis from conservation to restoration has also ushered in a new era for collaboration between ecologists and planners. Various questions have to be answered concerning suitability of locations, types of nature and their spatial distribution, management, connections to related natural areas and compatability with other land-use types. When tackling these questions we have to consider what kind of ecological information is needed and how this should be incorporated in the planning process. This chapter presents a planning procedure in which ecology contributes differently to the two stages of a planning process: plan design and plan evaluation (Figure 9.1). In the design stage, general ecological knowledge at landscape level contributes to a creative solution of the problem. The landscape structure as a whole has to be taken into account, not the attributes of the individual elements. A holistic approach predominates here and results in different scenarios for nature restoration. The next stage, the plan evaluation, is a more analytical activity. The plan has to be validated at the most concrete level of knowledge, i.e. the ecosystem or, even better, the individual species used as indicators for ecosystems. Deviations from a determined objective are assessed by measuring impacts on selected species. A computer model has been developed in order to evaluate the scenarios in this stage. Since the stages are not completed consecutively but may alternate cyclically the results of the evaluation are input in a new planning cycle in order to adjust the scenarios and to re-evaluate them. Eventually, a more comprehensive plan can be developed. The following case study illustrates how ecology is incorporated in landscape planning for nature restoration.

9.2 A REGIONAL CASE STUDY: THE CENTRAL OPEN SPACE PROJECT

The economic activities of The Netherlands are concentrated in the west of the country, the so-called Randstad, a loose conurbation consisting of Amsterdam, Rotterdam, The Hague, Utrecht and some smaller cities. Half of the country's population of 15 million people lives here. According to the Fourth Report on Physical Planning (Ministerie van VROM, 1988), this economic core should extend eastwards, to link up with the cities of Arnhem and Nijmegen. This Central City Belt contains several traffic axes connecting the Randstad by rail, road and waterways with the German conurbation 'Ruhrgebiet'. In the middle of this Central City Belt there is a large open agricultural area of about 400 000 ha, called the Central Open Space (Figure 9.2). Peatland, marine and fluviatile clay lowlands predominate in this Holocene area. In the early Middle Ages the area consisted of swamps and mud flats formed by sea and rivers, before it was reclaimed for farmland. It was gradually enclosed, diked and drained, enabling permanent settlements and agriculture to become established. The area is crossed by three parallel rivers (the Rhine, Waal and Meuse) and is bounded by coastal dunes in the west,

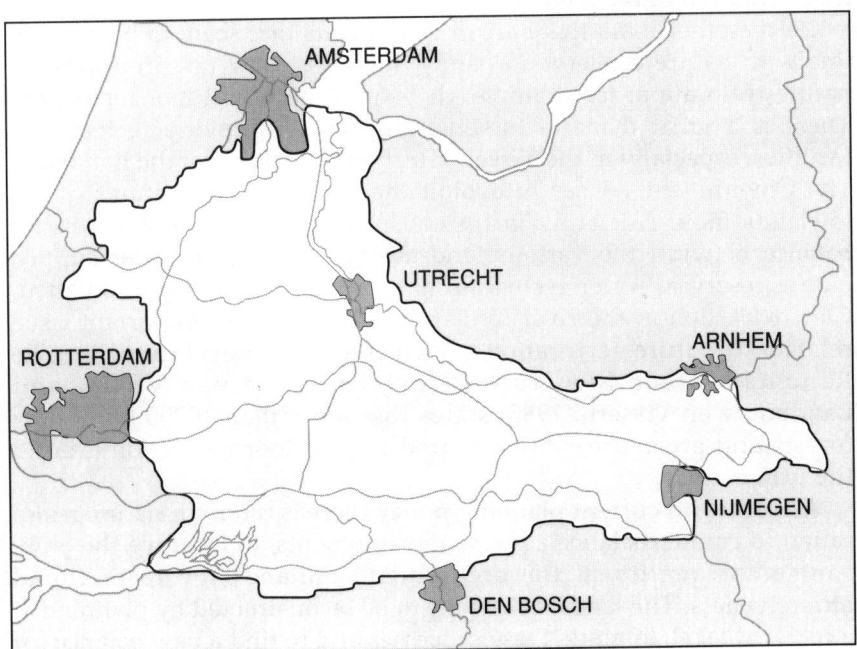

Figure 9.2 The study area The Central Open Space.

Pleistocene ice-pushed ridges in the north-east and aeolian sand deposits from the Pleistocene in the south. The slopes and the levees have long been favourite sites for settlements (Vos and Zonneveld, 1993, this volume).

Today the area is under stress. There is strong pressure towards urbanization: housing for one million people is planned for the extension of the Randstad. Glasshouse horticulture, one of the most important national industries, is expanding and threatening the open space. Most of the area is under grass, and dairy farming largely determines the appearance of the landscape. In the peat grassland areas a particularly drastic economic change in agriculture will be necessary because of the surpluses produced and the continuing costs of draining and land consolidation. Some economists expect that up to 20% of the land will be withdrawn from agriculture (Douw *et al.*, 1987). So the future of dairy farming on peat areas is very uncertain. However, the peat grasslands have great international value for the conservation of meadow birds. The Central Open Space is also affected by the growth of infrastructure: broadening of motorways, the construction of a new railway for the High Speed Train, widening of the important canal between Amsterdam and the Rhine. These developments will cause a further fragmentation of the landscape.

On the other hand there are developments that seem to be more in line with nature conservation and which give new opportunities for nature restoration: for example, the expansion of outdoor recreation. There is a great demand for enlarging and improving recreational facilities, especially in the 'Green Heart' in the middle of the Randstad. The Government wishes to exploit the characteristic features of the lowlands more efficiently in future. In large areas of water a better balance between the various land-use types, such as water supply, nature reserves, water recreation and transport is being stimulated. Clay extraction in the river forelands will also be an important issue related to nature restoration and recreational developments. The Randstad Green Structure Plan (Ministerie van VROM and Landbouw en Visserij, 1985) states that more than 10000 ha of new forests and green areas are required for outdoor recreation around the urban areas.

According to current planning policy there is room for nature restoration to counterbalance adverse developments, to improve the environmental quality of the area and to enhance the recreational attractiveness. The landscape needs to be reconstructed by planning at a regional level, to guide these processes and to find a new balance for all interests.

Four planning scenarios for nature restoration for the Central Open Space have been developed and evaluated in order to help solve these planning problems and to implement the national physical planning programme on a regional scale. A computer model (COSMO) has been constructed for the plan evaluation, to assess the ecological consequences (abiotic and biotic) of each scenario, and when and where these consequences will occur.

9.3 FOUR SCENARIOS FOR NATURE RESTORATION

To develop the scenarios we examined existing plans and schemes for landscape planning and nature restoration in particular. Most of the various concepts are ambiguous, a compromise or only partly useful. The underlying concepts about nature restoration often remain hidden. To understand the basic strategies in nature restoration it was necessary to reveal and clarify these concepts. Two questions are crucial here: what is the ecological objective and which sites are suitable?

9.3.1 ECOLOGICAL OBJECTIVES AND SPATIAL STRATEGIES

The first question concerning the ecological objectives is related to the ecosystem. Several different ecosystems can be developed on one site under different types of nature management. The choice of the system to be developed is a matter of nature restoration policy. One can focus on rare or endangered communities, on international significance or self-sustainability of the ecosystems involved, on seminatural species-rich communities sustained by human intervention or a non-intervention wilderness, probably with a lower species diversity.

The second question is related to the landscape as a whole, the spatial configuration of ecosystems. An existing arrangement or configuration is the result of an interaction of historical, social and landscape ecological patterns and processes. In this regard account is taken of other activities, like farming, infrastructure, recreation, water supply and mineral extraction which also require space in the study area. Nature restoration cannot be considered independently from those activities. They are sometimes interrelated in the sense of mutual support, but they can also be unconnected, or even incompatible. Nature restoration has

Table 9.1 Ecological objectives and spatial strategies of the four scenarios

Scenario	Ecological objective	Spatial strategy
GODWIT	Variety of ecosystems	Integration and zoning of land use
OTTER	Improvement of dispersal	Development of corridors and networks
ELK	Self-sustaining ecosystems	Segregation of land use
HARRIER	Variety of ecosystems	Segregation of land use optimal site selection

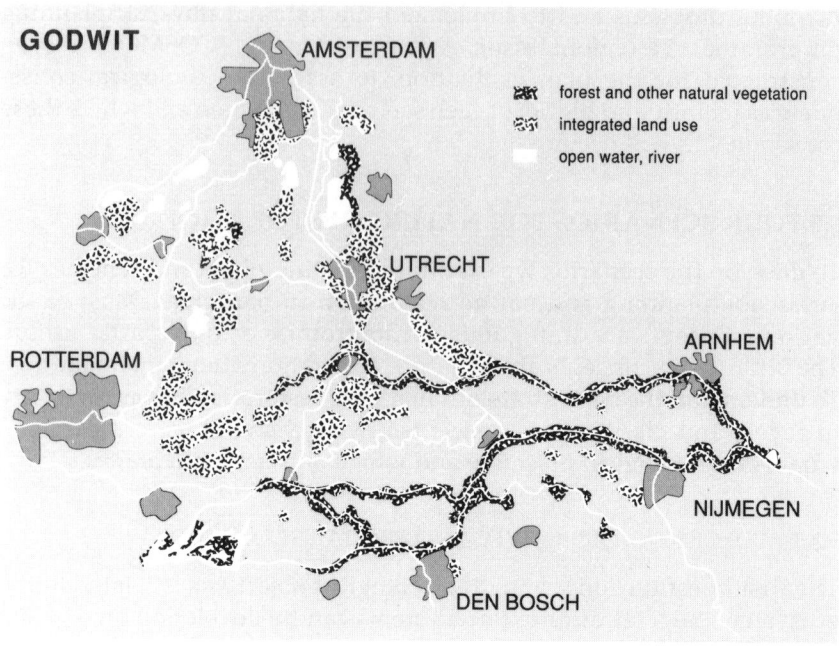

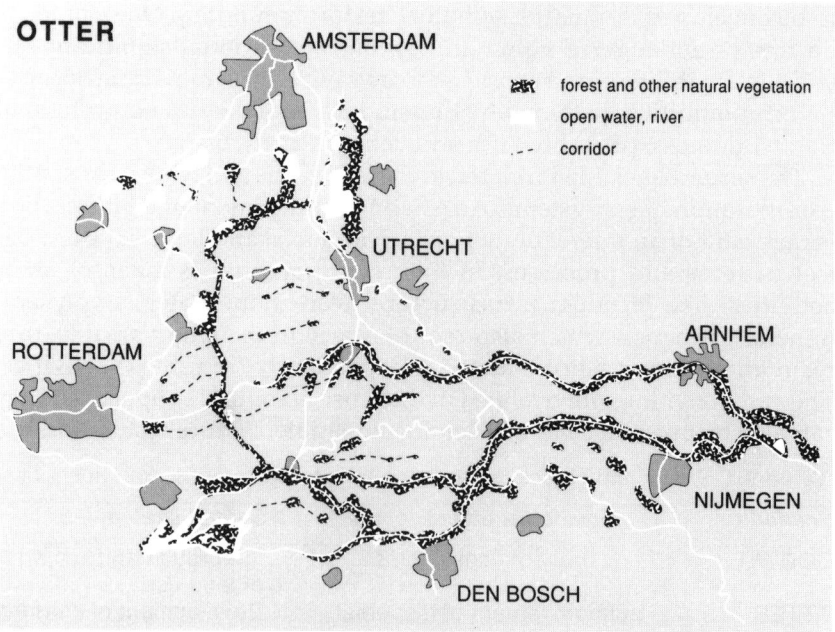

Figure 9.3 (a) Spatial layout of the four scenarios (simplified). (b)Visual impression of the four scenarios.

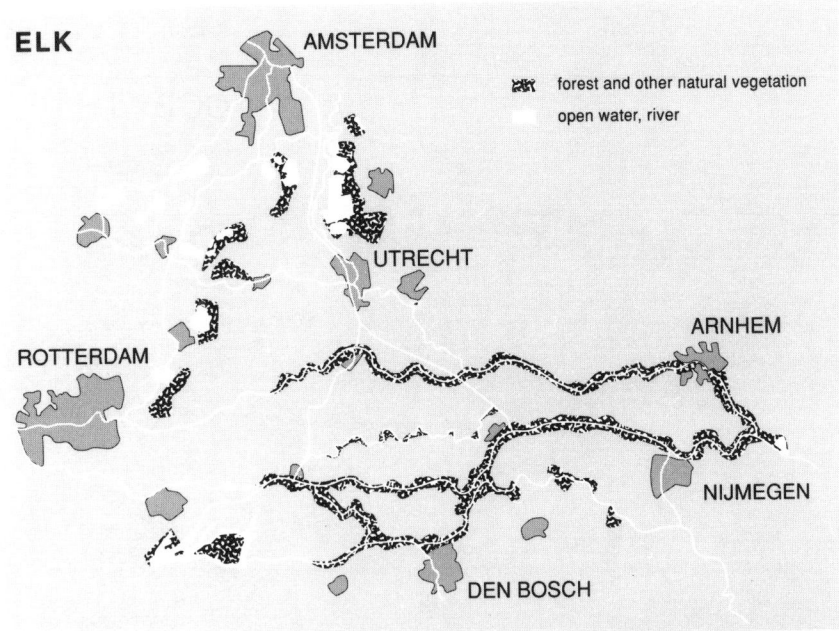

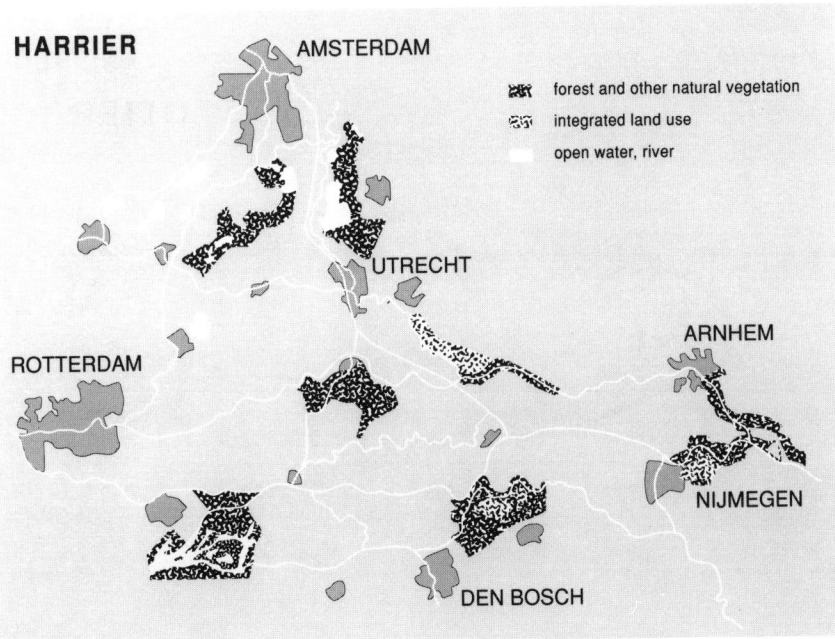

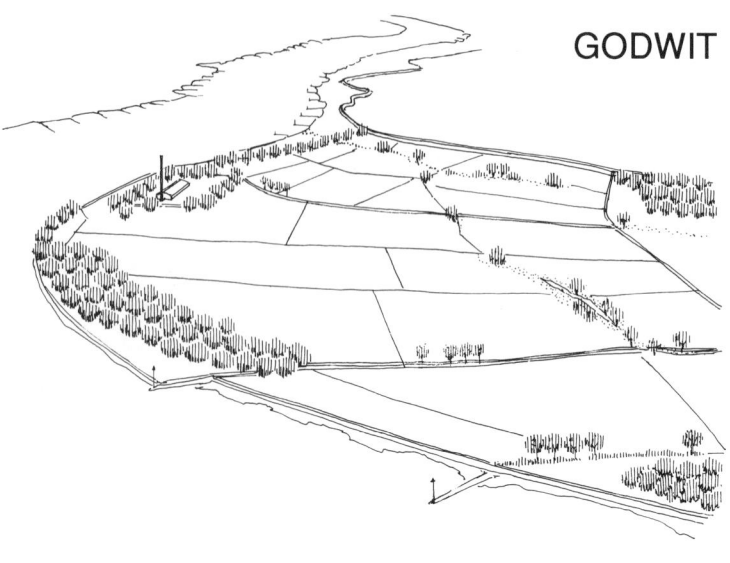

GODWIT

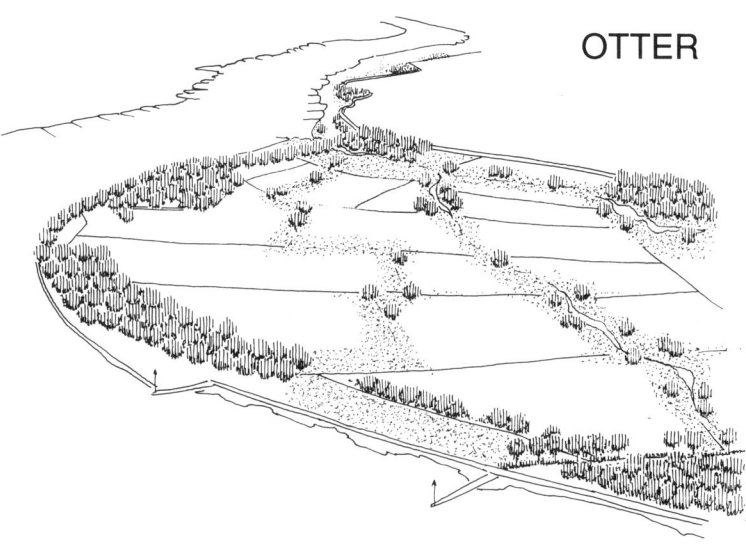

OTTER

ELK

HARRIER

to take note of these relations and fit new ecosystems into the existing order. So a spatial strategy is required. For instance: in order to sustain the ecological values of grasslands a less intensive form of agriculture often has to be integrated with well-defined nature conservation measures. As a consequence lower production levels have to be compensated by subsidies. Another strategy is the segregation of land uses. In areas with natural limitations to agricultural production such as flood plains, agriculture will not survive in the long run if unaided. Much of this land would probably be better used if nature areas and forests were located here. This would imply a segregation in space of incompatible land-use types. A set-aside programme could support this strategy.

We analysed existing plans and schemes in terms of the two sets of criteria: ecological objectives and spatial strategy. Using the elements thus gathered four different scenarios were developed for the Central Open Space. Since most of the existing material comprises only a part of the study area or is insufficiently detailed for a regional study, the ecological strategy often had to be extrapolated or elaborated to encompass the whole Central Open Space. The area designated for nature restoration was fixed at 10% of the study area. This percentage has been corrected for the planned nature areas integrated with other land uses. Each scenario was named after an animal species, that would benefit from its realization. The ecological objectives and spatial strategies of the concepts are summarized in Table 9.1. The spatial layout and a visual impression of the scenarios are shown in Figure 9.3a.

9.3.2 SCENARIO GODWIT

The ecological objective is to attain a maximum variety of species and species communities. In order to realize this objective, existing nature reserves will be improved and extended. In addition to the existing nature reserves the scenario exhibits a fine-grained pattern consisting of scattered and small natural elements like woodlots, shrubs, hedges, and ditch and dike verges. The spatial strategy achieves great diversity by integrating nature restoration with other land-use types. In buffer zones, among other locations, land-use types like ecological or integrated agriculture, nature-oriented recreation and forestry will be integrated. Around the cities forests and recreation areas are also planned as a buffer between urban and rural land. Much attention is paid to the development of park-like landscapes. To a large extent seminatural vegetation (mainly grassland) is planned to be interwoven with (subsidized) dairy farming. The wet grasslands on peat soils will provide ideal breeding conditions for meadow birds. The scenario corresponds with the trend in Dutch nature conservation policy during the last 15 years.

9.3.3 SCENARIO OTTER

This scenario aims at improving connectivity for species considered sensitive to habitat fragmentation. The concept of ecological infrastructure underlies this scenario. In this sense it is the opposite of the former scenario. The effects of isolation in fragmented landscapes have been demonstrated by many authors for a number of species (e.g. Oxley *et al.*, 1974; Storm *et al.*, 1976; Lynch and Whigham, 1984; Mader, 1984; Van Dorp and Opdam, 1987). Autonomous and planned developments aggravate the fragmentation of natural habitats.

The spatial strategy implies establishing a network of corridors and 'stepping stones'. Isolated fragments of woods, marches and oligotrophic grasslands have to be connected with larger similar nature areas (core areas). This means that in the waterlogged western part the lakes, reeds and marshes will be connected mainly in the south–north axis. This connection is also desirable for water tourism and recreation. A former study in the same area has indicated the efficacy of forest corridors for the expansion of forest-interior species (Harms and Knaapen, 1988). The results of this research were applied in the layout of this scenario.

9.3.4 SCENARIO ELK

Development of wilderness, self-sustaining 'complete' ecosystems is the objective here. This is a relatively new objective in the Dutch tradition of nature conservation. The emphasis is put on ecosystems with processes without human intervention or human management. In Dutch situations, sustainable systems in this sense can be supported by introducing or reintroducing wild grazers like the elk (*Alces alces*) (Van de Veen, 1985).

Although the course of development of ecosystems is uncertain the objectives are mainly climax vegetation like wet forests and marshes. To achieve these objectives requires a spatial strategy that can secure the ecological prerequisites over a long period of time. The scenario subscribes to the framework concept (Sijmons, 1990; Van Buuren and Kerkstra, 1993, this volume) as useful for nature restoration in this sense. In this concept land-use types that differ in the rate at which they change are separated in space. The less dynamic land-use types, like nature management, recreation and forestry, should be combined and linked in a stable and robust framework in order to protect sensitive ecological phenomena from surrounding influences. The more dynamic land-use types, like agriculture, industry and urbanization should be allowed maximum flexibility in large unfragmented areas with a potential for further development. Areas with natural limitations to dynamic and productive land use types, such as river forelands and peatland with a thick peat layer, should be designated as the less dynamic framework.

9.3.5 SCENARIO HARRIER

This scenario is the most drastic one. Like GODWIT the ecological objective is to maximize the variety of species and communities, but unlike GODWIT this scenario runs counter to current trends in Dutch nature conservation policy. New natural areas are not located adjacent to existing reserves but on optimal sites, i.e. places with suitable soil conditions, such as slopes, or sites where there is seepage of groundwater. All the expertise available in nature management will be exploited to achieve a maximum species diversity. A variety of vegetation like marshes, bogs, forests, vegetation of park-like landscapes, reeds, swamps, lakes, meadows, poor and rich fens etc. is planned. The spatial strategy is focused on a concentration of natural areas in large tracts. This will promote mutual interrelations between different communities. Moreover large areas are less vulnerable to external influences. Other land use types (e.g. forestry, agriculture) are allowed in the nature reserves insofar as they contribute to specific ecosystems, and to the overall variety of the area. Furthermore, segregation predominates, as in ELK. However, nature management actively intervenes in this scenario, unlike the non-intervention of ELK.

9.4 COSMO, A TOOL FOR EVALUATING THE SCENARIOS

A model, COSMO (Central Open Space MOdel), has been developed to evaluate the scenarios in terms of:

1. the suitability of the abiotic state factors;
2. the scenario's impact on vegetation and fauna (assessed by simulating the spatial and temporal development of the vegetation).

Essentially, COSMO is a knowledge-based system coupled with a grid-based GIS. The GIS used is MAP2 (Tomlin, 1983; Van den Berg *et al.* 1984). Both the data on the actual situation (vegetation, soil, land use) and the data on planned and expected development of site condition, vegetation and fauna are handled in a grid with cells of 1 km^2. The model provides a systematic way of using available ecological knowledge to evaluate the expected consequences of nature restoration plans. COSMO is based on a deterministic concept of vegetation dynamics that depends on physiotope, target vegetation and management (Figure 9.4) and of faunal habitat requirements which are also dependent on vegetation structure.

A physiotope is a spatial unit, characterized by independent state

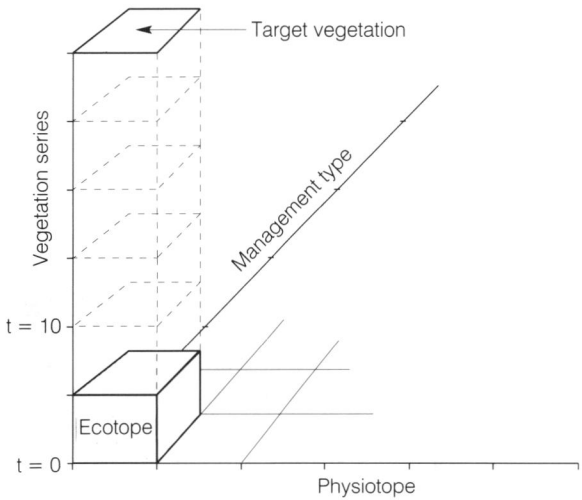

Figure 9.4 The concept of the GIS-based model COSMO.

factors (Jenny, 1980), primarily edaphic, hydrological and topographic. Class boundaries are drawn according to the response of the vegetation to those factors. Each type of physiotope can carry a limited number of vegetation types. Approximately 90 vegetation types are distinguished. This classification is sufficiently detailed to enable association with the physiotopes and with the habitat requirements of the faunal groups. A specific unique combination of vegetation and physiotope is called an ecotope (Neef, 1967; Zonneveld, 1972; Leser, 1976). Ecotopes are the basic units of the model. Development of ecotopes is modelled in discrete stages of five to thirty years. Each type of ecotope has its own determinate unidirectional successional development if no management is applied. The stages may be climax-like or cyclic with autogenous or management-dependent cycles.

Every five years it is possible to change the management. The kind of management chosen determines the further development of an ecotope type, i.e. the pathway to be taken through the ecotopes. This pathway is called a vegetation series (Harms and Damen, 1978; Harms and Kalkhoven, 1979). 'No management' is one of the management options. A further ten management types are distinguished. Examples of management types are: 'management of marsh–forest cycles' and 'development and management of oligotrophic species-rich fens'. The interface between the scenarios and the model is formed by a number of target vegetations. Each target vegetation provides a complete and logical set of choices for all the combinations of ecotope and management type that occur within the range of ecological and management parameters. Examples of target vegetations are: 'riparian forests on river forelands' and 'grassland vegetations not intensively used, suitable for meadow birds'.

The present-day situation was the starting point for the simulation. Data on soil type, hydrology and present-day vegetation were derived from a database of ecological variables on a national scale on a km^2 basis (De Veer and De Waal, 1988; Canters *et al.*, 1991). Only the dominant actual vegetation and soil type in a grid cell were considered.

The scenarios were translated into target vegetations and transferred on to a grid map. The first evaluation was the suitability of the state factors (physiotope types) for the chosen objectives of the scenario. If a target vegetation does not correspond to the prevailing physiotope type, the model can propose measures to modify the physiotope (e.g., raising the groundwater level or removing the topsoil). It is also capable of proposing alternative target vegetations. The planner can choose either solution, or both. Consequently, vegetation development was simulated in accordance with the target vegetation starting from the actual or adapted physiotope and vegetation. The results were evaluated at 10, 30 and 100 years after the hypothetical start of the development.

The effects on the fauna were accounted for in the model by calculating the area of potential habitat. Species of birds, mammals and butterflies considered typical of the Central Open Space, were selected for evaluation and were combined in ecological groups on the basis of an extensive literature survey. The areal requirements and home range area were determined for each group. Some of the groups distinguished use several spatially separated vegetation types to fulfil different needs. Most of these 'shuttle species' are birds, which breed in forests and forage in grasslands or marshes. Both foraging and breeding habitat (i.e. vegetation types) of these groups and area requirements were determined for both kinds of habitat. All vegetation types were classified for each group as optimal or non-optimal habitats. In order to evaluate the potential effects of the four scenarios on the fauna, vegetation maps of the scenario 10, 30 and 100 years after the starting point of the vegetation development were translated into maps of the potential habitat. Contiguous areas of potential habitat were aggregated, to calculate the population sizes that could be attained. Aggregates smaller than the individual home range area were rejected (too small). Larger areas were classified according to the expected population size.

9.5 RESULTS OF THE EVALUATION FOR NATURE RESTORATION

By and large the scenarios reach their objectives after adjustment. Most of the target vegetations develop as planned. Below we will focus on the time taken for the vegetation to develop and on the effects of the fauna.

The time taken for development of the target vegetations is indicated

Table 9.2 Development time of the scenarios. The percentage of the target vegetations completed in different stages is indicated.

Development time	GODWIT	OTTER	ELK	HARRIER
0	16	4	1	9
<10	37	20	5	10
10–30	10	16	0	12
30–100	30	29	50	50
> 100	2	29	44	14
Deviation	5	2	0	5

in Table 9.2. GODWIT was the fastest developing scenario. The target vegetation will be reached within 30 years on 65% of the area, because this scenario largely comprises grasslands and other seminatural vegetation. ELK has the longest development time as large areas of marshland woods and riparian hardwood forests are planned in this scenario. It takes more than 100 years to develop these kinds of natural vegetation. OTTER was a comparatively fast-developing scenario, because on a major part of the area grasslands and marshes are planned which develop within 30 years. HARRIER has a broad spectrum of vegetation types. As a consequence the development time of the areas differs: 50% of the area will take between 30 and 100 years to develop.

The evaluation of all ecological groups revealed large differences between the scenarios in terms of effects on the fauna. Only generalized results are given here. Table 9.3 summarizes the effects on the fauna expected after realizing the scenarios.

GODWIT enables a rather broad spectrum of animals to build up moderate to large populations. Meadow birds (including 'critical' species like ruff (*Philomachus pugnax*)), small mammals of rough pasture

Table 9.3 Effect of realization of the scenarios on ecological groups of birds and mammals with small and large home range areas. Ecological groups are classified according to biotope

Biotope	GODWIT	OTTER	ELK	HARRIER
Small home range area				
Open water	=	++	=	+
Grassland	++	=	—	-
Marsh/reed	+	+	=	+
Park-like	+++	-	+	++
Forest	++++	+++	+++	++++
Large home range area				
Open water	=	+	=	=
Grassland	++	-	-	=
Marsh/reed	=	+	+	+
Park-like	++	-	=	+++
Forest	+++	++	+++	++++

like *Microtus agrestis*, hare (*Lepus europaeus*) and butterflies of oligotrophic grasslands, such as *Pyrgus malva*, can be expected to benefit greatly from realization of this scenario, as will animals of the forest and park-like landscapes like bank vole (*Clethrionomys glareolus*), nuthatch (*Sitta europea*) and stoat (*Mustela putorius*). Even large herbivores like red deer (*Cervus elaphus*) may be able to develop populations albeit rarely exceeding 25 individuals. Animals of wet biotopes are hardly helped by this scenario.

OTTER does not favour any of the ecological groups more than the other scenarios, except birds of clear shallow waters like red-crested pochard (*Netta rufina*). This is the consequence of the scenario's objective, i.e. developing an ecological infrastructure for a limited number of ecological groups by constructing networks and corridors. For most groups, this does not lead to enlargement of potential habitat. The scenario does, however, have a considerable positive effect on forest animals like great spotted woodpecker (*Dendrocopos major*) and nuthatch and, to a lesser extent, on animals preferring grasslands or open water and marshland. It is the only scenario that improves the situation for the otter. Animals requiring large areas will only be able to form very small populations.

ELK will have a strong effect on a limited number of groups. This scenario has a very explicit objective. It aims at developing a kind of nature that will be able to persist with minimal management. This involves large tracts of primarily marshy woodland and riparian forest. As a result the area of potential habitat for animals restricted to these kinds of forests will increase greatly. Even the reintroduction of large herbivores like elk (*Alces alces*) can be successful. The elk will be able to form populations of considerable size. Also a typical animal of riparian biotopes like the beaver (*Castor fiber*), reintroduced recently, will find its area of potential habitat enlarged and less fragmented. However, because of the emphasis on riparian and marshy woodlands animals of drier forests and park-like biotopes are helped to a lesser extent. Because most of the forests will have an open structure due to grazing, small mammals and butterflies preferring wet or peaty grasslands (for instance *Clossiana selene*) will benefit from the realization of this scenario.

HARRIER will lead to a considerable increase of the area of potential habitat in virtually all ecological groups. This is because a broad spectrum of biotopes will develop. Each biotope is concentrated into large clusters of similar vegetation. The area of potential habitat increases greatly, especially for ecological groups preferring forest, marshy woodland and park-like biotopes. Because of the clustering of forests and park-like landscapes in this scenario wide-ranging animals such as pine marten (*Martes martes*) and red deer are able to form larger populations than in GODWIT. The only group that will suffer a decrease in area of potential habitat is the group of critical meadow birds (containing e.g. ruff). This is because of the afforestation of existing habitat (i.e. grasslands on river forelands).

The development of a considerable part of the target vegetations will take a century or more. In the planning of a particular policy a hundred years is a long time to wait for results. Therefore it is interesting to see how the fauna may develop within the next hundred years. In all the scenarios a considerable amount of forest or park-like landscape is planned. Apart from a minor part of the area considered, where existing young woods provide a good head start, these forests are planned on present-day grasslands. These grasslands are predominantly heavily exploited in dairy production. By being abandoned or by a change to much less-intensive grazing these grasslands will become more suitable for several small mammals like longtailed field mouse (*Apodemus sylvaticus*) and hare, and for musteloids like weasel (*Mustela nivalis*) and meadow birds like lapwing (*Vanellus vanellus*). As a result, the area of potential habitat for these animals will increase by approximately 80% in HARRIER and 50% in OTTER in comparison with the present situation. This increase is considerable, because the area designated for nature restoration is only 10% of the total study area. After the planned forests and park-like landscapes have grown there is a decrease in the area of potential habitat, resulting in an area of potential habitat equivalent to the present situation.

The succession of wet physiotopes proceeds via marsh and reed. Here, species like bearded reedling (*Panurus biarmicus*) and spotted crake (*Porzana porzana*) may find their area of potential habitat more than doubling.

Several of the shuttle species will also find an enormous increase of their foraging area. One such species is the purple heron (*Ardea purpurea*), which nests in reedland, marshland or in low shrubs on wet physiotopes. It forages in the breeding habitat as well as in moist or wet physiotopes and extensively farmed land. This species will find an increase of at least 100% in its potential breeding habitat after 10 to 30 years in all scenarios. However, because the purple heron requires large areas for foraging, the present situation does not allow a population of more than 10 individuals to build up anywhere in the study area. In three out of four scenarios the increase in potential foraging habitat lags far behind the increase in the area of potential breeding habitat. Thus development of larger populations is limited. In GODWIT, however, in addition to the increase in reedland and marshland on sites where wet forests are planned, there is a tremendous increase in nonintensively exploited grasslands. These are being developed partly towards oligotrophic grasslands, and partly towards drier forests and park-like land-

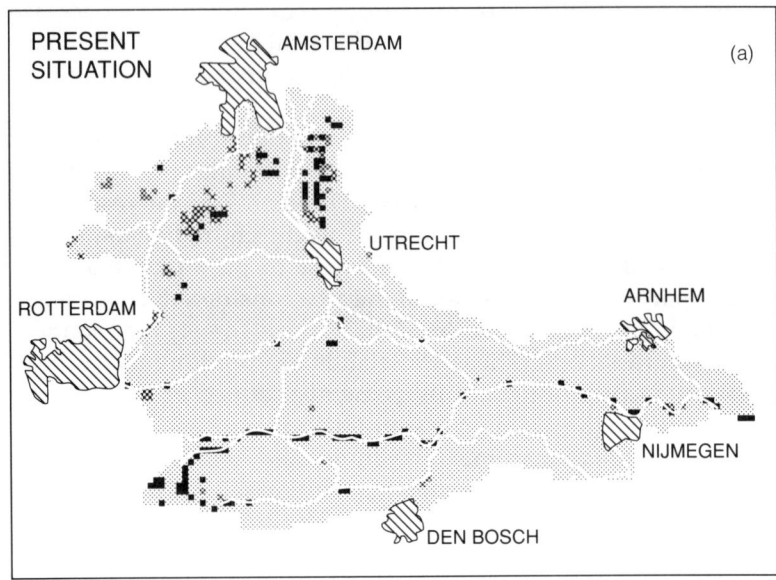

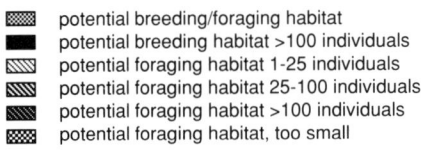
potential breeding/foraging habitat
potential breeding habitat >100 individuals
potential foraging habitat 1-25 individuals
potential foraging habitat 25-100 individuals
potential foraging habitat >100 individuals
potential foraging habitat, too small

Figure 9.5 Development of area of potential habitat for purple heron (*Ardea purpurea*) in scenario GODWIT, from the (a) present situation to that after (b) 30 and (c) 100 years.

scapes. Figure 9.5 shows that this involves doubling the area of potential breeding habitat after 30 years and at the same time an approximately eightfold increase in the area of potential foraging habitat. After 100 years the planned forests will have matured and the area of potential breeding habitat and the area of potential foraging habitat will have decreased, although it will still be larger than that at present.

9.6 DISCUSSION

The results of the evaluation help to explain the consequences of a chosen scenario. They do not, however, explicitly indicate which is the best scenario. That answer depends on the objectives of the policymakers. The four scenarios are exaggerated in the sense of consistent implementation of ecological objectives and spatial strategy. The scenarios and their ecological consequences have to be understood as an abstract frame delin-

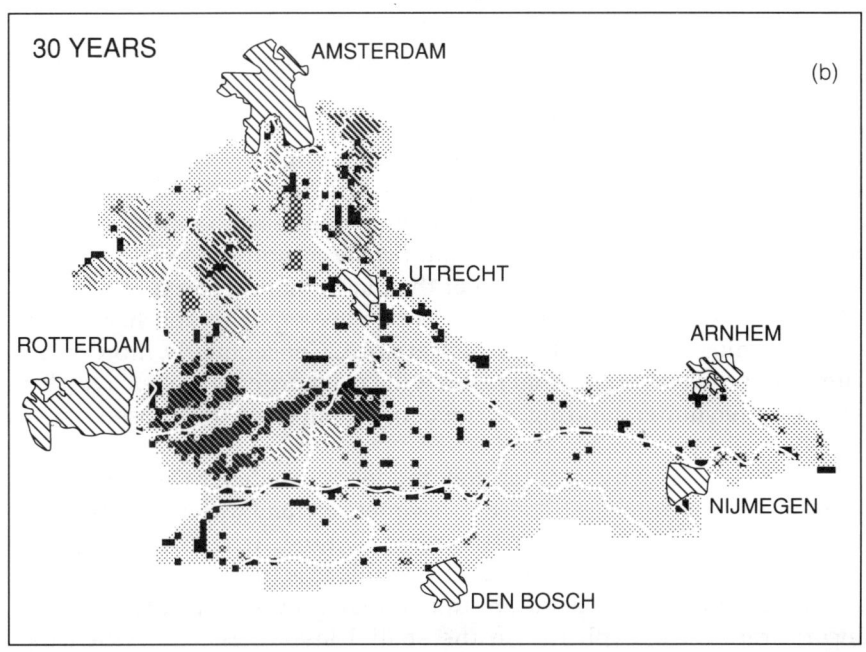

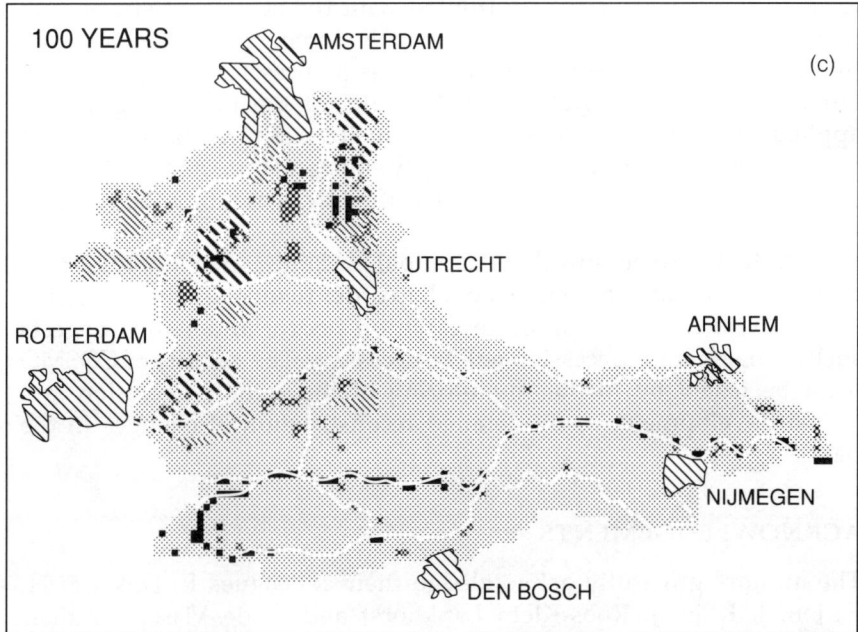

eating the possibilities for nature restoration, in which only the corners of the frame have been surveyed. The projected consequences of these extremes enable policymakers to choose a feasible compromise.

As described, the planning procedure used is cyclical, with alternating stages of design and evaluation. In the design stage the scenarios have been developed and implemented for the Central Open Space. The ecological approach to this stage was holistic, requiring knowledge at the level of the landscape as an ecosystem, which is expressed in the ecological objectives and the spatial strategies of the scenarios. In contrast, the evaluation stage needs knowledge at the most detailed level, which is expressed by the results of the computer model COSMO. In our study, only one cyclic run of the planning procedure was made. The results of the evaluation should be new input to another run in order to adjust and re-evaluate the scenarios. Consequently, more cyclic runs will be needed. The final result will be a more comprehensive plan and a better understanding of the consequences of the chosen objectives.

One may ask why a computer model is needed to evaluate the ecological consequences. These consequences may differ greatly from the aims of the scenario, for two reasons. First, when designing the spatial layout of the scenario, it is impossible to scrutinize the ecological requirements of all vegetation and faunal groups considered, or to incorporate these explicitly in the spatial layout. Second, when the ideas of the scenarios are confronted with the actual pattern of land occupation many choices have to be made, forcing one to deviate from the optimal layout. Therefore, the consequences for vegetation and fauna are not the straightforward results of the target vegetations applied. In particular, the extent to which certain faunal groups are helped or hindered by a scenario largely emerges from the scenario, and this needs to be demonstrated and clarified by evaluating the scenario with the model.

COSMO has to be considered as a tool for landscape planners rather than as an accurate predictive model. In order to increase the predictive power of such models, more research on poorly understood aspects such as marshland succession will be necessary. The validity of COSMO must be tested by further monitoring of the actual vegetation and faunal development, or by detailed reconstruction of changes in the past.

ACKNOWLEDGEMENTS

The authors gratefully acknowledge their colleagues E. Dijkstra, H. Farjon, J. Klijn, J. Roos-Klein Lankhorst and R. de Visser for their contributions to the underlying study of this chapter or for constructive criticism of the manuscript.

REFERENCES

Canters, K.C., Den Herder, C.P., De Veer, A.A., *et al* (1991) Landscape-ecological mapping of the Netherlands. *Landscape Ecology* **5** (3), 145–62.

De Veer, A.A. and De Waal, R.W. (1988) Landscape ecological mapping of the Randstad area, the Netherlands, in *Connectivity in Landscape Ecology* (ed. K.F. Schreiber), Münsterische Geographische Arbeiten 29. Münster, pp. 169–73.

Douw, L., Van der Giessen, L.B. and Post, J.H. (1987) *De Nederlandse landbouw na 2000, een verkenning.* Landbouw Economisch Instituut, The Hague.

Harms, W.B. and Damen, M. (1978) Die Anwendung von Vegetationskomplexen in einem Landschaftsplanungsprojekt, in *Tuxen (red): Assoziationskomplexe (Sigmeten) und ihre praktische Anwendung.* Ber. Int. Symp. Int. Ver. f. Vegetationskunde 1977 Cramer, Vaduz, pp. 435–53.

Harms, W.B. and Kalkhoven, J.T.R. (1979) *Landschapsecologie en natuurbehoud in Midden-Brabant.* Rijksinstituut "De Dorschkamp", rapport nr. 208, Wageningen; RIN Rapport nr. 79/15, Leersum.

Harms, W.B. and Knaapen, J.P. (1988) Landscape planning and ecological infrastructure: the Randstad study, in *Connectivity in Landscape Ecology* (ed. K.F. Schreiber), Münsterische Geographische Arbeiten 29. Münster, pp. 163–8.

Jenny, H. (1980) The soil resource: origin and behaviour. *Ecological Studies: Analysis and Synthesis.* Vol. 37. Springer-Verlag, New York.

Leser, H. (1976) *Landschaftsoekologie.* UTB 521 Ulmer, Stuttgart.

Lynch, J.F. and Whigham, D.F. (1984) Effects of forest fragmentation on breeding bird communities in Maryland, USA *Biological Conservation,* **28** 287–324.

Mader, H.J. (1984) Animal habitat isolation by roads and agricultural fields. *Biological Conservation,* **29**, 81–96.

Ministerie van Landbouw, Natuurbeheer en Visserij (1990) *Natuurbeleidsplan regeringsbeslissing,* SDU, The Hague.

Ministerie van Volkshuisvesting, Ruimtelijke Ordening en Milieubeheer en Ministerie van Landbouw en Visserij (1985) *Nota ruimtelijk kader Randstadgroenstructuur,* SDU, The Hague.

Ministerie van Volkshuisvesting, Ruimtelijke Ordening en Milieubeheer (1988) *Vierde Nota over de Ruimtelijke Ordening.* SDU, The Hague.

Neef, E. (1967) *Die theoretischen Grundlagen der Landschaftslehre.* Haach, Gotha/Leipzig, GDR.

Oxley, D.J. Fenton, M.B. and Carmody, G.R. (1974) The effects of roads on populations of small mammals. *Journal of Applied Ecology* **11**, 51–9.

Storm, G.L., Andrews, R.D., Philips, R.L. *et al.* (1976) Morphology, reproduction, dispersal and mortality of mid-western red fox populations. *Wildlife Monograms*, **49**, 5–82.

Sijmons, D. (1990) Regional planning as a strategy. *Landscape and Urban Planning*, **18** 3/4, 265–73.

Tomlin, C.D. (1983) *Digital Cartographic Modelling Techniques in Environmental Planning*. Yale University.

Van Buuren, M. and Kerkstra, K. (1993) The Framework Concept: a new perspective in the design of a multifunctional landscape, in *Landscape Ecology of a Stressed Environment* (eds C.C. Vos and P. Opdam), Chapman & Hall, London. pp. 219–43.

Van Den Berg, A., Van Lith, A.J., and Roos J. (1984) MAP: a set of computer programs that provide for input, output and transformation of cartographic data, in *Methodology in Landscape Ecological Research and Planning*. Proc. 1st Int. Sem. IALE Vol. III. RUC, (eds J. Brandt and P. Agger), Roskilde. pp. 101–13.

Van De Veen, H.A. (1985) Natuurontwikkelingsbeleid en bosbegrazing. *Landschap* **1** (2), 14–28.

Van Dorp, D. and Opdam, P.F.M. (1987) Effects of patch size, isolation and regional abundance on forest bird communities. *Landscape Ecology* **1**, 59–73.

Vos, C.C. and Zonneveld, J.I.S. (1993) Patterns and processes in a landscape under stress: the study area, in *Landscape Ecology of a Stressed Environment* (eds. C.C. Vos and P. Opdam), Chapman & Hall, London. pp. 1{29.

Zonneveld, J.I.S. (1972) *Land Evaluation and Land(scape) Science*. ITC Textbook of Photointerpretation. ITC Enschede, The Netherlands.

The framework concept and the hydrological landscape structure: a new perspective in the design of multifunctional landscapes

10

Michaël van Buuren and Klaas Kerkstra

10.1 INTRODUCTION

Processes with a conflicting nature, expressing incompatible goals in society, cause fundamental problems in multifunctional landscapes (Kerkstra and Vrijlandt, 1990; Sijmons, 1990; Van Buuren *et al.*, 1991). Some types of land use, such as modern agriculture, demand rapid adjustments of the physical landscape structure to technical innovation and changing market conditions. Other types require sustainable environmental conditions, continuity in space and time and hence long-term planning (e.g. nature development and water supply). The traditional landscape structure is too tight to host intensive and rapid agricultural developments. Consequently the ecological integrity of the landscape is threatened (De Molenaar, 1980; Anonymous, 1985; Anonymous, 1987; Weinreich and Musters, 1989; Brouwer *et al.*, 1991) dis-

Landscape Ecology of a Stressed Environment
Edited by Claire C. Vos and Paul Opdam
Published in 1993 by Chapman & Hall London. ISBN 0 412 44820 3

rupting the relationship between the landscape pattern and the underlying natural physical structure (Buitenhuis *et al.*, 1986; Kerkstra and Vrijlandt, 1988; Meeus *et al.*, 1988; De Regt, 1989; Van Buuren *et al.*, 1991).

In this perspective the landscape planning issue is how to create ecologically sound landscapes with possibilities for the development of conflicting types of land use (Kerkstra and Vrijlandt, 1990; cf Forman, 1990). This involves the (re)allocation of land-use types in a way that mutual negative impacts are prevented. At the same time the allocation of land-use types should correspond to the potentials of the natural physical structure. Landscape relations, especially those with a chorological character (Vos *et al.*, 1982; Forman and Godron, 1986; Zonneveld, 1989), are crucial. These relations determine how and to what extent the various parts of the landscape are either connected or isolated from each other. The movement of water is one of the most important phenomena creating landscape relations (Farjon, 1982; Roelofs *et al.*, 1982; Grootjans, 1985; Pedroli, 1989; Everts and De Vries, 1991).

This chapter focuses on the way the knowledge of landscape relations induced by flows of water may contribute to allocate land use in multifunctional landscapes. A case study is discussed, illustrating problems and possibilities.

10.2 THE HYDROLOGICAL LANDSCAPE STRUCTURE

Water moving through the landscape transports matter, energy and organisms. Flows of groundwater and surface water and related chemical and physical processes result in chorological relations between different parts of the landscape (Farjon, 1982; Roelofs *et al.*, 1982; Grootjans, 1985; Pedroli, 1989; Everts and De Vries, 1991). Figure 10.1 demonstrates that hierarchically ordered and superimposed groundwater flow systems may be found (Tóth, 1963; Freeze and Witherspoon, 1966, 1967, 1968; Domenico, 1972; Engelen and Jones, 1986; Engelen *et al.*, 1989; Ophori and Tóth, 1990). Each flow system has an infiltration zone and one or more exfiltration zones. Qualitative and quantitative characteristics of the water differ within and between the flow systems (Van Wirdum, 1979; Engelen and Jones, 1986; Kemmers, 1986; Pedroli, 1989).

The pattern of groundwater flow causes a hierarchically structured differentiation of environmental conditions throughout the landscape. In the infiltration zones the chemical composition of the groundwater resembles rainwater ('atmotrophic'; Van Wirdum, 1979). During transportation through the (sub)soil, mineralization and other processes occur, resulting in a different chemical composition of water in exfiltration zones: 'lithotrophic' (Van Wirdum, 1979). Shallow, local ground-

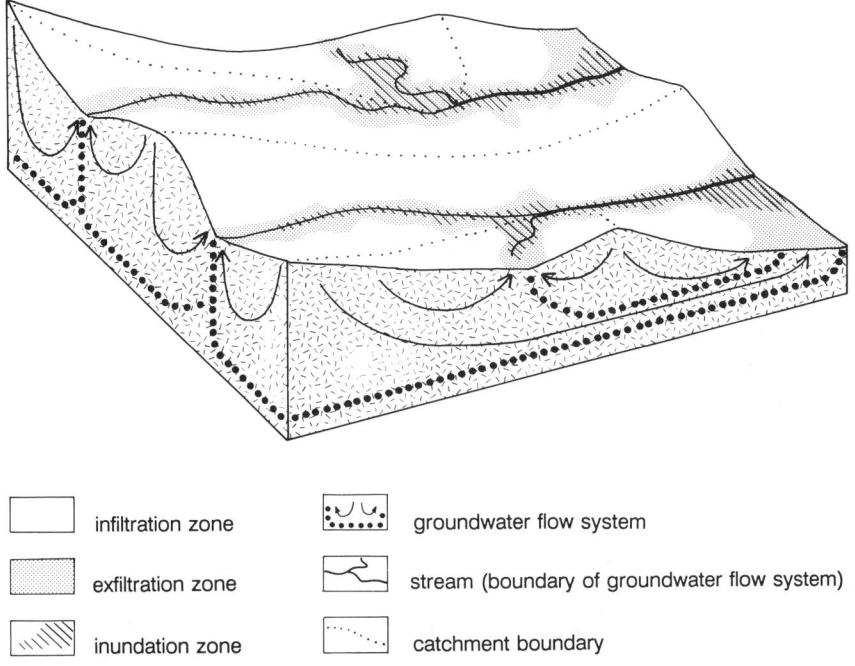

infiltration zone	groundwater flow system
exfiltration zone	stream (boundary of groundwater flow system)
inundation zone	catchment boundary

Figure 10.1 Environmental differentiation in the landscape resulting from flows of water; a schematic representation.

water flow systems appear and disappear depending on meteorological conditions, creating ecologically rich transition zones (Van Wirdum, 1979; Pedroli, 1989). Deep(er) groundwater flow systems and their exfiltration zones have a (more) permanent character.

Surface water flow is closely related to the groundwater. Generally, in north-west European landscapes, streams rise in flat areas were rainwater and local groundwater converge (De Vries, 1974). Going downstream the amount of water from deep(er) groundwater flow systems increases. As a result a hydrochemical gradient from atmotrophic to lithotrophic conditions is found (Van der Aart *et al.*, 1988). Sediment load and other materials (seeds, organisms), inundation and erosion in streams may contribute to environmental differentiation in inundation zones (Jongman, unpublished; Knaapen and Rademakers, 1990). In upper reaches quantitative properties (amount of water, velocity) change frequently, resulting in dynamic aquatic environments (Higler, 1983; Tolkamp, 1981). More stable conditions appear when the influence of deeper groundwater flow systems dominates.

The description of the flows of groundwater and surface water in the landscape shows that these processes cause specific patterns of more or less related units in the landscape, defined here as the **hydrological**

landscape structure. The different environmental conditions within these units lead to specific land-use potentials (Van Buuren, 1991).

Exploiting the qualities of the hydrological landscape structure in making landscape plans is called here a hydrological approach to land-

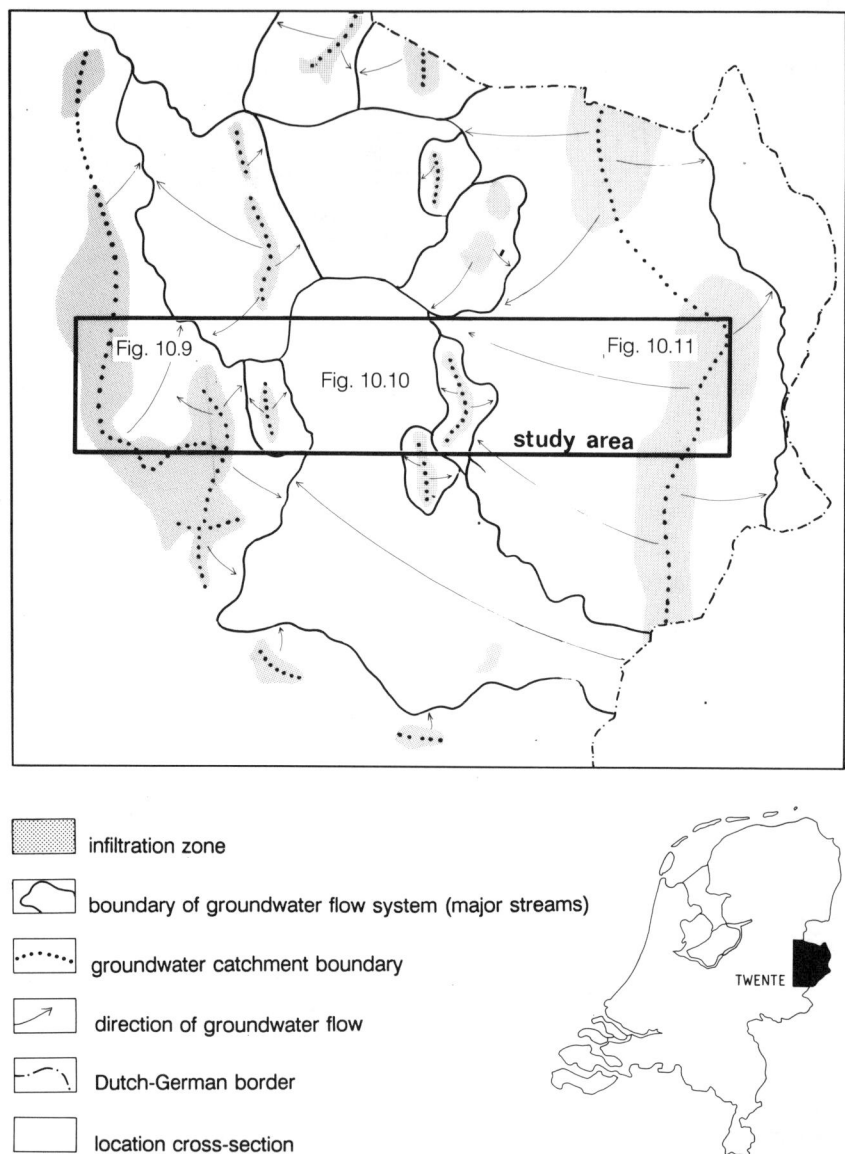

Figure 10.2 The Twente region is situated in the eastern part of The Netherlands. The major groundwater flow systems of Twente are depicted. The study area is situated in the centre of Twente.

scape planning (Van Buuren, 1991; cf. Dunne and Leopold, 1978; Ferguson, 1985; Toth, 1990). The central components of the approach are:

1. a description of the hydrological landscape structure;
2. an analysis of the way the spatial distribution of land use is related to the hydrological landscape structure;
3. the reallocation of land use corresponding to the units within the hydrological landscape structure.

The hydrological approach to landscape planning is illustrated below by discussion of a case study in the Twente region (Figure 10.2).

10.3 THE HYDROLOGICAL LANDSCAPE STRUCTURE AND LAND USE IN THE TWENTE REGION

The hydrological landscape structure and related land use of Twente are characteristic for an agricultural landscape in the sand region (Vos and Zonneveld, 1993, this volume). They were analysed at a regional scale (1:50.000; Van Buuren, unpublished). This analysis is a qualitative interpretation of existing maps and data on geology, geohydrology, geomorphology, soil types, groundwater tables, relief, drainage patterns, water quality, land use and vegetation, applying the method of the regional hydrological systems analysis (Engelen and Jones, 1986; Engelen *et al.*, 1989). Computer simulations with FLOWNET (Van Elburg *et al.*, 1989) – a two-dimensional model for steady-state saturated groundwater flow in rectangular inhomogeneous anisotropic sections of the subsoil – were used to evaluate this interpretation. The description of the hydrological landscape structure offers fundamental understanding of the direction of processes at work in the landscape and the existence of landscape relations, but gives no information on exact quantities and intensities of these phenomena. Furthermore, the restrictions in input data are reflected in the interpretation.

These factors have to be kept in mind, when reading the next sections. The analysis results in a reconstruction of the historic (around 1850) and a characterization of the contemporary situation (Figures 10.3–10.7). By doing this, both the processes in the landscape over the last century and actual problems are illustrated. The reconstruction of the hydrological landscape structure of the 19th century gives an indication of environmental conditions that occur in less human dominated landscapes. Figures 10.3 and 10.4 represent surfacewater and groundwater characteristics. The crosssections of Figure 10.5 schematize hydrogeological conditions and patterns of groundwater flow, and the way land use is superimposed on the physical structure. In Figures 10.6 and 10.7 maps of wet areas and land use are presented.

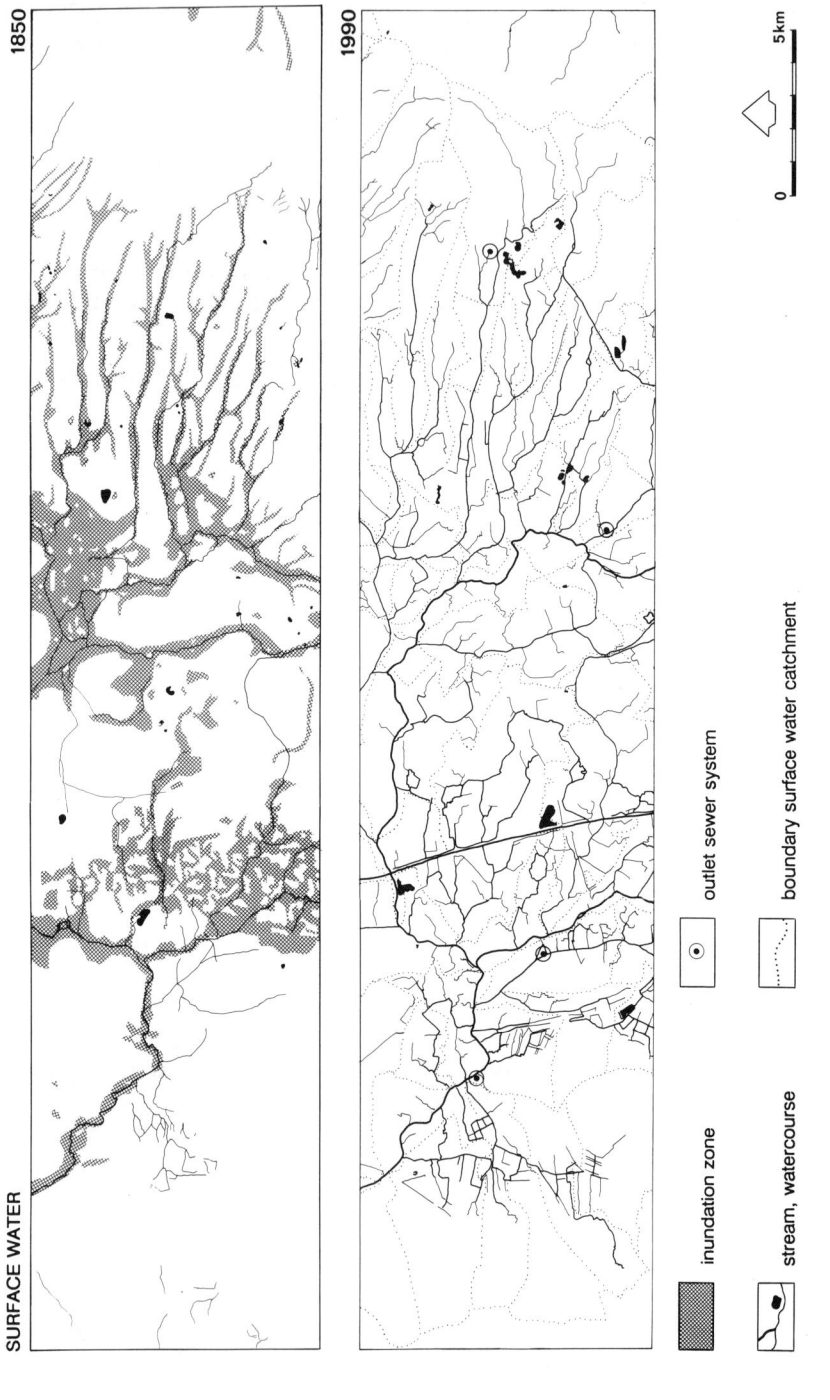

Figure 10.3 A reconstruction of the historical and the contemporary surface water characteristics in the study area. Notice the intensification of the drainage pattern, and the loss of inundation zones.

GROUNDWATER

1850

1990

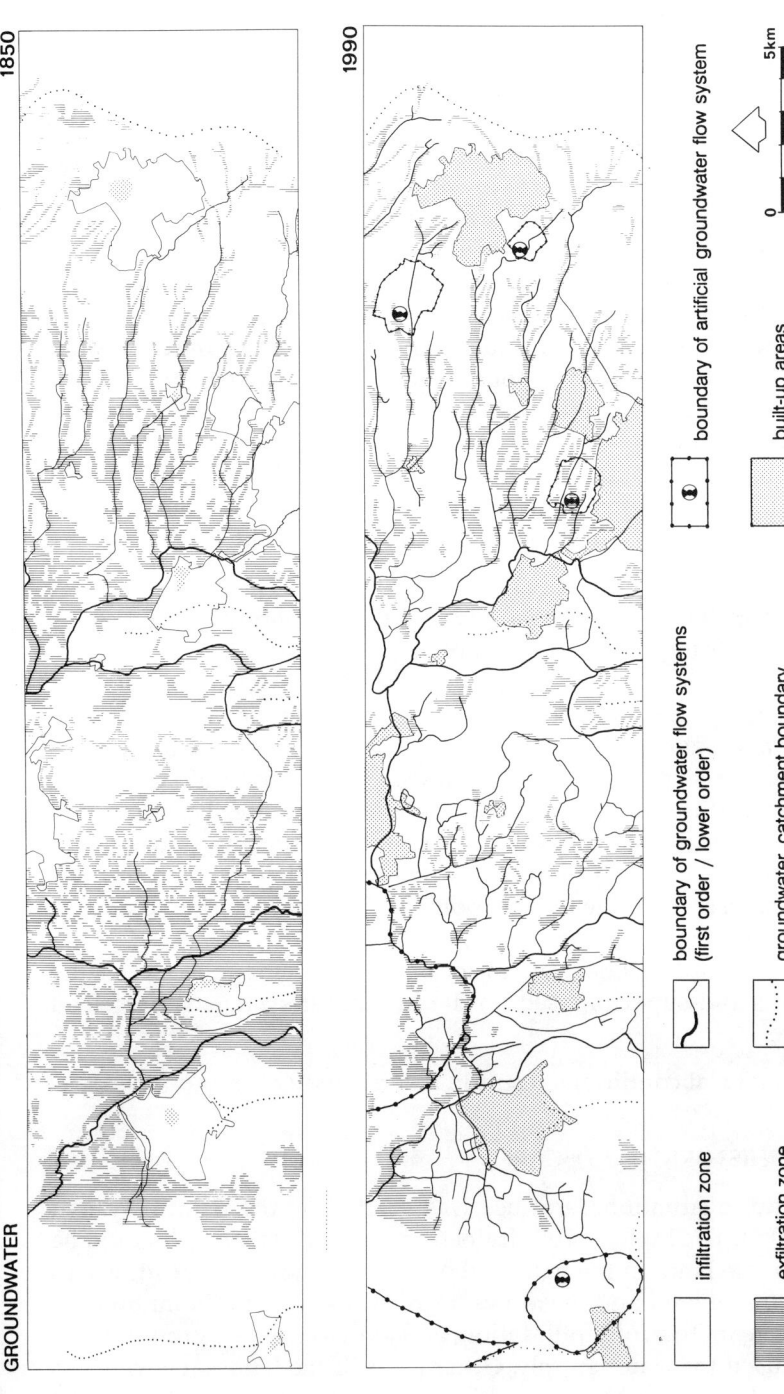

infiltration zone

exfiltration zone

boundary of groundwater flow systems
(first order / lower order)

groundwater catchment boundary

boundary of artificial groundwater flow system

built-up areas

0   5km

Figure 10.4 A reconstruction of the historical and the contemporary groundwater characteristics in the study area. Notice the decrease of exfiltration zones and the appearance of artificial groundwater flow systems. For the pattern of groundwater flow, see Figure 10.5.

10.3.1 THE HYDROLOGICAL CONDITIONS

The hydrological situation in Twente is dominated by two ice-pushed hills, with a flatter, hummocky relief in between (Figure 10.5). In the western hills sandy material was pushed up. The impermeable formations lay at great depths, resulting in a deep aquifer. Precipitation surpluses are predominantly transported through groundwater flow. Consequently drainage density is low. The groundwater exfiltrates in lowlands in between the hills and along stream valleys (cf. Figure 10.9).

In the flatter central part of the study area little ridges cause local groundwater flow systems, appearing in wet periods only (cf. Figure 10.10). Here streams used to inundate large areas, depositing clay. More to the east higher ridges are found; the lateral extent of the related groundwater flow systems is limited by the shallowness of the aquifers.

The eastern hills consist of clay formations, limiting groundwater flow to a shallow layer of sandy material. At the flank of the hill (almost) impermeable formations lay at greater depths, extending the flow domain of groundwater. Going from the top of the hill downstream, the following zones are found (cf. Figure 10.11):

1. a zone of shallow, local groundwater flow systems. These systems result in springs, feeding small brooks;
2. the infiltration zones of a first order groundwater flow system at the flank of the hill. From here groundwater flows to different exfiltration zones and in stream valleys. A first exfiltration zone is found at a step in the relief, marked by the rising of streams. A second exfiltration zone is found where the aquifer becomes very shallow and its transmissivity becomes insufficient to accommodate the incoming groundwater from uphill. In this zone new streams rise. From the highest point at the flank, groundwater flow to the drainage base may occur;
3. the lower order relief causes local flow systems in between the streams and superimposed on the first order system as described before.

The hydrological conditions result in a dense pattern of brooks.

10.3.2 THE HISTORIC LANDSCAPE

In the historic situation, land use was grafted on the natural system (Figure 10.5). The higher, well drained, moist and richer soils could be used as arable land. In the valleys the wet soils provided meadows. In between arable land and meadows, farmsteads were built, far enough from the stream to avoid inundation, and at places where groundwater could be used for water supply. Other parts of the landscape were too dry, too wet or too steep to be cultivated. A striking difference may be observed between the sand hills and the clay hills. The weathered clay soils could be cultivated up to the top of the hills, whereas the sandy soils appeared to be too dry for this. Characteristically, the historic landscape of Twente included extensive wet areas like marshes, bogs and meadows (Figure 10.6). The land use pattern (Figure 10.7) reflected the spatial structure of the historic landscape. Fine grained, enclosed patches of agricultural land surrounded by hedges, alternated with open fields with heathland and marshes.

10.3.3 THE PRESENT LANDSCAPE

The industrialization which started in the second half of the 19th century caused an extensive urban development. This together with the intensification of agriculture, resulted in a dramatic change of the landscape and its hydrological features (Figures 10.3–10.7; cf. Vos and Zonneveld, 1993, this volume).

The reclamation of wetlands and the intensified agricultural land use is reflected by a dense drainage pattern of regulated streams and watercourses (Figure 10.3). The extraction of groundwater for the supply of drinking water affects patterns of regional groundwater flow (Figure 10.4). The result is a decrease in the amounts of (deeper) groundwater flow and a drop in groundwater tables (Figure 10.5; Van Amstel et al., 1989; Gieske, 1990). The reduction of potentials in deeper groundwater flow systems may cause the increase of flow domains of local flow systems that are generally more polluted.

The intensification of land use has major effects on water quality. Natural hydrochemical gradients have disappeared (Anonymous, 1989b; Verdonschot, 1990). In the urban areas the effluent from sewer systems causes poor water quality in streams. Heavy rain storms in the hilly urban areas of east Twente produce high discharge peaks and inundation in downstream areas, even in summer periods. The high input of nutrients and biocides in agriculture affects groundwater and surface water quality. Especially in wet periods, when shallow groundwater flow systems start to flow, large amounts of these chemical agents are transported (cf. Stuurman et al., 1989).

The processes as described before result in a serious decrease of environmental differentiation: there is an overall tendency towards dry and eutrophic conditions (Verhoeven et al., 1993, this volume). The decline of wet ecosystems is obvious (Figure 10.6). The remaining wetlands are threatened by drought and or pollution. The development of urban and agricultural land use, and the concurrent loss of nature areas and areas with low land-use intensity (Figure 10.7), marks the increased nutrient levels in the landscape. Most streams and rivers are polluted and regulated. Watercourses with an artificial physical struc-

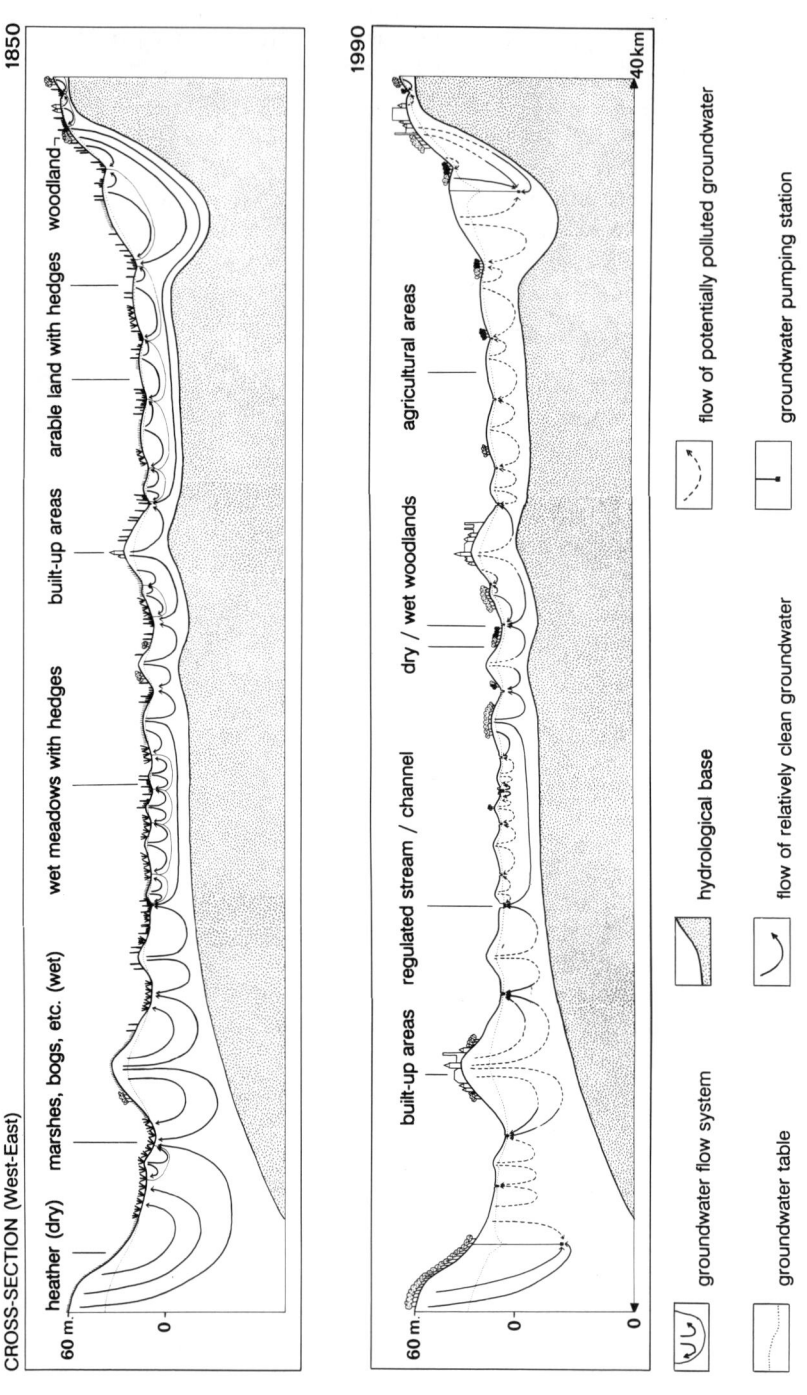

CROSS-SECTION (West-East)

1850

heather (dry) marshes, bogs, etc. (wet) wet meadows with hedges built-up areas arable land with hedges woodland

60 m.

0

1990

built-up areas regulated stream / channel dry / wet woodlands agricultural areas

60 m.

0

40 km

0

⌐ groundwater flow system

⌐ groundwater table

⌐ hydrological base

⌐ flow of relatively clean groundwater

⌐ flow of potentially polluted groundwater

⌐ groundwater pumping station

Figure 10.5 The hydrological landscape structure and related land use and landscape patterns in a historic and a contemporary situation; a cross-section. Notice the changes in the flowpattern and the quality of the groundwater.

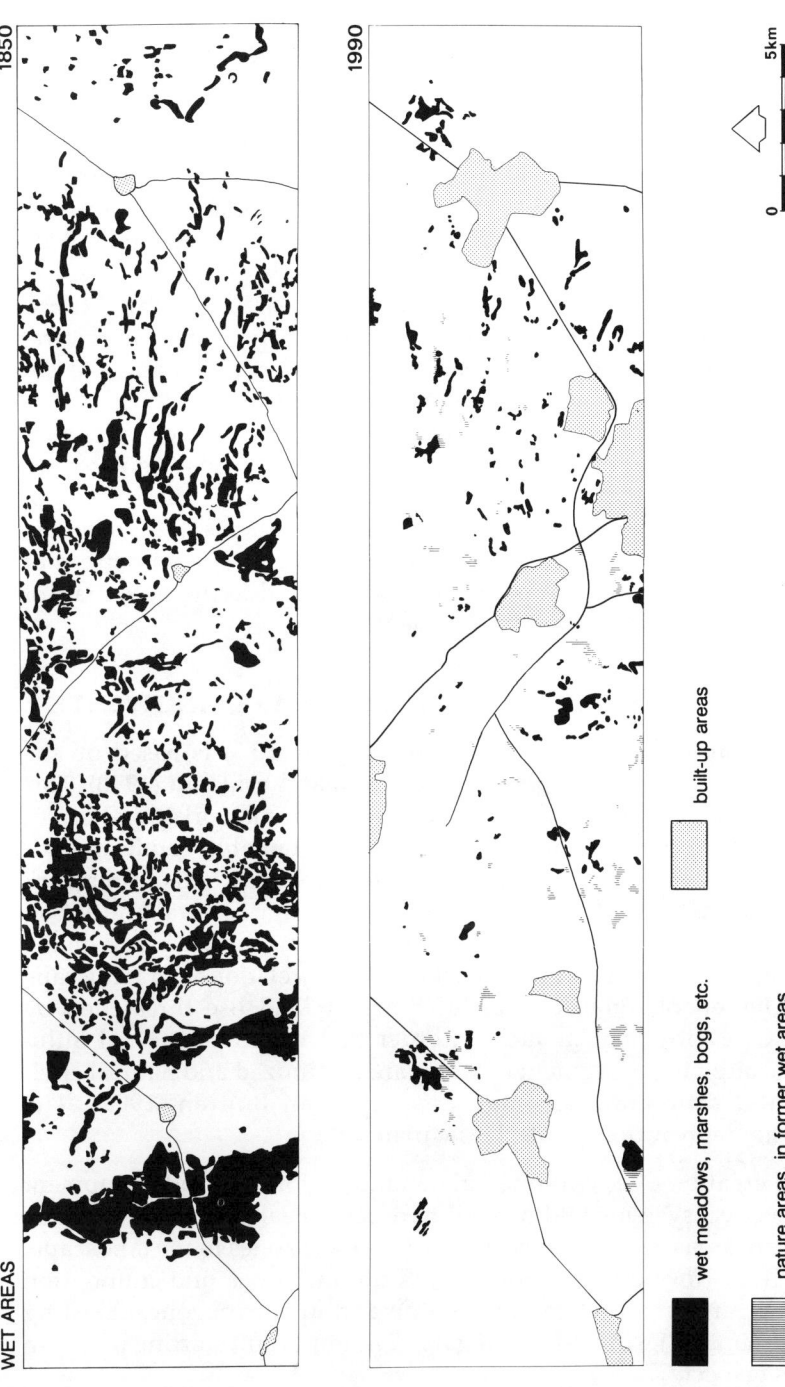

WET AREAS

1850

1990

wet meadows, marshes, bogs, etc.

built-up areas

nature areas, in former wet areas

0   5km

Figure 10.6 A reconstruction of the historical and the contemporary marshy areas in the study area, showing the decrease in the area of wetlands.

ture and regimen remain (Heijdeman and Peters, 1981; Verdonschot, 1990). Small landscape elements such as hedges and woodlots disappear and are fragmented (Figure 10.7). These developments have caused the extinction of many flora and fauna species (Grootjans *et al.*, 1993, this volume). The isolated spatial position of the remaining habitat patches, the wetlands in particular, presumably does not permit the recolonization of habitat patches once a local population became extinct, at least for species with small dispersal capacity (Opdam *et al.*, 1993, this volume).

The historic landscape patterns have disappeared, leaving a transformed and fragmented landscape no longer expressing the natural structure. The differences between (former) open fields and the enclosed agricultural areas have vanished (Figure 10.7).

At the same time drainage intensity and groundwater extraction causes water shortages for agriculture during dry periods (Anonymous, 1989b). The pollution of groundwater is threatening the supply of drinking water, whereas the demand for drinking water is still expanding.

Many of the problems are due to an inadequate spatial pattern of land use types, in relation to spatial relations and to the potentials of the physical structure of the landscape.

10.4 A LANDSCAPE PLAN FOR THE STUDY AREA IN TWENTE

The landscape plan for the study area (Figure 10.8) is based on the framework concept as proposed by Kerkstra and Vrijlandt (1990). This concept involves the planning of a landscape framework, a pattern of interconnected zones in which long-term, sustainable conditions for nature development and water supply are provided. This framework envelopes expanses of land in which dynamic agricultural and urban development is allowed. The location of the framework is based on the hydrological landscape structure and its (inter)relations at a landscape level. This enables the location of types of land use threatened by dynamic development in such an order that negative external influences through the movement of water are minimized and natural environmental differentiation may develop (Van Buuren, 1991). The following deliberations underly the plan.

1. The creation of environmental conditions for the (re)development of wet ecosystems and natural hydrochemical gradients. For this, nature areas are located in (potentially) wet zones in the landscape, such as (former) lithotrophic exfiltration zones and inundation zones, and in all related infiltration and upstream zones. Existing (remnants of) wetlands are regarded as important starting points in this respect;

2. The management of nature areas will be aimed at the development of self-supporting, native communities (Baerselman and Vera, 1989). This may imply the (re)introduction of ecological key-processes such as browsing and grazing by large herbivores and erosion and sedimentation in streams (Hynes, 1970; De Bruin *et al.*, 1987);

3. The preservation of high quality water resources within the framework for drinking water supply. To avoid conflicting interests with nature development a shift from groundwater to surface water as the major source for drinking water is necessary. The retention of precipitation surplusses by hampering run off may enhance this;

4. The designation of the areas with the best soils for modern agricultural production. To protect vulnerable types of land use these agricultural areas are located at hydrologically isolated positions. Environmental prerequisites are necessary to maintain good production conditions in future.

By planning the landscape framework it is assumed that a network of wetlands and streams may enhance the development and dispersal of related plant and animal species. However, in the design process here this aspect is not elaborated.

In the plan of Figure 10.8 the spatial characteristics that may develop within the framework are indicated. It is assumed that grazing and browsing by free ranging cattle and horses may stimulate the development of dense woodlands in the wettest parts, large open tracts in dry situations and a mosaic pattern in the transition zone in between.

The way the hydrological landscape structure is used in locating the framework will be discussed in more detail for three representative situations (Figures 10.9–10.11). At the same time an indication will be given of the environmental conditions that may appear within the framework. In the Twente case the following typology for environmental conditions is used (Westhoff and Den Held, 1969; Kalkhoven *et al.*, 1976; Van Beusekom *et al.*, 1990). Descending from the hills into the valley the following zones are encountered:

1. Permanent infiltration conditions. Here a (very) dry and atmotrophic environment will be found.

2. Infiltration combined with capillary rise. Periodically, plant roots are, through capillary rise, in contact with the groundwater table. This results in moist environmental conditions with an atmotrophic character.

3. Alternation of infiltration and stagnation in poor sandy soils. These conditions are found in flat, relatively high areas where a sandy layer was deposited on top of clay formations. After rain storms the groundwater flow is hampered by the clay resulting in the stagna-

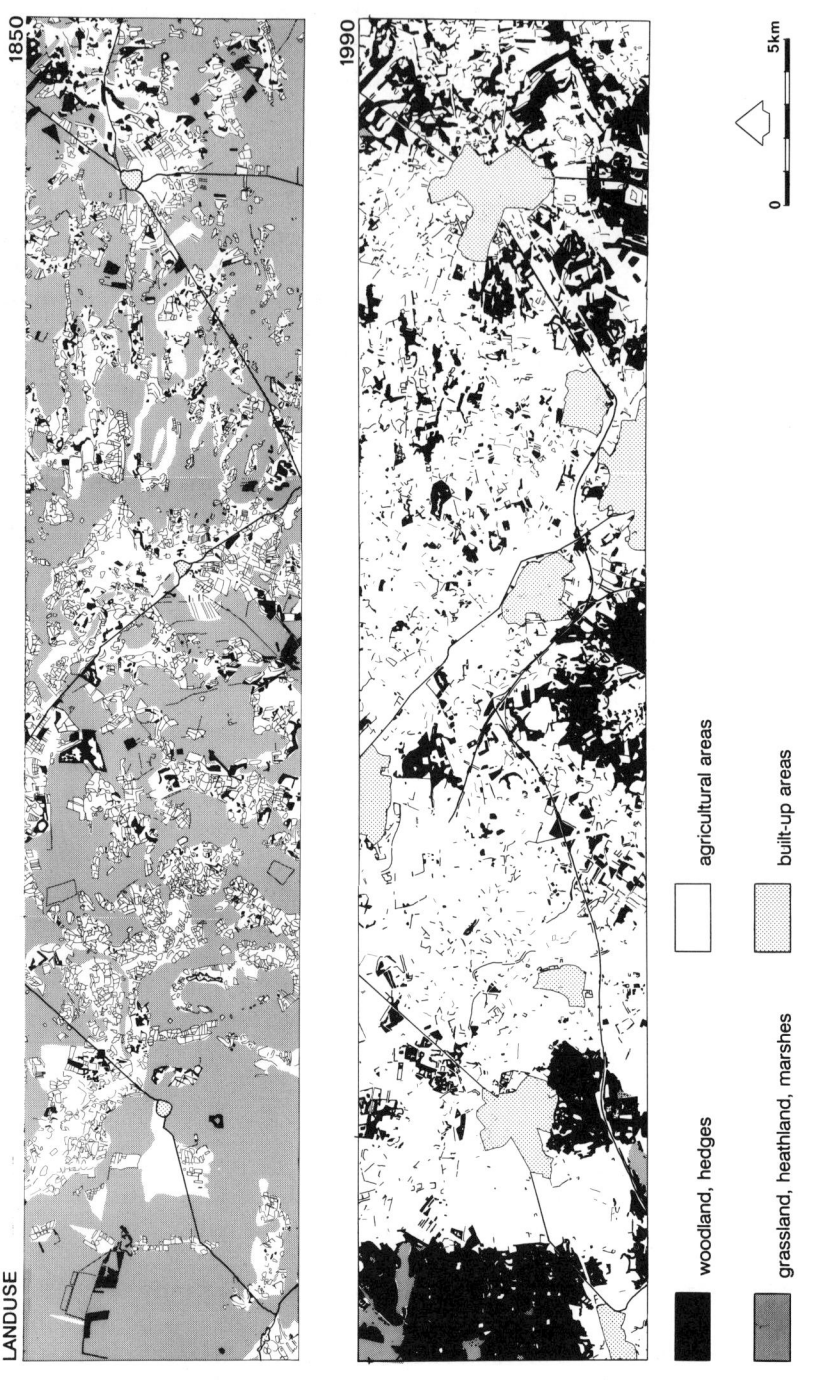

Figure 10.7 A reconstruction of the historical and the contemporary land-use patterns in the study area. The historic landscape patterns disappeared, leaving a transformed and fragmented landscape no longer expressing the natural physical structure.

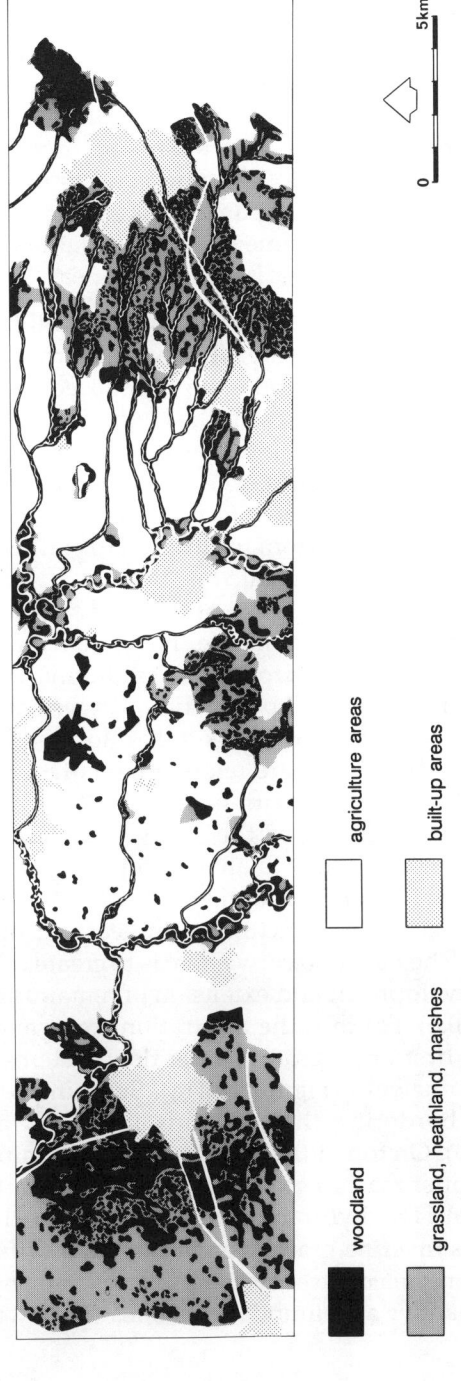

woodland

grassland, heathland, marshes

agriculture areas

built-up areas

0   5km

Figure 10.8 A landscape plan for the study area: a landscape framework, based on the hydrological structure.

tion of water. In drier periods infiltration takes place. As a result moist to wet, atmotrophic conditions may be found.

4. Alternation of infiltration and stagnation in weathered clay soils. The hydrological conditions are comparable to zone 3. However, the weathered clay soils cause nutrient-rich conditions.

5. Exfiltration. Here permanent moist to wet conditions of a lithotrophic character are found.

6. Combination of exfiltration and inundation. This combination occurs in the lowest parts of the stream valleys. Periods of inundation increase environmental differentiation. As a result wet lithotrophic to eutrophic conditions may appear.

Figure 10.9 represents the western hills. The proposed framework covers the groundwater flow system from the catchment boundary to the drainage base and a superimposed flow system of a lower order (Figure 10.9a). At present the major part of the related infiltration zone is covered by woodlands, resulting in a high quality of groundwater in the exfiltration zones. The plan proposes to exploit this situation by changing the rather wet and marginal agricultural lands into nature areas. A gradient from infiltration to exfiltration and inundation zones may develop (Figure 10.9b).

In the flat central part of the study area agriculture will remain the dominant type of land use. Here, many water management schemes have produced hydrological conditions that satisfy the demands of modern agricultural production. Larger nature areas are maintained as elements of the framework only if located in infiltration zones (Figure 10.10). These areas are relatively unaffected by agriculture because of their 'up-stream' position. The small and isolated (remnants of) wetlands in this part of the study area offer, due to their position as hydrological 'sinks' in an agricultural environment, little prospect for conservation or development of nature.

The complex structure of the eastern hills is illustrated in Figure 10.11. The major objective here is to create conditions for the protection, (re)development and extension of remaining wetlands (cf. Figures 10.4 and 10.6). For this, the infiltration zones and related exfiltration zones of the first-order groundwater flow systems are designated as a part of the framework (Figure 10.11a). Superimposed groundwater flow systems bordering the wetlands involved, are also included (Figure 10.11b). On top of the hills, the areas draining towards the springs and the upper reaches of the brooks become part of the framework (Figure 10.11b). The hydrological complexity in this part of the study area results in a fine-grained environmental differentiation (Figure 10.11c).

Along major streams belts of trees and shrubs (in upstream reaches) or meander and inundation zones (in downstream reaches) establish

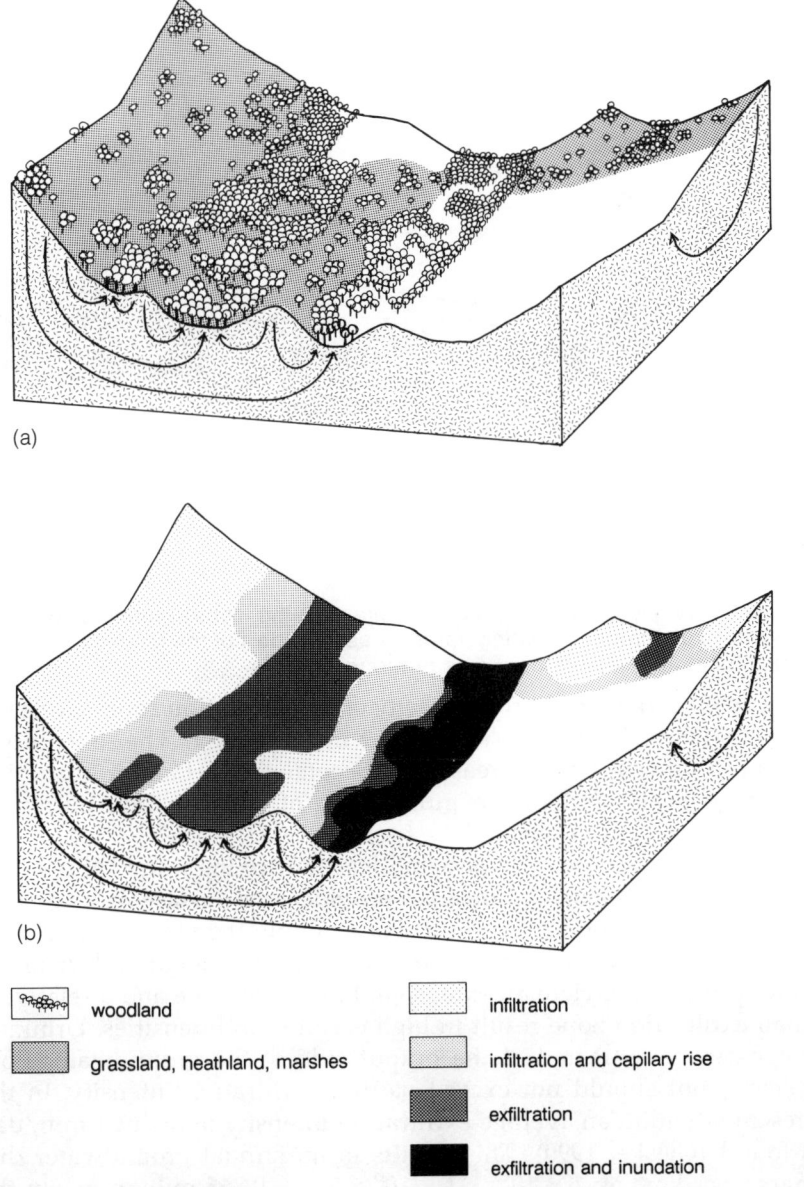

(a)

(b)

woodland	infiltration
grassland, heathland, marshes	infiltration and capilary rise
	exfiltration
	exfiltration and inundation

Figure 10.9 Detail of the landscape plan for the western ice-pushed hills. (a) Location of the landscape framework. A complete gradient from groundwater catchment boundary to drainage base may develop. (b) Environmental conditions within the framework.

corridors between the different parts of the framework. Here a good water quality is enhanced by concentrating watercourses from agricultural areas on specific locations through purification marshes (Van der

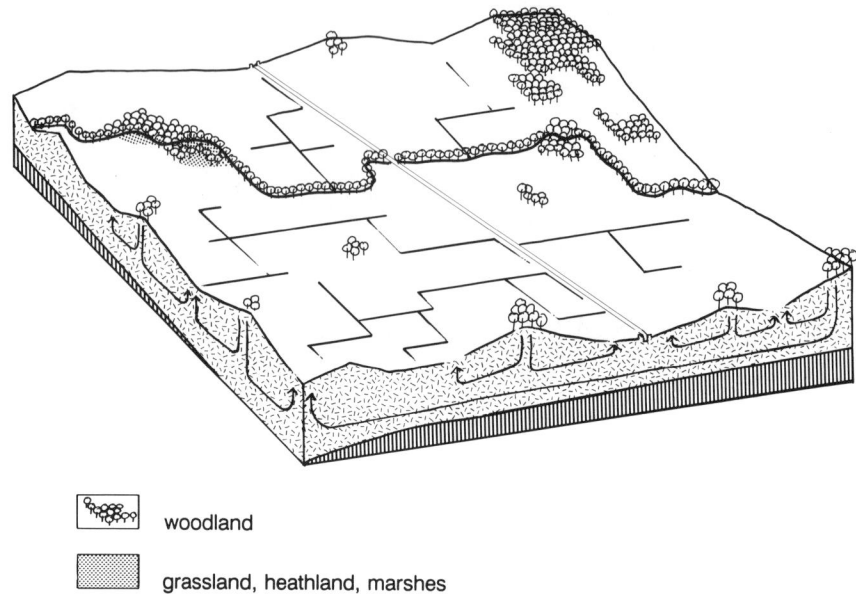

 woodland

grassland, heathland, marshes

Figure 10.10 Detail of the landscape plan for the central hummocky area. Agriculture dominates. Existing nature areas are part of the landscape framework if located in infiltration zones or along streams.

Aart, 1990) and by the buffer capacity of the vegetation zones along the streams (Peterjohn and Correll, 1984; Kuenzler, 1989). The zones along streams draining urban areas may become important in absorbing discharge peaks and in the purification of waste water. Eutrophic environmental conditions will be found here.

The framework provides possibilities for drinking water supply (van Beusekom *et al.*, 1990). Nature areas provide good surface water quality and a relatively constant run off. For example the western hills offer perspectives for a combination of nature development and drinking water production. Here an extensive infiltration zone and a relatively small exfiltration zone result in high exfiltration intensities. Drinking water can be extracted at the output side of the groundwater flow systems, but should not exceed natural exfiltration intensity. In the present situation an average exfiltration intensity of about 1 mm/day is found (Gieske, 1990). This results in an annual groundwater discharge per km^2 of: $1 \times 10^{-3} \times 1 \times 10^{6} \times 365 = 0.365$ million m^3. In the approximately 10 km^2 of exfiltration area this may result in a theoretical yield of about 3.65×10^{6}m^3 of high quality drinking water. Under conditions as proposed by the plan this quantity may increase when exfiltration rates are stimulated by:

1. conversion of groundwater to surface water as a major source of drinking water;

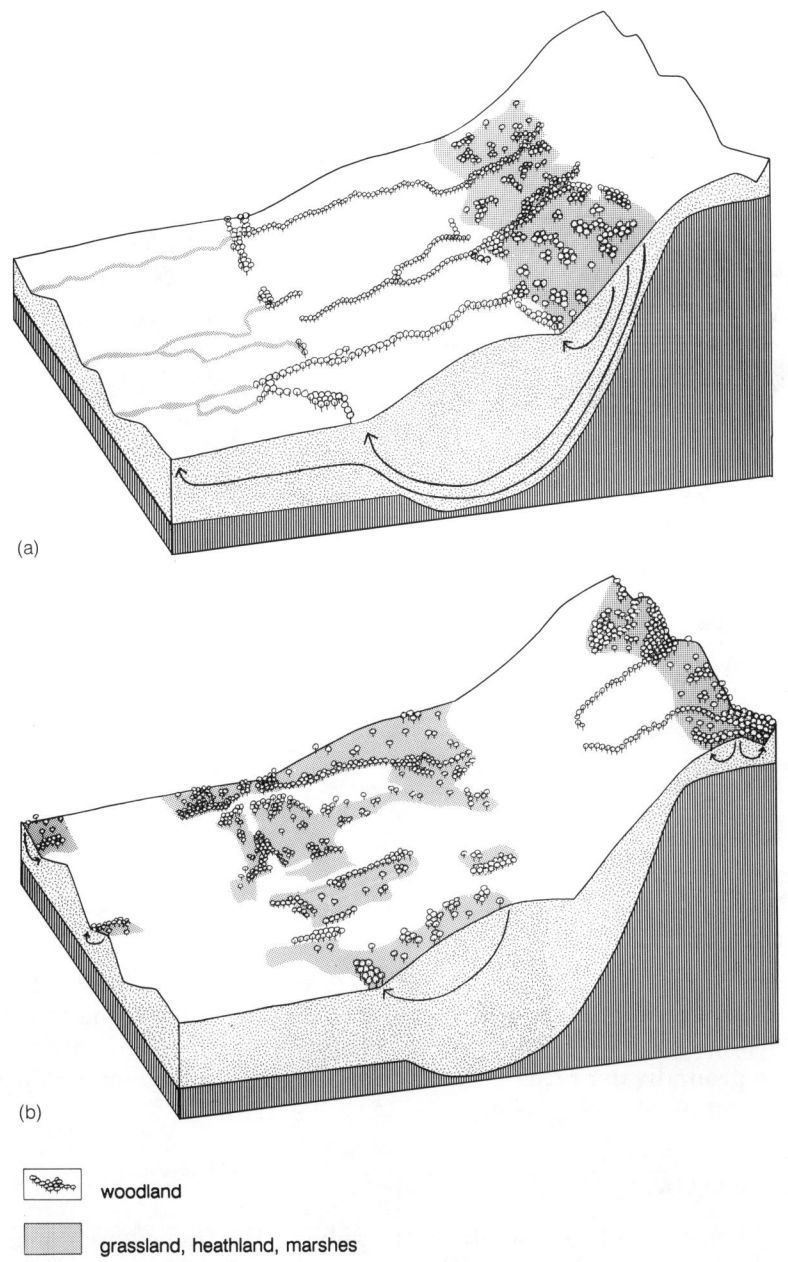

woodland

grassland, heathland, marshes

Figure 10.11 Detail of the landscape plan for the eastern ice-pushed hills. (a) Location of the landscape framework based on first-order infiltration zones and related exfiltration zones. (b) Location of the landscape framework based on superimposed, lower-order infiltration and related exfiltration zones. (c) Environmental conditions within the framework.

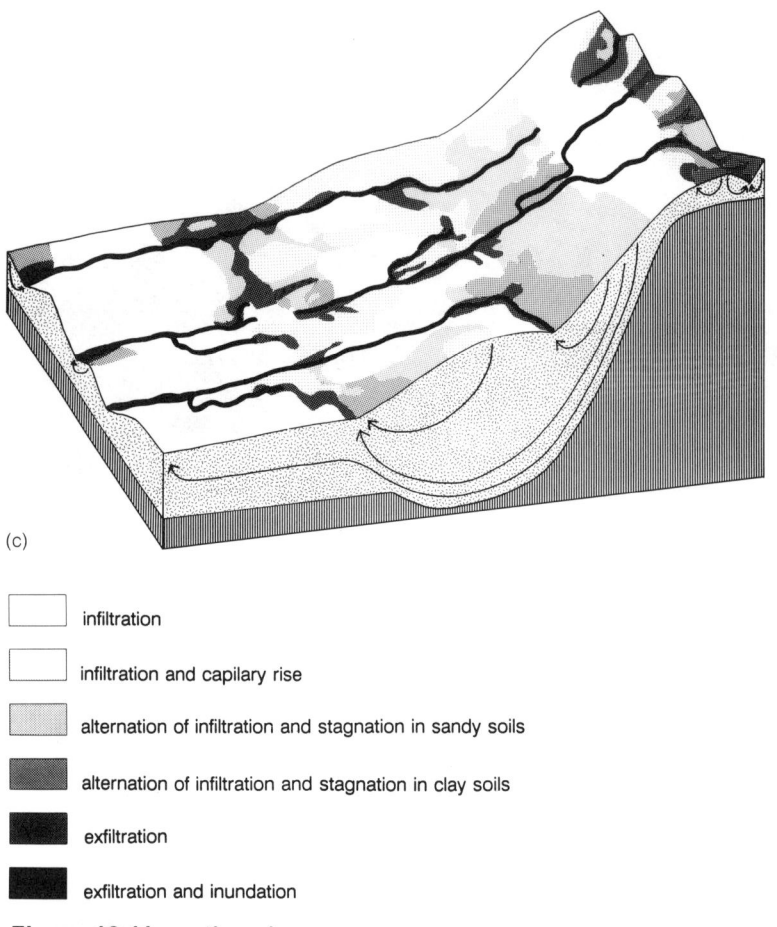

(c)

☐ infiltration

☐ infiltration and capilary rise

▨ alternation of infiltration and stagnation in sandy soils

▨ alternation of infiltration and stagnation in clay soils

■ exfiltration

■ exfiltration and inundation

Figure 10.11 continued.

2. transformation of present coniferous vegetation (transpiration of 612 mm annually) into an open native deciduous forest (half forest-half grassland; transpiration of 439 mm per year) which contributes to groundwater recharge (data from Jongman and Souer, 1990);
3. retention of surface water.

10.5 EVALUATION

The Twente case is a first effort of applying the framework concept based on the hydrological landscape structure. The case represents a transitory phase in an iterative research process. Here, design is regarded as an experiment in which theoretical concepts are translated into images of future landscapes. The evaluation of resulting plans may produce new questions, demanding further research. Design can thus serve as an exploration of the consequences of applying ecological and hydrological theories (Lang, 1987).

Consequently the landscape plan of Figure 10.8 is not meant to be implemented. Nevertheless, it is in line with national and provincial policies to develop more offensive strategies for nature protection (De Bruin *et al.*, 1987; Anonymous, 1989a; Baerselman and Vera, 1989). The plan proposes a new landscape, expressing the underlying natural physical structure. The key-point of the plan is a reallocation of land use, predominantly based on infiltration–exfiltration relations. This reallocation is expected to contribute to a sustainable development of land use in the landscape. A landscape framework is designed, consisting of large nature areas, connected by corridors along streams. To restore natural groundwater flow patterns, the conversion of groundwater into surface water as a major source for the supply of drinking water is required. It is proposed that the area for nature development be extended.

Further research will be necessary to evaluate and elaborate the landscape plan and the underlying concepts. The qualitative description of hydrological processes may be sufficient to plan at a regional scale. The elaboration of the plan at a more detailed scale however, demands more quantitative hydrological methods. Issues that need further consideration are, for example, the combination of nature development and drinking water production, the redevelopment of wetlands and the restoration of regulated streams into corridors across intensive agricultural areas.

From an ecological perspective it is necessary to evaluate the biogeographical aspects of the plan. Important issues are the locations and the character of corridors for different species and the way the effects of major infrastructure on the dispersal of species may be minimized. Further ecological research is necessary to give a better indication of the ecosystems that may develop within the framework, of the (recovery) times needed for (re)development of ecosystems and the environmental prerequisites that have to be fulfilled from this perspective.

REFERENCES

Anonymous (1985) *Environmental Data Compendium*, Organisation of European Cooperation and Development, Paris.

Anonymous (1987) *The State of the Environment in the European Community 1986*, Office for Official Publication of the European Communities, Luxemburg.

Anonymous (1989a) *Natuurbeleidsplan*, Ministerie van Landbouw, Natuurbeheer en Visserij, SDU Uitgeverij, 's-Gravenhage.

Anonymous (1989b) *Waterbeheersplan 1989–1994*, Waterschap Regge en Dinkel, Almelo.

Baerselman, F. and Vera, F.W.M. (1989) *Natuurontwikkeling; een verkennende studie*, Achtergrondreeks Natuurbeleidsplan nr. 6, Ministerie van Landbouw, Natuurbeheer en Visserij, SDU Uitgeverij, 's-Gravenhage.

Brouwer, F.M., Thomas, A.J. and Chadwick, M.J. (eds) (1991) *Land Use Changes in Europe. Processes of Change, Environmental Transformations and Future Patterns*, Kluwer Academic Publishers, Dordrecht.

Buitenhuis, A., Van de Kerkhof, C.E.M., Van Randen, I.J., and De Veer, A. (1986) *Schaal van het landschap*, Stichting voor Bodemkartering, Wageningen.

De Bruin, D., Hamhuis, D., Van Nieuwenhuijze, L. *et al.* (1987) *Ooievaar; de Toekomst van het Rivierengebied*, Stichting Gelderse Milieufederatie, Arnhem.

De Molenaar, J.G. (1980) *Bemesting, Waterhuishouding, Intensivering in de Landbouw en het Natuurlijk Milieu*, Rijksinstituut voor Natuurbeheer, Leersum.

De Regt, A.L. (1989) Kleinschalig landschap in een grootschalig Europa, in *Ruimtelijke Verkenningen 1989*, Jaarboek Rijksplanologische Dienst; Ministerie van V.R.O.M., 's-Gravenhage, pp. 12–44.

De Vries, J.J. (1974) *Groundwater Flow Systems and Stream Nets in the Netherlands*. A groundwater-hydrological approach to the functional relationship between the drainage system and the geological and climatical conditions in a Quaternary accumulation area, Proefschrift, Rodopi NV, Amsterdam.

Domenico, P.A. (1972) *Concepts and Models in Groundwater Hydrology*, McGrawHill, New York.

Dunne, T. and Leopold L.B. (1978) *Water in Environmental Planning*, W.H. Freeman, San Francisco.

Engelen, G.B. and Jones, G.P. (eds) (1986) *Developments in the Analysis of Groundwater Flow Systems*, International Association of Hydrological Sciences No. 163.

Engelen, G.B., Gieske J.M.J. and Los, S.O. (1989) *Grondwaterstromingsstelsels in Nederland; een landsdekkend beeld van de grondwaterstromingsstelsels,* Achtergrondreeks Natuurbeleidsplan nr. 2, Ministerie van Landbouw, Natuurbeheer en Visserij, SDU uitgeverij, 's-Gravenhage.

Everts, F.H. and De Vries, N.P.J. (1991) *De vegetatieontwikkeling van beekdalsystemen; een landschaps-ecologische studie van enkele Drentse beekdalen,* Historische Uitgeverij, Groningen.

Farjon, J.M.J. (1982) *Landschapsecologische relaties via het oppervlaktewater op nationaal en regionaal niveau,* Rijksinstituut voor Onderzoek in de bos- en landschapsbouw De Dorschkamp Rapport nr. 316, Wageningen.

Ferguson, B.K. (1985) Land environments of water resource management. *Journal of Environmental Systems,* **14** (3), 291–312.

Forman, R.T.T. (1990) Ecologically sustainable landscapes: the role of spatial configuration, in *Changing Landscapes: an Ecological Perspective* (eds I.S. Zonneveld and R.T.T. Forman), Springer Verlag, New York, pp. 261–78.

Forman, R.T.T. and Godron, M. (1986) *Landscape Ecology,* John Wiley and Sons, New York.

Freeze, R.A. and Witherspoon, P.A. (1966) Theoretical analysis of regional groundwater flow: 1. analytical and numerical solutions to the mathematical model. *Water Resources Research,* **2** (4), 641–56.

Freeze, R.A. and Witherspoon, P.A. (1967) Theoretical analysis of regional groundwater flow: 2. effect of water-table configuration and subsurface permeability variation. *Water Resources Research,* **3** (3), 623–34.

Freeze, R.A. and Witherspoon, P.A. (1968) Theoretical analysis of regional groundwater flow: 3. quantitative interpretations. *Water Resources Research,* **4** (3), 581–90.

Gieske, J.M.J. (1990) *Effecten van ingrepen op de grondwatersystemen in het Reggedal,* TNO-rapport OS 90-27-B, TNO/DGV, Oosterwolde.

Grootjans, A. (1985) *Changes of Groundwater regime in Wet Meadows,* Thesis University of Groningen.

Grootjans, A.P., Van Diggelen, R., Everts, H.F., Schipper, P.C., Streefkerk, J. De Vries, N.P.J. and Wierda, A. (1993) Linking ecological patterns to hydrological conditions on various spatial scales in *Landscape Ecology in a Stressed Environment* (eds. C.C. Vos and P. Opdam), Chapman & Hall, London, pp.60–78.

Heijdeman, B. and Peters, A. (1981) *Twente, een Hydrobiologisch Onderzoek van de Beken,* Provinciale Planologische Dienst van Overijssel, Zwolle.

Higler, L.W.G. (1983) De Nederlandse beken. *Natura,* **80** (1), 4–8.

Hynes, H.B.N. (1970) *The Ecology of Running Waters,* Liverpool, University Press, Liverpool.

Jongman, R., and Souer, M. (1990) *Landscape Ecological and Spatial Impacts of Climatic Change in Two Areas in the Netherlands,* Agricultural University, Wageningen.

Kalkhoven, J.T.R., Stumpel, A.H.P. and Stumpel-Rienks, S.E. (1976) *Environmental Survey of The Netherlands,* Research Institute for Nature Management, Verhandeling 9, SDU, 's-Gravenhage.

Kemmers, R.H. (1986) Perspectives in modelling of processes in the root zone of spontaneous vegetation at wet and damp sites in relation to regional water management, in *Proceedings and Information* nr. 34, (ed. J.C. Hooghart), TNO Committee on Hydrological Research, The Hague.

Kerkstra, K. and Vrijlandt, P. (1988) *Het Landschap van de Zandgebieden; Probleemverkenning en Oplossingsrichting*, Directie Bos en Landschapsbouw, Ministerie van Landbouw en Visserij/Vakgroep Landschapsarchitectuur Landbouwuniversiteit, Utrecht/Wageningen.

Kerkstra, K. and Vrijlandt, P. (1990) Landscape planning for industrial agriculture: a proposed framework for rural areas. *Landscape and Urban Planning*, **18**, 275–87.

Knaapen, J.G. and Rademakers, J.G.M. (1990) *Rivierdynamiek en Vegetatieontwikkeling*, Staring Centrum, Rapport 82, Wageningen.

Kuenzler, E.J. (1989) Value of forested wetlands as filters for sediments and nutrients, in *The Forested Wetlands of the Southern United States*, Proceedings of the symposium, Orlando, Florida, 12–14 July 1988. United States Department of Agriculture, Forest Service, Southern Forest Experiment Station, General Technical Report SE-50, pp. 85–96.

Lang, J. (1987) *Creating Architectural Theory; the Role of the Behavioral Sciences in Environmental Design*, Van Nostrand Reinhold, New York.

Meeus, J., Van der Ploeg, J.D. and Wijermans, M. (1988) *Changing Agricultural Landscapes in Europe: Continuity, Deterioration or Rupture?*, IFLA-Conference The European landscape, Changing Agriculture, Changing Landscapes, Rotterdam.

Opdam, P., Van Apeldoorn, R., Scholman, A., and Kalkhoven, J. (1993) Population responses to landscape fragmentation, in *Landscape Ecology of a Stressed Environment* (eds C.C. Vos and P. Opdam), Chapman & Hall, London, pp.143–71.

Ophori, D. and Tóth, J. (1990) Relationships in regional groundwater discharge to streams: an analysis by numerical simulation. *Journal of Hydrology*, **119**, 215–44.

Pedroli, G.B.M. (1989) *The nature of landscape; a contribution to landscape ecology and ecohydrology with examples from the Strijper Aa landscape, Eastern Brabant, The Netherlands*, Thesis Free University Amsterdam.

Peterjohn, W.T. and Correll, D.L. (1984) Nutrient Dynamics in an agricultural watershed: observations on the role of a riparian forest. *Ecology*, **65**, 1466–75.

Roelofs, H.J., Beukeboom, Th. J., Ebregt, A. and Vos, W. (1982) *Landschapsecologische relaties via het grondwater op nationaal en regionaal niveau*, Rijksinstituut voor Onderzoek in de bos- en landschapsbouw De Dorschkamp Rapport nr. 317, Wageningen.

Sijmons, D. (1990) Regional planning as a strategy. *Landscape and Urban Planning*, **18**, 265–73.

Stuurman, R.J., Van der Mey J.L., Engelen, G.B. *et al.* (1989) *De hydrologische systeemanalyse van westelijk Noord-Brabant en omgeving*, Dienst Grondwaterverkenning TNO/Vrije Universiteit Amsterdam. DGV-TNO rapport OS 89-59, Delft.

Tolkamp, H.H. (1981) *Organism–Substrate Relationships in Lowland Streams*, Agricultural Research Reports No. 907, Pudoc, Wageningen.

Tóth, J. (1963) A theoretical analysis of groundwater flow in small drainage basins. *Journal of Geophysical Research*, **68**, 4795–812.

Toth, R.E. (1990) *Hydrologic and Riparian Systems: the Foundation Network for Landscape Planning*. International Conference on Landscape Planning, University of Hannover, 6–8 June 1990.

Van Amstel, A.R., Braat, L.C. and Garritsen, A.C. (1989) *Verdroging van natuur en landschap in Nederland: beschrijving en analyse*, Min. van Verkeer en Waterstaat, 's-Gravenhage.

Van Beusekom, C.F., Farjon, J.M.J., Foekema, F. *et al.* (1990) *Handboek Grondwaterbeheer voor Natuur, Bos en Landschap*, Studiecommissie Waterbeheer Natuur, Bos en Landschap, SDU Uitgeverij, 's-Gravenhage.

Van Buuren, M. (1991) A hydrological approach to landscape planning. The framework-concept elaborated from a hydrological perspective. *Landscape and Urban Planning*, **21**, 91–107.

Van Buuren, M., Kerkstra, K. and Vrijlandt, P. (1991) Kleinschalig, verweven of casco? Het landschap van de zandgebieden nader beschouwd. *Landinrichting*, **31**, 1–18.

Van der Aart, P.J.M. (ed.)(1990) *Wetlands for the Purification of Wastewater*, Post Academic Course, Plant Ecology News Report 11, Department of Plant Ecology, University of Utrecht, Utrecht.

Van der Aart, P.J.M., Verhoeven, J.T.A. and Kemmers, R.H.(1988) Eutrofiering in het landelijk gebied; probleemschets en mogelijke oplossingen vanuit een landschaps-ecologisch kader, in *Integraal Waterbeheer, een nieuwe aanpak* (ed. A. Bijlsma), NIBI, Utrecht, pp. 17–55.

Van Elburg, H., Engelen, G.B. and Hemker, C.J. (1989) *FLOWNET version 5.1*, Institute of Earth Sciences at the Free Reformed University of Amsterdam, Amsterdam.

Van Wirdum, G. (1979) Dynamic aspects of trophic gradients in a mire complex, in *Proceedings and Information 25*, TNO Committee on Hydrological Research, The Hague, pp. 66–82.

Verdonschot, P.F.M. (1990) *Ecological characterization of surface waters in the Province of Overijssel (The Netherlands)*, Thesis Wageningen Agricultural University; Province of Overijssel; Research Institute for Nature Management, Leersum.

Verhoeven, J., Kemmers, R.M. and Koerselman, W (1993) Nutrient enrichment of freshwater wetlands, in *Landscape Ecology of a Stressed Environment* (eds. C.C. Vos and P. Opdam), Chapman & Hall, London, pp.33–59.

Vos, C.C. and Zonneveld, J.I.S. (1992) Patterns and processes in a landscape under stress: the study area, in *Landscape Ecology of a Stressed Environment* (eds. C.C. Vos and P. Opdam), Chapman & Hall, London, pp. 1–28.

Vos, W., Harms, W.B. and Stortelder, A.H.F. (1982) *Vooronderzoek naar landschapsecologische relaties tussen ecosystemen*, Rijksinstituut voor Onderzoek in de bos- en landschapsbouw De Dorschkamp Rapport nr. 246, Wageningen.

Weinreich, J.A. and Musters, C.J.M. (1989) *Toestand van de natuur; veranderingen in de Nederlandse natuur*, Ministerie van Landbouw, Natuurbeheer en Visserij, Achtergrondsreeks Natuurbeleidsplan nr. 4, SDU uitgeverij, 's-Gravenhage.

Westhoff, V. and Den Held, A.J. (1969) *Plantengemeenschappen in Nederland*, W.J. Thieme, Zutphen.

Zonneveld, I.S. (1989) The land unit – a fundamental concept in landscape ecology, and its applications. *Landscape Ecology*, **3** (2), 67–86.

Nature conservation and extraction of drinking water in coastal dunes: the Meijendel area

11

Theo W. M. Bakker and Pieter J. Stuyfzand

11.1 INTRODUCTION

The coastal dunes of The Netherlands are an area rich in nature. Although they cover only 1% of the total Dutch surface, some 800–900, i.e. about 64%, of all plant species found in The Netherlands occur here. Among them are very rare species such as 20 species of orchid and 30 species of sedge. Wet slacks contribute an important part of these high natural values. Here the vegetation, with many rare species, is influenced by groundwater. In the natural situation about one-third of the total dune surface consists of wet slacks.

Large parts of these coastal dunes are not only a nature reserve, but also supply drinking water to the western part of The Netherlands, one of the most densely populated areas of the world.

Due to the activities of the drinking water companies a lot of geohydrological and hydrochemical research has been carried out. The abiotic, civil engineering approach has dominated since the beginning of the twentieth century. Only during the last decade, has the increased concern about nature conservation triggered the study of biotic features, especially vegetation, and ecohydrological research has been undertaken.

On the basis of these researches it is possible to extract drinking

Landscape Ecology of a Stressed Environment
Edited by Claire C. Vos and Paul Opdam
Published in 1993 by Chapman and Hall, London. ISBN 0 412 44820 3

water from the dunes in such a way as to retain natural values. To demonstrate this coexistence, we will focus on the Meijendel dunes, north of The Hague.

11.2 COASTAL DUNES IN THE NETHERLANDS, LANDSCAPE ECOLOGICAL STRUCTURE

The 400 km^2 coastal dunes to be found in The Netherlands are a part of a long and mainly narrow range of dunes stretching from Calais (northern France) to the northern tip of Denmark. Within this range the dunes of The Netherlands are exceptional because of their large extent, high biological diversity and undisturbed conditions (Bakker *et al.*, 1979; Van der Maarel, 1989).

According to Londo (1971) there are four key factors that determine the high spatial variation in dune landscapes.

1. The coastal climate: The first 1–2 km bordering the sea suffer from blowing sand and a high deposition of chloride because of seaspray. These two factors determine the open character of the vegetation in this zone. Here trees and scrub play a minor role. The proportion of trees and scrub in the vegetation cover increases with distance from the sea. At the inner dune fringe, some 4 km from the sea, real woodlands may occur. For this reason dunes show a strong zonation parallel to the coast.
2. The differences in lime content: The lime content of the primary dune sand changes markedly along the Dutch coast (Figure 11.1). The northern dunes are poor in lime, the southern ones are rich in lime. In the former acid soils may easily develop. Each of these zones has its specific vegetation.
3. The great abundance of relief: Because of the pronounced relief of the dunes there is a hydrological gradient from wet slacks to dry slopes and a gradient in the amount of direct sunlight, north-facing slopes differing greatly in this aspect from south-facing slopes.
4. The groundwater: Groundwater plays a major role in the big difference between wet slacks and dry slopes.

These four factors produce a very fine grained, mosaic-like vegetation pattern, with many species.

A further differentiation occurs because of differences between the three main types of coast, which can be distinguished in The Netherlands (Figure 11.1):

1. The estuary coast in the south
2. The mainland coast of northern and southern Holland
3. The wadden coast in the north

Figure 11.1 Main coastal types along the Dutch coast.

Each of these coastal types has its own characteristics. In the estuary and wadden coast, which are far more dynamic than the mainland coast, there are a number of islands, separated by wide inlets. In some places islands are eroded and dunes have disappeared. In other places growing sandflats enlarge the island and new dune-areas are formed. After the disastrous storm surges of 1953, it was decided within the scope of the so-called Deltaplan, to close the inlets of the estuary coast, causing the typical characteristics of this type of coast to disappear.

The turbulent historical development of the estuary and wadden coast has given these parts of the Dutch coastal dunes a varied geomorphological composition which is reflected in the vegetation. Although in some parts, because of the extraction of drinking water, a considerable disturbance of the groundwater situation has taken place, large unaffected areas with wet slacks and even dune lakes still remain.

The mainland coast forms a continuous and almost straight line, along which erosion and sedimentation are basically in balance. Changes in the coastline have therefore been relatively insignificant over the past century. This area is characterized by dune ridges and slacks which run parallel to the coast. These dunes have been strongly affected by water extraction and artificial recharge of the groundwater (Bakker *et al.*, 1979). Some remnants of wet slacks are found here on only a very small scale. Although the groundwater table has risen in the neighbourhood of the spreading basins, used for the production of water, the eutrophic water and the high rate of groundwater flow pre- vented the formation of natural wet slack environments (Van Dijk, 1984).

11.3 THE COASTAL DUNES OF THE NETHERLANDS AS AN AREA FOR THE EXTRACTION OF DRINKING WATER

Around 1850 AD groundwater in the dunes was discovered to be of excellent quality and well-suited as a source of drinking water for the cities in the western part of the Netherlands, such as Amsterdam, The Hague and Haarlem. Canals were dug in the dunes close to this urban zone and drinking water was extracted. The first obvious consequences of this action were a lowering of the groundwater table and drying up of wet slacks.

Van Zadelhoff (1981) describes the effects of a lowering of the groundwater table by 0.5–1.5 m on the vegetation of the wet slacks in the neighbourhood of Bergen. Within some decades the number of phreatophytes had decreased by 56% and the most vulnerable species by 75%. Species which disappeared included: *Sagina nodosa, Samolus valerandi, Drosera rotundifolia, Eleocharis quinqueflora, Scirpus setaceus, Centaurium erythraea, Centunculus minimus, Hierochloe odorata, Juncus acutiflorus, Gentianella campestris, Orchis incarnata, Orchis majalis, Herminium monorchis* and *Liparis loeselli*.

11.3.1 THE GEOHYDROLOGICAL APPROACH

Because of the activities of the drinking water companies there has been a great deal of geohydrological and, later on, geohydrochemical re- search. The scope of this research was purely abiotic, the dunes being studied from a civil engineering point of view. The main question was what quantity of fresh water could be extracted from the dunes, and, to answer this question the geohydrological system had to be understood. The first important principle in understanding the geohydrology of coastal dunes was formulated by Badon Ghyben in 1889 and Herzberg in 1901 (Bakker, 1990).

Figure 11.2 gives a cross-section of the coastal dunes with a rough

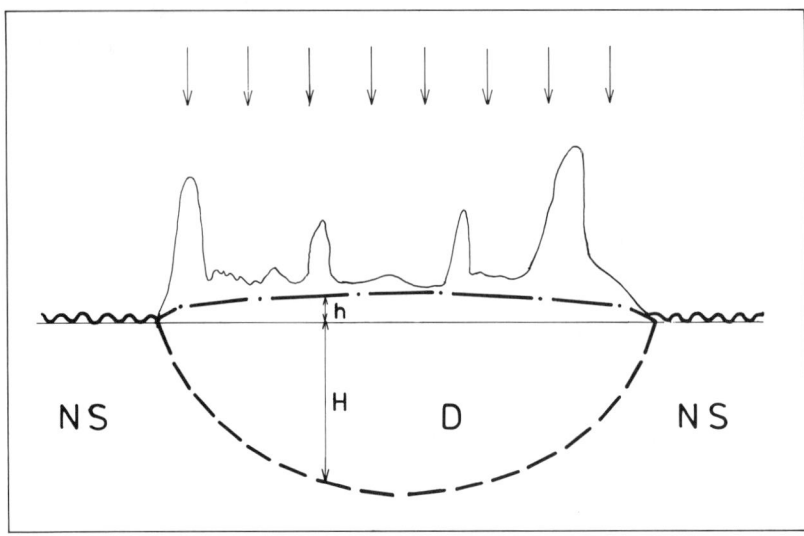

Figure 11.2 Quantities in the Ghyben–Herzberg relation (see text) within a cross-section with a schematic groundwater situation. (h, height of groundwater above sea-level (m); H, Depth of fresh water below sea-level (m); D, dune water; NS, North Sea water)

scheme of the flow of groundwater and shows the specific geohydrological circumstances of coastal dunes with a hinterland below sea level. The fresh water is present in a so-called lens of fresh water which 'floats' on the salt water. At the fresh–salt interface the pressure of the fresh water above must equal the pressure of the salt water below:

$$(h + H) * \rho_f = H * \rho_s \text{ or: } H = \frac{\rho_f}{\rho_s - \rho_f} * h$$

Where h = Height of groundwater table above sea-level (m)

H = Depth of fresh water below sea-level (m)

ρ_f = Density of fresh water (kg/m³)

ρ_s = Density of salt water (kg/m³)

According to this so-called Ghyben–Herzberg relation and applying $\rho_f = 1000$ kg/m³ and $\rho_s = 1025$ kg/m³, the depth of the lens of fresh water below sea-level becomes 40 times the height of the lens above it. In reality, due to the occurrence of semi-impervious layers, like peat and clay, this value is never reached. Values of 15–20 are normal (Bakker, 1981). In the above scheme geohydrology is assumed to be static. Figure 11.3 shows a more realistic view. The dynamic character of the groundwater is demonstrated. An everlasting input of fresh water by precipitation is counterbalanced by evapotranspiration and flow of groundwater to the sides of the dune area.

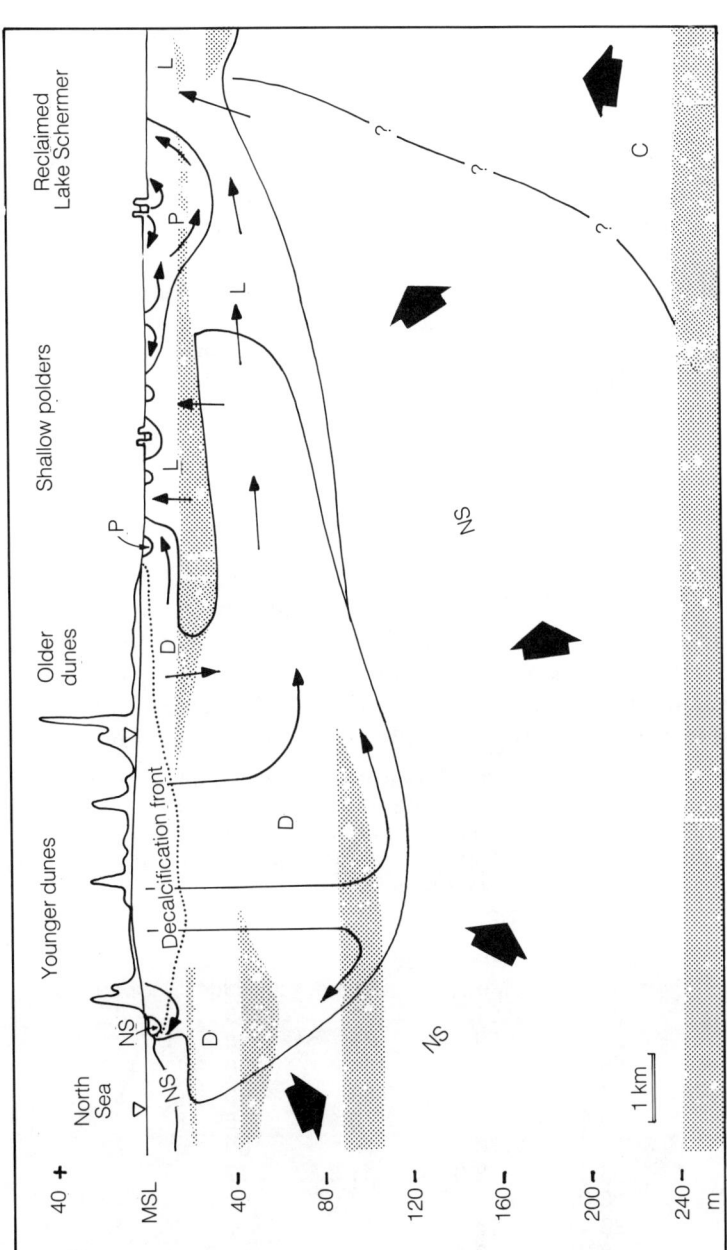

Figure 11.3 The spatial distribution of groundwater types and groundwater flow in a cross section over the dunes north of Bergen primarily poor in lime and without artificial recharge (simplified from Stuyfzand, 1990). (D, dune water; L, relict Holocene transgression water; P, polder water; NS, North Sea water; C, connate Lower Pleistocene sea water; MSL, mean sea level.)

11.3.2 ARTIFICIAL RECHARGE THROUGH SPREADING

With the increasing population in the Western Netherlands there was a rapid increase in the amount of extracted water. This caused a strong mining of the freshwater lens and locally brackish groundwater, unfit as drinking water, was pumped up.

A second phase in the history of the extraction of drinking water from the Dutch coastal dunes began. Plans were developed to bring water from the river Rhine to the dunes. In the 1950s this new technique was started. Pretreated water from the river Rhine replenished the groundwater via spreading basins and was forced towards the extraction wells. Both nature-conservation boards and drinking water companies were optimistic: a nearly unlimited amount of water was available. It was anticipated that groundwater tables would rise, dried slacks would become wet again, and there would be a return of the vulnerable vegetation. However, as the great importance of the water quality was not yet understood, the results were disappointing from a nature-conservation point of view. Although slacks became wet again the typical vegetation did not reappear.

Several types of surface water are used for artificial recharge. The mean composition of some of these types is shown in Table 11.1. The extremes are: pretreated water from the river Meuse, which exhibits the best quality in many respects and which is used in the Meijendel dunes, and untreated water from a polder, which is flushed by Rhine water. As compared to mean shallow dune water, they are clearly more polluted.

Table 11.1 Mean concentrations of pretreated water from Meuse and Rhine and polder water just before spreading in the dunes and of water used for deep-well infiltration. (As a reference also the quality of shallow groundwater of 'natural' dunes, below a mixture of bare sand, mosses and sea buckthorn, is presented.)

	Natural Dune	Pretreated Meuse	Rhine	Polder	Deep-well
Cl (mg/l)	31	61	161	150	82
SO$_4$ (mg/l)	28	59	70	110	91
HCO$_3$ (mg/l)	166	190	154	250	135
Na (mg/l)	17	35	86	89	50
K (mg/l)	0.6	4.8	6.1	15	5.9
Ca (mg/l)	71	76	79	100	72
Mg (mg/l)	2.3	8.9	11.3	17	8.9
NH$_4$ (mg/l)	0.03	0.07	0.07	1.2	<0.01
NO$_3$ (mg/l)	23	15	19	7	11
PO$_4$ (mg/l)	0.05	0.14	0.09	1.8	<0.03
As (μg/l)	<0.8	1.4	1	6.3	<0.5
Cd (μg/l)	<0.1	0.13	0.1	0.25	<0.1
Pb (μg/l)	2	2.7	1	3.6	<1

Table 11.2 Fluxes of atmospheric deposition at 2 km from the high water line of the North Sea and for dune areas with artificial recharge which are infiltrated with pretreated water from Meuse and Rhine and raw polder water.

	Atmospheric deposition	Pretreated Meuse		Pretreated Rhine		Polder basin
		Basin	Whole	Basin	Whole	
Cl ($mgm^{-2}d^{-1}$)	41	7400	900	36225	7250	16500
SO$_4$ ($mgm^{-2}d^{-1}$)	36	7100	900	15750	3150	12100
HCO$_3$ ($mgm^{-2}d^{-1}$)	1	22800	2900	34650	7000	27500
Na ($mgm^{-2}d^{-1}$)	23	4250	400	19350	3900	10000
K ($mgm^{-2}d^{-1}$)	0.8	600	70	1373	275	1650
Ca ($mgm^{-2}d^{-1}$)	3	9200	1100	17775	3500	11000
Mg ($mgm^{-2}d^{-1}$)	3	1100	130	2543	600	1900
NH$_4$ ($mgm^{-2}d^{-1}$)	11.5	8	1	16	3	132
NO$_3$ ($mgm^{-2}d^{-1}$)	18	1800	200	4275	850	770
PO$_4$ ($mgm^{-2}d^{-1}$)	0.2	17	2	20	4	198
As ($\mu gm^{-2}d^{-1}$)	2.7	168	20	23	45	693
Cd ($\mu gm^{-2}d^{-1}$)	0.8	16	2	2	5	28
Pb ($\mu gm^{-2}d^{-1}$)	52	324	40	23	45	396
H$_2$O (mm/d)	1.3	120	15	225	45	110

However, low concentrations in combination with a high flow velocity may yield high fluxes, which are also problematic. Table 11.2 gives fluxes for ions from the atmosphere (atmospheric deposition) and fluxes for three artificial recharge water types: pretreated water from the rivers Rhine and Meuse and raw polder water. Fluxes in the spreading basins are a few hundred to a thousand times higher than those caused by atmospheric deposition. When we consider the artificial recharge area as a whole, including the area in between the basins and the pumping wells, the difference between fluxes by precipitation and fluxes by artificial recharge decreases but still remains very high NH$_4$ and Pb excluded.

Because of these high fluxes the vegetation of artificial recharge basins and slacks under the influence of recharged river water differs greatly from the vegetation of natural wet slacks. Van Dijk (1984) stated that in wet slacks under the influence of infiltration water, species such as Urtica dioica, Epilobium hirsutum, Mentha aquatica, Eupatorium cannabinum, Lycopus europaeus, Circium arvense, Calamagrostis epigejos and Phragmites australis cover 50–80% of the surface. The indigenous species such as Epipactus palustris, Carex panicea, Schoenus nigricans and Parnassia palustris are found in only a few places. As will be shown, one of the benefits of the geohydrochemical and ecohydrological approach is that we understand the mechanisms behind the existence of these few 'natural' places within a dune-area with artificial recharge (Figure 11.5 and matching text).

Nowadays the total amount of artificially recharged water in the Dutch coastal dunes is in the order of 175 million m^3 per year. The total supply by precipitation is some 300 million m^3 per year of which some 50% returns to the atmosphere by evapotranspiration. This means that the net yearly supply of groundwater by precipitation is slightly less than the artificially recharged amount. Annually some 225×10^6 m^3 of water is extracted for drinking water.

11.3.3 DEEP-WELL RECHARGE

In 1990, with the introduction of deep-wells, a third phase in the history of the extraction of drinking water from the dunes was initiated. With this technique strongly pretreated river water (Table 11.1) is pumped into deeper soil layers by means of so-called deep-wells, and is extracted at the same deep level. In this way the upper groundwater is hardly influenced and keeps its natural quality and behaviour. Wet slacks may therefore develop under almost natural circumstances.

Within the Dutch coastal dunes two plants, each with a capacity of 4 million m^3/year, are now in full operation. One of them is situated in the Meijendel dunes.

11.3.4 THE GEOHYDROCHEMICAL APPROACH

The geohydrological approach, discussed above, mainly was concerned with the quantity of groundwater; quality aspects of the water were not considered important. However, since the 1980s the quality of the groundwater has received much more attention – the geohydrochemical approach was introduced (Stuyfzand, 1990).

A first important division within the fresh water (Cl < 300 mg/1) in the dunes is between true dune water, which originates from the local precipitation excess, and artificially recharged surface water. The spatial distribution of both types of groundwater in a longitudinal section (Figure 11.4), reveals that dune water is still the predominant water type and that recharged Rhine water has reached the fresh–brackish interface at one recharge plant only: the extensive Scheveningen plant in the Meijendel dunes to the north of The Hague. The fresh and salt (Cl > 10 000 mg/1) waters, are separated by a 8–80 m thick brackish zone. This zone reaches its maximum thickness around the North Sea Canal, which connects the port of Amsterdam with the North Sea and where dune water was extracted for drinking and industrial water supply. Another area with a thick transition zone is situated between Castricum and Bergen, where the expulsion of relict Holocene transgression water is strongly retarded by aquicludes and the local flow pattern (Stuyfzand, 1990).

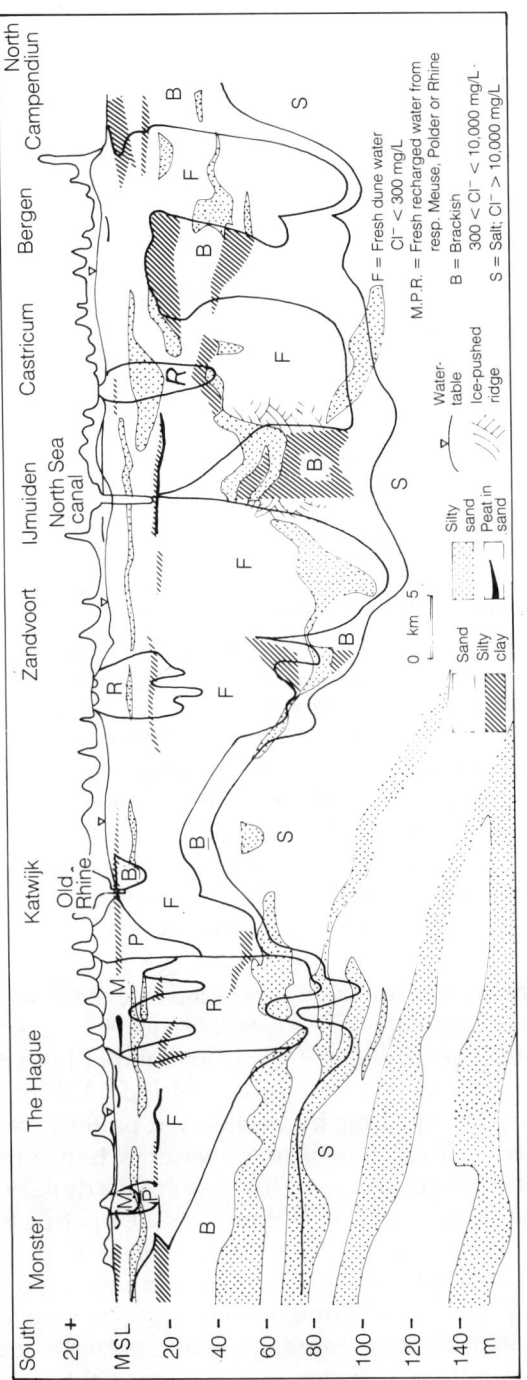

Figure 11.4 Longitudinal section over coastal dunes of the western part of The Netherlands, with the spatial distribution of groundwater types (simplified from Stuyfzand, 1989). (MSL, mean sea level.)

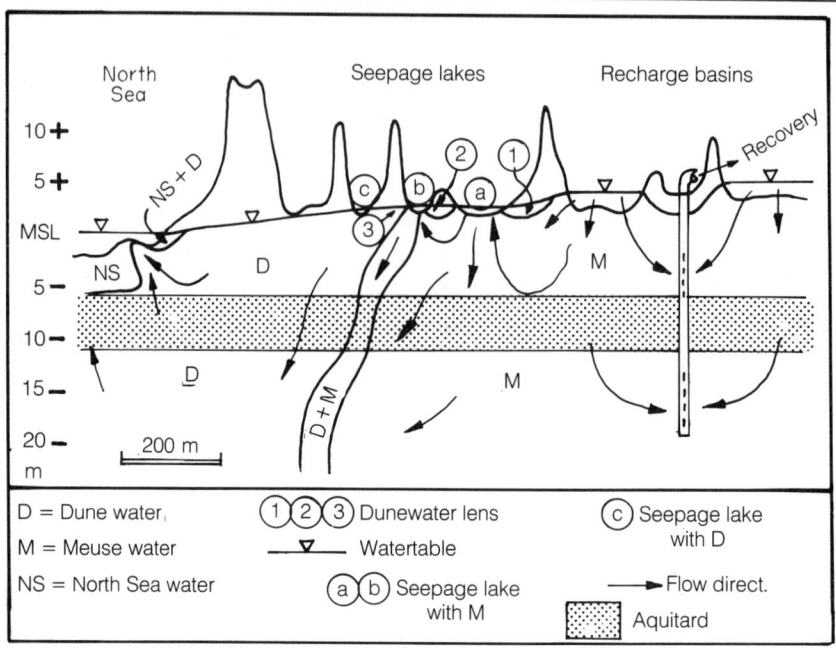

Figure 11.5 The spatial distribution of groundwater types and groundwater flow in a cross section over the calcareous Meijendel dunes north of The Hague with artificial recharge (simplified from Stuyfzand, 1991). (MSL, mean sea level; 1–3, dune water lenses; a–c, seepage lakes.)

In dune areas without artificial recharge, the quality of the groundwater depends, among other things on the quality of the precipitation, the variations in the content of primary $CaCO_3$ (calcareous versus non-calcareous dunes) and a strong gradient in sea-spray deposition and vegetation covers perpendicular to the coastline (Stuyfzand, 1987).

A part of this variation is shown in a cross section over dunes primarily poor in lime, north of Bergen (Figure 11.3). The dune water is present amidst connate Lower Pleistocene sea water, intruded North Sea water (Late Holocene to recent), relict Holocene transgression water and recent polder water. The reclamation of lakes in the coastal plain behind the dunes, mainly in the period 1400–1860, has a dominant impact on the dynamic hydrochemical pattern and also on the actual flow pattern. Acid dune groundwater is transformed at about 20 m below MSL (mean sea level) into calcareous dune water by the dissolution of shell fragments in Holocene and Upper Pleistocene formations.

In dunes with artificial recharge, thanks to precipitation, lenses of dune water may develop on top of laterally migrating recharged river water (Figure 11.5). Hydrodynamic dispersion prevents their formation in areas with closely spaced spreading basins and extraction systems. In the other areas, sufficiently remote from the spreading basin, lenses may contain pure dune water, with a thickness up to several metres (point 3 in Figure 11.5). Both water types are well mixed in seepage lakes with- in 100–250 m from spreading basins without extraction on that side.

11.3.5 THE ECOHYDROLOGICAL APPROACH

The ecohydrological approach developed more or less in parallel to the geohydrochemical approach. Grootjans who did research on the Wadden island of Schiermonnikoog played a crucial role in this development (Grootjans *et al.*, 1988, 1989; Lammerts *et al.*, 1992).

Ecohydrological research deals primarily with the occurrence of plant species in relation to biotic and abiotic features of the environment. The main question in this research is: How can the occurrence (or disappearance) of plant species be explained from patterns (or changes in patterns) in soil conditions and the quality and quantity of the groundwater? Ecohydrological research is of importance in the field of nature conservation and nature management. Once we understand the reason why certain plants occur on certain places we can manipulate the circumstances and, by doing so, stimulate the occurrence of vulnerable plant species. An example of this approach is the understanding of the occurrence of lenses of dunewater in a dune area with artificial recharge as is demonstrated in Figure 11.5. These lenses are very important from a nature conservation point of view. Here, within an area with artificial recharge, the characteristic and vulnerable wet slack vegetation may develop. In the Meijendel dunes, species such as *Parnassia palustris, Gentianella amarella, Dactylorhiza incarnata, Sagina nodosa* and *Carex trinervis* are found in these places.

By this pattern-oriented way of nature management, focusing on these water lenses and their specific vegetation, a multifunctional land-use can be practised, and nature conservation and extraction of drinking water may be combined.

The vegetation types aimed for in nature conservation, the so called 'goal type' vegetations for wet slack communities are: the Samolo-Littorelletum, the Centaurio-Saginetum, moniliformis, the Parnassio-Juncetum atricapilli and the Junco balticischoenetum nigricantis.

In the course of the last century some 70% of the total wet dune area of The Netherlands has lost this type of vegetation. This decline is associated with processes of 'ageing' and of 'drying out'. The former is related to processes such as decalcification (acidification) and accumulation of organic matter (succession), the latter with the draw-down of the groundwater table (desiccation).

To start with the last process, Noest and Van der Maarel (1989) (see also Noest *et al.*, 1989; Noest, 1990) related the occurrence of some 100 plant species of the 'goal type' communities to the groundwater table. They found that the presence of vulnerable plant species of wet slacks is mainly determined by the following factors:

1. Period of inundation of a slack during the growing season
2. The total range of the yearly fluctuation of the groundwater table (under 'natural' circumstances this range of fluctuation is the order of 0.5–1 m; Bakker, 1981).

Van Zadelhoff (1981) reviews the research on the drying out of wet slacks. In many cases a draw-down of the average groundwater table by 10 cm causes a disturbance of the vegetation. A drawdown of one metre is enough to cause a strong desiccation of the slacks and the disappearance of many vulnerable plant species.

Lammerts *et al.* (1992) consider the process of acidification. They summarize the research done on the Dutch Wadden islands in the field of the quality of groundwater and soil in relation to the plant species of wet slacks. They determine the following main factors for the occurrence of 'goal type' plant species of wet slacks:

3. The lime content of the top soil. If this content exceeds 0.25% important elements of the 'goal type' communities are often present.
4. The groundwater composition. When the lime content of the top soil layer is below 0.25% the quality of the groundwater becomes the predominant factor. The vegetation 'goal types' are only found when the calcium concentration of the groundwater exceeds 40 mg/1.
5. Management activities like sod cutting, mowing and grazing which prevent a rapid accumulation of organic matter.

These five factors determine to a high degree the presence of wet slack vegetations.

Figure 11.6 shows an example of the way these factors, especially 3 and 4, contribute to the distribution of plant species (Esselink *et al.*, 1989). In a cross-section over a wet slack some types of groundwater are shown in relation to the surface. The lime content of the soil in this slack is well below 0.25%. So the calcium concentration of the groundwater is the prime factor to determine the occurrence of calciphilous plant species. Under the small and low dune ridges within the slack, small lenses of (acid) rainwater are present. Groundwater rich in calcium reaches the root zone of the vegetation in only a few places. Here the last remnants of a calciphilous vegetation, with species such as *Schoenus nigricans, Epipactus palustris* and *Parnassia palustris,* are still found. Acid circumstances prevail in this slack and plant species such as *Potentilla palustris* and *Eriophorum angustifolium* are common.

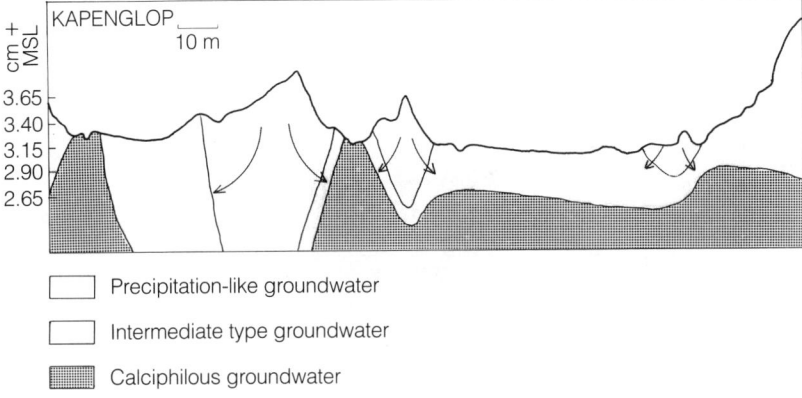

Figure 11.6 Distribution of three types of groundwater in the upper metre of the wet slack Kapenglop on the Frisian island of Schiermonnikoog (simplified from Esselink *et al.*, 1989). (MSL, mean sea level.)

Therefore, by using classical approaches like geohydrology and geohydrochemistry and by paying special attention to biotic features like species composition, ecohydrological research achieves an integrated approach to the landscape and provides tools for better nature management.

11.4 THE MEIJENDEL DUNES NEAR THE HAGUE, A CASE STUDY

The Meijendel dunes (Figure 11.7) cover an area of about 2000 hectares. The sand is predominantly calcareous. In the south-eastern part, where older dune deposits crop out, the lime content may have dropped below 0.25% and acid soil conditions prevail. The area consists, from the sea inland, of regularly ridged fore dunes parallel to the coast, parabolic dunes alternating with depressions, extensive slacks with abandoned 19th century farmland, and a broad inner dune ridge with a steep leeward inner dune face. These main landscape types have developed almost parallel to the coastline.

The most important utility functions of this area are the extraction of drinking water, nature conservation, recreation and coastal defence.

The extraction of drinking water in this area started in 1874. In 1955 artificial recharge with pretreated water from the river Rhine was introduced. Since 1976 water from the less-polluted river Meuse has been used for artificial recharge. The pretreated water is pumped into spreading basins which have a lake-like character. From there the water

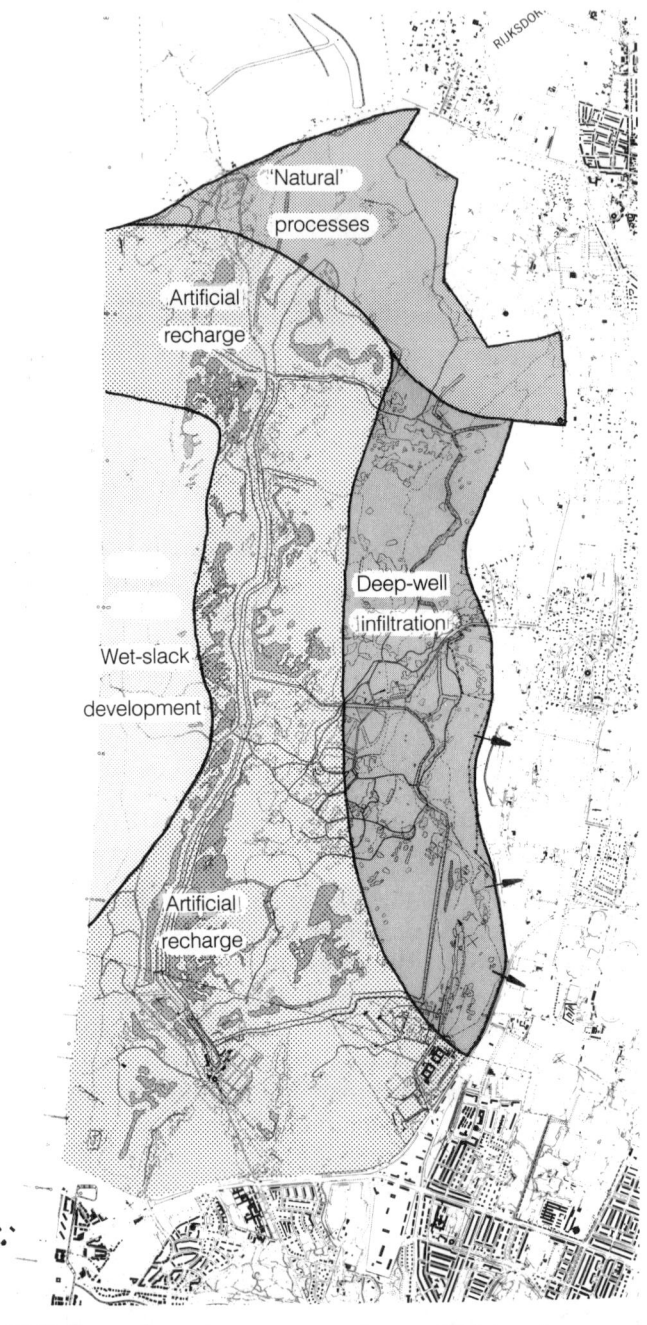

Figure 11.7 Zones for nature-conservation, artificial recharge and extraction of drinking water within the Meijendel dunes near The Hague. (MSL, mean sea level.)

flows underground to shallow extraction devices such as wells and a network of horizontal drains. In 1990 the first deep-well infiltration plant, with a capacity of four million m^3/year was implemented. Nowadays the yearly amount of drinking water produced in the Meijendel dunes is approximately 50 million m^3.

The Dune Water Company of Southern Holland not only deals with the supply of drinking water but also with the conservation and management of the Meijendel dunes. One of the major management aims is to enlarge the area of the 'goal type' vegetation of wet slacks. The following strategy has been scheduled.

Figure 11.5 shows that in the zone directly behind the coast, lenses of natural dunewater are present. As the Meijendel dunes are rich in lime, and because of these lenses, the possibilities for the 'goal type' vegetation are favourable. The problem which remains to be solved, is that of the required groundwater table. The so-called HYVEG-model was developed for this purpose (Noest, 1990), with which it is possible to determine levels of watertables in the basins for artificial recharge in such a way that in the boardering slacks the desired groundwater tables are established. Together with calciphilous soil and groundwater conditions vulnerable vegetation types may then develop.

These more or less artificial wet slacks have developed within an area of about 50 ha in the last decade and it is expected that another 50 ha will be created before the year 2000. To do so the extraction of drinking water in this part of the Meijendel dunes has to be abandoned and substituted by deep-well infiltration elsewhere.

Another aim for nature conservation in the next decade is to improve the quality of the spreading basins. This is done by giving the recharge water from the Meuse a further pretreatment and by reshaping the borders of the spreading basins to gently sloping banks. The latter will provide better opportunities for various plant and bird species.

As was shown in Table 11.2, the fluxes of some major ions within the basins of dune-areas with artificial recharge are a hundredfold and more higher than those under 'natural' circumstances. Also after further pretreatment fluxes will remain high. Therefore, the vegetation of the spreading basins will change only a little and will keep its nutrient-rich character. The greatest benefit for nature-conservation will be an increase in the number of rare species of breeding birds (Hoekstra and Van der Hagen, 1990; Slings and Schekkerman, 1990).

The land side part of the Meijendel dunes, where the freshwater lens reaches its maximum range, is best suited for deep-well infiltration and here the new deep-well infiltration plants are situated. Thereby groundwater tables will rise and favourable conditions for the development of wet slacks and brooklets may be created.

Finally, in the northern part of Meijendel, some hundreds of hectares of dune area are much less affected by the activities of the drinking water supply. Here hydrological circumstances are more or less natural and nature management may be more process orientated. By stimulating characteristic dune-forming processes like the blowing of sand, large-scale blowout developments can go their way and wet slacks may develop. In order to enlarge this area and to give blowouts better chances, the extraction of drinking water and artificial recharge in the bordering areas must be terminated and substituted by deep-well infiltration elsewhere.

11.5 CONCLUSIONS

The Dutch coastal dunes are rich in nature and belong to the most vulnerable dune-landscapes of the world. Yet there is an intensive use of these dunes, as water catchment area for the supply of drinking water for the densely populated hinterland. Wet slacks have suffered strongly from the extraction of drinking water and have lost their vulnerable vegetation.

Ecohydrological research has proved that five main factors of the environment are important for the existence of 'goal type' wet slack vegetations: the period of inundation, the total range of the yearly fluctuation of the groundwater table, the lime content of the top soil, the quality of groundwater, and management activities such as sod cutting, mowing and grazing. Further ecohydrological research showed ways to undertake a pattern-oriented way of nature management in which artificial recharge and extraction of drinking water combine with the development of wet slacks. This approach may be well practised in dune areas where several functions, such as extraction of drinking water, nature conservation, coastal defence and recreation, must be served.

In more remote places, where nature conservation may be the only or major type of land use, a more process-oriented way of nature management is better suited. The stimulation of large-scale blowing within an area with 'natural' hydrological circumstances will result in the development of wet slacks with 'goal type' vegetation.

ACKNOWLEDGEMENT

We wish to thank Professor Peter Jungerius and Dr Larry Boorman for their comments on the text.

REFERENCES

Bakker, T.W.M. (1981) *Nederlandse kustduinen; Geohydrologie.* Thesis, Pudoc, Wageningen.

Bakker, T.W.M. (1990) The geohydrology of coastal dunes, in *Dunes of the European Coasts, Geomorphology, Hydrology, Soils,* (eds T.W.M. Bakker, P.D. Jungerius and J.A. Klijn), Catena supplement 18, Cremlingen - Destedt, pp. 109–19.

Bakker, T.W.M., Klijn, J.A. and Van Zadelhoff, F.J. (1979) *Duinen en duinvalleien, een landschapsecolgische studie van het Nederlandse duingebied.* Pudoc, Wageningen.

Esselink, H., Grootjans, A.P., Hartog, P.S. and Jager, T.D. (1989) Kalkrijke vegetaties in een duinvallei op Schiermonnikoog. *Duin,* **2,** 75–80.

Grootjans, A.P., Hendriksma, P., Engelmoer, M. and Westhoff, V. (1988) Vegetation dynamics in a wet dune slack I: rare species decline on the Waddeneiland of Schiermonnikoog in the Netherlands. *Acta Botanica Neerlandica,* **37** 265–78.

Grootjans, A.P., Esselink, H., Van Diggelen, R. *et al.* (in press). Decline of rare calciphilous dune slack species in relation to decalcification and changes in local hydrological systems. *Proceedings Coastal Dune Congress Seville, 1989.* EUCC, Leiden.

Hoekstra, A.C. and Van der Hagen, H.G.J.M. (1990) Infiltratiepannen in Meijendel en de slibverwijdering in pan 13, in: *Natuurwaarden en Waterwinning in de Duinen* (eds W. Koerselman, M.A. Den Hoed, A.J.M. Jansen and W.H.O. Ernst), KIWA-Nieuwegein, mededeling nr. 114, pp. 65–89.

Lammerts, E.J., Sival, F.P., Grootjans, A.P. and Esselink, H. (1992). Hydrological conditions and soil buffering processes controlling the occurrence of dune slacks species on the Dutch Wadden Sea islands in *Coastal Dunes. Geomorphology, Ecology and Management for Conservation* (eds R.W.G. Carter, T.F.G. Curtis and M.J. Skeffington) A.A. Balkema, Rotterdam/Brookfield 533 pp.

Londo, G. (1971) *Patroon en Proces in Duinvalleivegetaties Langs een Gegraven meer in de Kennemerduinen.* Verhandelingen Rijks Instituut voor Natuurbeheer, Leersum.

Noest V. (1990) *HYVEG, een interaktiemodel hydrologie – vegetatie voor jonge vochtige duinvalleien.* Dunewatercompagny of Southern Holland, Den Haag.

Noest, V. and Van der Maarel, E. (1989) A new dissimilarity measure and a new optimality criterion in phytosociological classification. *Vegetatio,* **83,** 157–65.

Noest, V., Van der Maarel, E., Van der Meulen, F. and Van der Laan, D. (1989) Optimum-transformation of plant species cover-abundance values. *Vegetatio*, **83**, 167–78.

Slings, Q.L. and Schekkerman, H. (1990) Vogels van de infiltratiegebieden in het Noordhollands Duinreservaat, in *Natuurwaarden en Waterwinning in de Duinen* (eds W. Koerselman, M.A. Den Hoed, A.J.M. Jansen and W.H.O. Ernst), KIWA–Nieuwegein, mededeling nr. 114, pp. 89–103.

Stuyfzand, P.J. (1984) Groundwater quality evolution in the upper aquifer of the coastal dune area of the western Netherlands, in *Hydrochemical Balances of Freshwater Systems. Proceedings of the Uppsala Symposium Sept. 1984* (ed. E. Eriksson), IAHS Publication 150, pp 87–98.

Stuyfzand, P.J. (1987) Effects of vegetation and air pollution on groundwater quality in calcareous coastal dunes near Castricum the Netherlands. Lysimeter observations. *Proceedings International Symposium Acidification and water pathways*. Bolkesjo Norway 4–8 May 1987. Norwegian National Commission for Hydrology, **2**, 115–25.

Stuyfzand, P.J. (1989) Hydrochemical evidence of fresh and salt water intrusions in the coastal dunes aquifer system of the Western Netherlands. *Natuurwetenschappelijk Tijdschrift*, **70**, 9–29.

Stuyfzand, P.J. (1990) Hydrochemical facies analysis of coastal dunes and adjacent low lands: The Netherlands as an example, in *Dunes of the European coasts, Geomorphology, Hydrology, soils* (eds. T.W.M. Bakker, P.D. Jungerius and J.A. Klijn), Catena supplement 18, Cremlingen - Destedt, pp. 121–32.

Stuyfzand, P.J. (1991) Een hydrochemische facies analyse voor hydro-ecologisch onderzoek: theorie en toepassing op Hollands kustduinen en aangrenzende polders, in *Natuurwaarden en Waterwinning in de Duinen*. (eds. W. Koerselman, M.A. Den Hoed, A.J.M. Jansen and W.H.O Ernst) KIWA-Nieuwegein, mededeling 114, pp 191–213.

Thijse, J.P. (1927) *Texel*, Verkade-album, Zaandam.

Udo de Haes, H.A., Drijver, C.A. and Vertegaal, C.T.M. (1980) Is duininfiltratie wel zo gunstig voor de vogelstand? *Duin*, **3**, 3–11.

Van der Maarel, E. (1989) De betekenis van de Nederlandse kust in nationaal en Europees verband. *De Levende Natuur* **90**, 131–4.

Van der Meulen, F., Jungerius, P.D. and Visser, J.H. (eds) (1989) Perspectives in coastal management. *Proceedings of the European Symposium Leiden*, 7–11 September 1987. SPB Academic Publishing, Leiden.

Van Dijk, H.W.J. (1984) *Invloeden van oppervlakte-infiltratie ten behoeve van duinwaterwinning op kruidachtige oevervegetaties*. Thesis, LU Wageningen.

Van Zadelhoff, F.J. (1981) *Nederlandse kustduinen Geobotanie*. Thesis, Wageningen.

Artificial wetlands: a device for restoring natural wetland values

12

Harm Duel, Roel During and Cees Kwakernaak

12.1 EUTROPHICATION OF WETLANDS

In The Netherlands, many oligotrophic and mesotrophic surface waters have been eutrophied. Eutrophication of rivers, streams and lakes is also a major problem in other West European countries (Melzer, 1976; Whitton, 1984; Hoffmann, 1985). The influx of phosphorus into Dutch inland waters can be attributed to three main sources. In 1985, inland sources accounted for 40% of the total influx of phosphorus, while the rivers Rhine and Meuse accounted for 45% and 12% respectively. In 1985, the main sources of nitrogen were inland locations (33%), river Rhine (51%), river Meuse (3%) and other transboundary rivers (11%). Table 12.1 gives the nutrient loading figures in tons per year. Drainage and runoff from manured agricultural areas is the main inland source of eutrophication. Due to high manuring levels, the average nutrient loading of the discharge from the agricultural lands to the surface waters is 2.0 kg P/ha/year and 20 kg N/ha/year. Nutrient emissions are highest in areas of intensive agriculture, situated on sandy soils in the eastern and southern parts of The Netherlands (Vos and Zonneveld, 1993, this volume). Also, in the lowland peat areas used for agriculture nutrient emissions are high.

The eutrophication of water systems has affected aquatic vegetations and aquatic wildlife. The quantity of submerged waterplants has decreased dramatically in the formerly oligotrophic and mesotrophic

Landscape Ecology of a Stressed Environment
Edited by Claire C. Vos and Paul Opdam
Published in 1993 by Chapman and Hall, London. ISBN 0 412 44820 3

Table 12.1 The nutrient loading of surface waters in The Netherlands (1985) and the predicted effects of the national nutrient reduction programme(1995), including the expected results of the international nutrient reduction programmes (Ministerie van Verkeer en Waterstaat, 1989)

Source	Phosphorus		Nitrogen	
	1985	1995	1985	1995
Sewage untreated	2300[a]	600	8000	5000
Sewage treated	11000	3500	38000	24000
Industry	13500	7600	16000	10000
Agriculture (runoff and drainage	6000	6700	169000	155000
Atmospheric deposition	-	-	20000	14400
Contribution Rhine	36600	18300	378000	189000
Contribution Meuse	9600	4800	2500	12500
Contribution other transboundary rivers	3000	1500	82000	41000
Total	82000	43000	736000	450900

[a]Discharges in tons/year.

surface waters (Best *et al.*, 1984; De Nie, 1987). External phosphorus loading of formerly clear-water ecosystems led to increased production levels of phytoplankton, especially blue–green algae. As a consequence of enhanced algal productivity, clear-water species disappeared and were replaced by other species. The abundance of bream, smaller zoo-plankton species and blue–green algae is indicative of the prevailing eutrophic and hypertrophic conditions in many standing surface waters in Western Europe (Deufel *et al.*, 1986; Lammens, 1986; De Nie, 1987; Van Liere *et al.*, 1990).

Hence, eutrophication control is one of the major issues in West European environmental policy. Eutrophication control policy in Western Europe has been determined in two international programmes: the Rhine Action Programme and the North Sea Action Programme. The programmes aim at a 50% reduction of the 1985 level of phosphorus and nitrogen emissions from point sources by 1995. However, model studies reveal that for most lakes in The Netherlands this reduction will not be sufficient to solve the eutrophication problems (Hosper and Jagtman, 1990). Additional measures are required, for example, dredging lake sediments, lake flushing and chemical phosphorus fixation. Artificial wetlands have been mentioned as a promising measure to reduce the problem of eutrophication locally. Additionally, landscape planners promote the application of artificial wetlands as a welcome introduction of characteristic and useful landscape elements. As a consequence, artificial wetlands are becoming a familiar part of many re-allotment plans in The Netherlands.

This chapter discusses the applicability of artificial wetlands for ecological restoration of eutrophied natural wetlands. The significance of applying artificial wetlands within the framework of nature restoration strategies in the peatland area Green Heart of Holland is worked out in more detail.

12.2 ARTIFICIAL WETLANDS FOR EUTROPHICATION CONTROL

During the last decades in many countries considerable attention has been paid to the application of artificial wetlands for wastewater treatment (Tourbier and Pierson, 1976; Godfrey *et al.*, 1985; Van der Aart, 1985; Reddy and Smith, 1987). Because of the large surface area required for wastewater purification by artificial wetlands, this application is generally not possible in the densely populated Netherlands. Therefore, this kind of application is not dealt with here.

In The Netherlands, the feasibility of applying artificial wetlands is focused on treating eutrophied surface water systems (Duel and Saris, 1986; Duel and During, 1990). Agricultural runoff and drainage water treatment by artificial wetlands has been studied in other European countries and in the United States (Hoffmann, 1985; Goldstein, 1987; Kadlec, 1987a; Hammer. 1989).

12.2.1 REMOVAL PROCESSES

Artificial wetlands may act as efficient purification systems for water with high nitrogen and phosphorus concentrations (Nichols, 1983; Kadlec, 1987a; Brix and Schierup, 1989). Nitrogen and phosphorus can be removed by a complex variety of biological, physical and chemical processes, including plant uptake, bacterial transformation, sedimentation, adsorption and precipitation (Kadlec, 1987a; Good and Patrick, 1987; Patrick, 1990).

Removal of nutrients is most efficient during the growing season. In wintertime, nitrogen and phosphorus removal processes are considerably less active (Reed *et al.*, 1984; Brix and Schierup, 1987; Kadlec, 1987a). Although artificial wetlands can be applied to reduce nutrient concentrations, they can not alter significantly the macro-ionic composition (Duel and During, 1990).

12.2.2 NITROGEN REMOVAL

Uptake by helophytes and denitrification can be regarded as key processes in nitrogen removal. Maximum nitrogen storage capacity in aboveground biomass by helophytes proved to be between 250 and 530 kg N/ha/year (Duel and During, 1990). The daily uptake of nutrients by wetland vegetation depends on numerous factors. In Western Europe, the estimated average daily uptake in aboveground biomass in the summer is 1.5–2.5 kg N/ha (Duel and During, 1990; Koerselman, 1990). In the winter period, nutrient uptake by aboveground biomass does not occur. Harvesting is required to remove nitrogen stored in biomass from the wetland system. In non-harvested wetlands the litter will release its nutrients into the water by rapid decomposition.

Denitrification rates in wetland soils depend on the nitrate concentration level in the overlying water body (Andersen, 1977, 1981; Reddy and Patrick, 1984). Linear relationships were found between the denitrification rates in reed swamp sediments in shallow eutrophic Danish lakes and the nitrate concentration levels in the overlying lake water (Andersen, 1981). Duel *et al.* (1991) estimate denitrification rates in artificial wetlands at 0.2 kg N/ha/day per mg N/l nitrate in the overlying water.

12.2.3 PHOSPHORUS REMOVAL

Key processes in phosphorus removal are sedimentation of suspended load, uptake by helophytes, adsorption and precipitation. The highest amounts of phosphorus storage in aboveground biomass of helophytes that have been found range from 45 to 110 kg P/ha/year (Duel and During, 1990). During summer, the average daily uptake in aboveground biomass is estimated for Western Europe at 0.15–0.25 kg P/ha (Duel and During, 1990; Koerselman, 1990). Storage of phosphorus in aquatic soils should be regarded as a temporary phenomenon because adsorption and precipitation are reversible processes and because the adsorption capacity of soil particles is necessarily limited. The maximum amount of phosphorus that can be fixed in wetland soils is unknown.

12.2.4 CONDITIONAL FACTORS

Water depth, water residence time and the hydraulic loading rate are key variables in assessing the performance of artificial wetlands as water treatment systems. The maximum water depth of artificial wetlands covered by helophytes should be between 20 and 50 cm, depending on the type of emergent macrophytes (Tchobanoglous, 1987). The water residence time determines the period available for nutrient removal from surface waters (Kadlec, 1987b). For a nutrient reduction of more than 50% the water residence time should be generally over 10 days (Duel *et al.*, 1991).

Nutrient removal efficiency in artificial wetlands proved to be correlated to hydraulic loading rates (Figure 12.1). A more than 50% nutrient removal efficiency can be reached in the summer if hydraulic loading

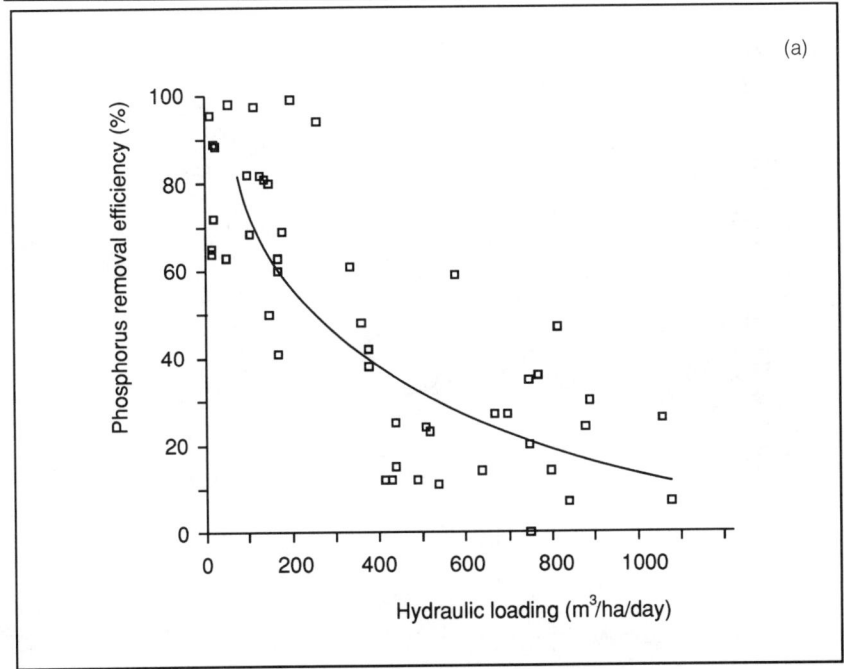

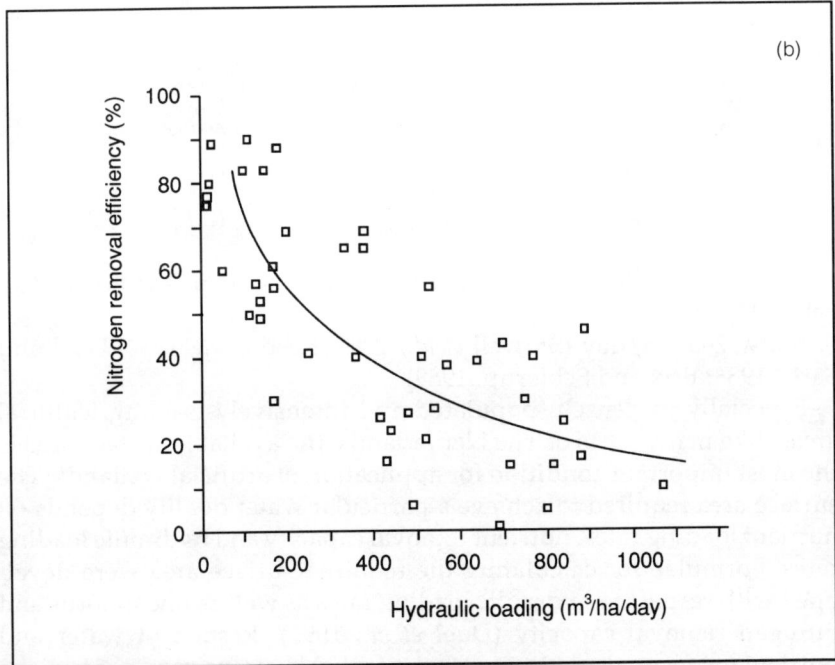

Figure 12.1 The nutrient removal efficiency of artificial wetlands used for municipal and agricultural wastewater treatment (a) phosphorus; (b) nitrogen. Information is based on data from artificial wetlands in North America and Western Europe.

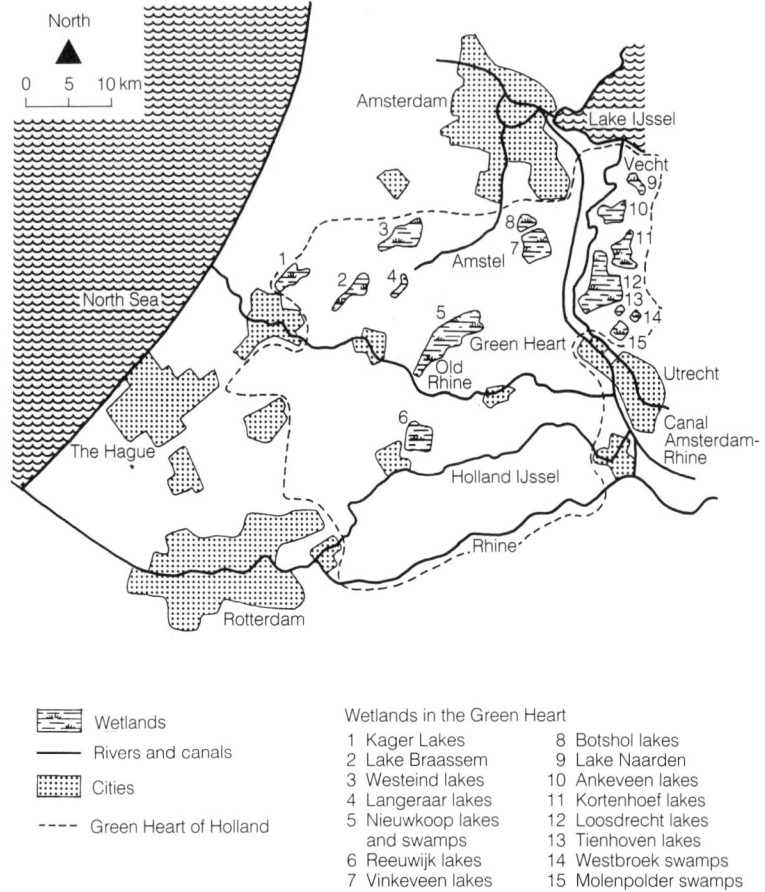

Figure 12.2 Location of the Green Heart: the peatland area in the western part of The Netherlands.

is below 2–3 cm/day (Stowell *et al.*, 1981; Reed *et al.*, 1984; Duel and Saris, 1986; Brix and Schierup, 1989).

Especially in densely populated and intensively used agricultural areas, like many parts of The Netherlands, the available surface area is the most important condition for application of artificial wetlands. The surface area required to achieve a particular water quality depends on nutrient loading rates, nutrient removal capacity and hydraulic loading rates. Formulae for calculating the required surface area were developed with respect to hydraulic loading rates as well as phosphorus and nitrogen removal capacity (Duel *et al.*, 1991). Results of water and nutrient balance calculations are an essential basis for good estimations of the required surface area.

12.3 EUTROPHICATION OF WETLANDS IN THE GREEN HEART OF HOLLAND

The Green Heart of Holland is an open peatland area situated in the western part of The Netherlands, enclosed by the urban areas of Amsterdam, The Hague, Rotterdam and Utrecht (Figure 12.2). The average soil surface level in this peatland area is below average sea level. The Green Heart is representative of polder landscapes. It is characterized by numerous extended meadows separated by small ditches, shallow lakes, small rivers and small remnants of originally vast mesotrophic and oligotrophic mires. The land is mainly used for agriculture (grazing, greenhouse cultures), recreation (boating) and nature conservation (wetlands). Villages and smaller towns in the Green Heart tend to annex peatland areas for building suburban districts. Nature reserves and connecting landscape structures have been dissected by highways and railroads. The many ground claims in the Green Heart make it necessary to make choices in order to give room to the several types of land use in a sustainable way.

An integrated approach to physical planning, environmental management, ecological restoration and water management on a regional scale has been developed for the Green Heart. Table 12.2 shows that

Table 12.2 The quality of the surface waters in the Green Heart of Holland in 1986. The Dutch basic water quality standards concerning nitrogen and phosphorus concentrations are set at 0.15 mg P/l and 2.2 mg N/l

Surface water system	Ammonium concen- tration	Nitrate concen- tration	Dissolved phosphor- us concen- tration	Total phosphor- us concen tration	Chloride concen- tration
	(mg N/l)	(mg N/l)	(mg P/l)	(mg P/l)	(mg/l)
River Rhine	0.57	4.35	0.28	0.47	160
River Vecht	2.2	3.43	0.57	1.08	127
River Amstel	0.54	2.24	0.22	0.29	147
River Old Rhine	1.4	3.2	0.29	0.48	156
Canal Amsterdam– Rhine	0.75	3.89	0.41	0.57	138
Molenpolder Swamps	0.32	0.28	0.44	0.58	72
Westbroek Swamps	0.39	0.12	0.08	0.29	21
Loosdrecht Lakes	0.21	0.26	0.05	0.15	59
Kortenhoef Lakes	0.1	0	0.02	0.26	224
Ankeveen Lakes	0.1	0	0.02	0.15	97
Lake Naarden	0.1	0	0.03	0.09	197
Botshol Swamps	0.2	0.38	0.07	0.1	497
Vinkeveen Lakes	0.18	0.37	0.16	0.05	249
Nieuwkoop Lakes and Swamps	0.2	0.16	0.05	0.22	103
Westeinder Lakes	0.25	0.64	0.16	0.23	145

eutrophication of surface waters is a serious environmental problem. An important goal of the integrated approach of the Green Heart is the rehabilitation of former ecological wetland values. Sustainable conservation and ecological rehabilitation are to be achieved mainly by taking measures for eutrophication control and for water conservation. Quantitative water management is linked largely with the eutrophication problem of peatland areas. This is explained in the following sections.

12.3.1 HYDROLOGY AND WATER MANAGEMENT

The Green Heart peatland area is a good example of a so-called inversion landscape. Originally low-lying landscape elements such as streams and lakes have become the most elevated parts in the present landscape, situated up to 5 m above surrounding reclaimed areas and former lakes. This phenomenon is caused by peat excavations and reclamation of the lakes on the one hand, and draining and cultivating of peat swamps, turning these into agriculturally used polders on the other hand. After the lowering of the groundwater table, the peat layer shrinks as a result of mineralization and oxidation. At present, this leads to a mean surface lowering of 0.3–1.0 cm/year, depending on the depth of the drained top layer (Kwakernaak *et al.*, 1991). Moreover, mineralization of the drained peat layer is an important source of nutrients in surface waters.

In the polders of the Green Heart, surface water levels are regulated to keep the land suitable for agriculture. Water levels in polders and wetlands are controlled by water boards. To maintain the water depth in the ditches and lakes at a constant level, the water surplus in periods of heavy rainfall is almost immediately pumped out into a network of canals and lakes which constitute the main water reservoir, the so-called boezem, of the Green Heart. This water surplus in the boezem is discharged into the sea. In periods when evapotranspiration and downward seepage to the groundwater exceed the amounts of rainfall and seepage, water from the boezem is pumped in. To compensate for the freshwater deficits in the boezem and to counteract the intrusion of brackish groundwater into the agricultural areas, eutrophic water (mainly from the river Rhine) is supplied.

River water supply influences the ecological quality of wetlands in several ways. The river water carries a high loading of nutrients. It has a macro-ionic composition completely different from natural wetland waters. Consequently, river water supply has resulted in a levelling out of the natural diversity of water types and in a decline of highly valued aquatic species and ecosystems in the Green Heart (Kwakernaak *et al.*, 1991).

Table 12.3 Vegetation types as nature management objectives and required water quality. The Dutch basic water quality standard concerning nitrogen and phosphorus concentration in surface water is set at 0.15 mg P/l and 2.2 mg N/l

Vegetation type	phosphorus concentration (mg P/l)	nitrate concentration (mg N/l)	ammonium concentration (mg N/l)
Chara-type	<0.05	<0.5	<0.2
Potamogeton-type	<0.15	<1.0	<0.2
Stratiotes-type		basic nutrient level	
Reed swamp		basic nutrient level	
Quagfen (floating mire)	<0.05	<1.0	<0.2
Forested swamp		basic nutrient level	

12.3.2 ECOLOGICAL AND WATER QUALITY OBJECTIVES

In recent nature rehabilitation plans in the Green Heart, ecological objectives have been set to lakes, fens, reed swamps, forested swamps and wet grasslands in extensive agricultural use. Depending on the desired vegetation types, water quality has to be improved and the groundwater level to be raised in order to achieve the required environmental conditions.

In the boezem, water quality objectives are based on the Dutch standards for nitrogen and phosphorus. Standards for total phosphorus and total nitrogen levels are set at 0.15 mg P/l and 2.2 mg N/l. In smaller lakes and ponds originating from peat excavation, water quality objectives have been derived from ecological objectives. The presence of *Chara* and *Potamogeton* species characterizes submerged waterweed beds in lakes. Nutrient concentrations found in these vegetation types are given in Table 12.3. The succession of waterweed beds results in a free-floating mass of plants with *Stratiotes aloides* (Verhoeven *et al.*, 1988; Van Wirdum, 1991). *Stratiotes aloides* is a characteristic plant of the ditches of virtually the whole Green Heart.

Vegetation in areas of groundwater discharge can be very susceptible to penetration of alien boezem water. Formerly, the groundwater discharge was equal to or exceeded the evapotranspiration rate. Examples of this type of vegetation are the quagfens in the eastern part of the Green Heart. These vegetations are extremely species-rich. Increasing the area and abundance of quagfens is one of the ecological objectives.

12.4 MANAGEMENT STRATEGIES FOR ECOLOGICAL RECOVERY OF THE WETLANDS IN THE GREEN HEART OF HOLLAND

In the Green Heart, ecological recovery of natural wetlands can only be achieved by following two main strategies:

1. reducing the eutrophication problem
2. reducing the loss of (clean) regional groundwater and surface water (autochtonous water).

The first strategy is mainly followed when the aim is to achieve the basic nutrient level. The second strategy has to be followed if specific ecological goals are to be achieved.

These strategies have to be worked out in sets of measures, directed at an integrated approach of the water quality and water quantity problems. Water conservation is only relevant if water quality is sufficient for sustainable conservation and development of nature values in wetlands. In Figure 12.3 the various sources of phosphorus within the management area of the water board of Rijnland are presented. This area (ca. 114 000 ha) covers a large part of the Green Heart area. Therefore, these figures can be regarded as indicative for the Green Heart. It is clear that manuring and municipal wastewater treatment are the main sources of nitrogen and phosphorus. Environmental management for agricultural land use aims at a balanced manuring level in the year 2000, when input of nutrients should no longer exceed the output. Also, compared with the situation in 1985, a 50% reduction of nutrient loading in wastewater treatment effluents should be achieved by 1995; this should be a step towards a targeted reduction of 75% for both nitrogen and phosphorus (Ministry of Transport and Public Works, 1989). Until sustainable solutions for the eutrophication problem in surface waters and wetlands in the Green Heart are found, measures directed at mitigating the ecological impact of nutrient loading have to be taken. One of these measures is the application of artificial wetlands. If structural solutions are found, the functional use of the artificial wetlands can be altered. The purification function can

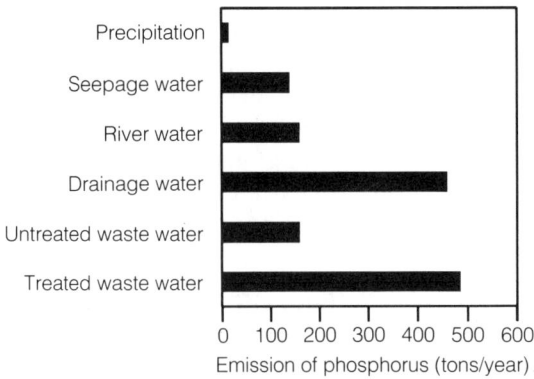

Figure 12.3 Calculated phosphorus emissions from several sources in the management area of the waterboard of Rijnland, situated in the Green Heart. (From Hoogheemraadschap van Rijnland, 1991.)

be abandoned in favour of the development of a eutrophic natural swamp.

12.5 PERSPECTIVES FOR APPLICATION OF ARTIFICIAL WETLANDS IN THE GREEN HEART

In a number of water management projects in the Green Heart, there have been measures to improve the water quality. These measures aim at a reduction of the emissions of nutrients from agriculture and municipal wastewater, in order to reach the basic water quality standards with regard to nitrogen and phosphorus. In The Netherlands, the standards for total phosphorus and total nitrogen levels are set at 0.15 mg P/l and 2.2 mg N/l. These standards are sufficient for developing reed and forested swamps and *Stratiotes* vegetations. To create good conditions for vegetation types such as quagfens and *Chara* and *Potamogeton* waterweed beds, additional measures are required with respect to the nutrient level (Table 12.3). In such cases, water quality improvement by artificial wetlands should be carried out in combination with changes in hydrology, which are directed at the conservation of unpolluted water with a certain regional hydrochemical character. The possible role of artificial wetlands as a means to achieve the basic nutrient level, or even a lower nutrient level, is discussed here.

In the Green Heart the feasibility analyses of artificial wetlands have focused on two types of utilization:

1. purification of eutrophic supplementary water;
2. purification combined with hydrological alterations in the surrounding areas.

12.5.1 PURIFICATION OF EUTROPHIC SUPPLEMENTARY WATER

During summer, eutrophic river water is supplied to lakes and wetlands to compensate for waterlosses by evapotranspiration and recharging of groundwater (Table 12.4). Artificial wetlands can be used to reduce phosporus and nitrogen concentrations in this supplementary water. If this supplementary water consists mainly of Rhine water, it needs to be purified to avoid eutrophication problems. Table 12.4 indicates the required surface area for purification of supplementary water for different natural wetlands in the eastern part of the Green Heart to a level that corresponds to the Dutch standards.

12.5.2 PURIFICATION COMBINED WITH HYDROLOGICAL ALTERATIONS

For wetland vegetation types such as quagfens, the macro-ionic composition of the water is important. These vegetation types are situated

Table 12.4 The artificial wetland area required to obtain the basic nutrient levels in supplementary water for some wetlands in the Green Heart (Duel, 1991)

Wetland area	Size (ha)	Average water supply in summer (m^3)	Artificial area required (ha)
Molenpolder Swamps	260	370 000	20
Westbroek Swamps	440	100 000	5
Loosdrecht Lakes	3150	13 000 000	300
Kortenhoef Lakes	710	2 800 000	125
Ankeveen Lakes	415	2 500 000	85
Lake Naarden	700	1 100 000	35
Botshol Swamps	300	1 250 000	50
Vinkeveen Lakes	945	7 200 000	175
Nieuwkoop Lakes and Swamps	2270	5 400 000	130

in areas where in the past groundwater discharge exceeded evapotranspiration (Verhoeven *et al.*, 1993, this volume). The chemical composition of this water is extremely constant. Because of reclamation of former lakes and because of drainage of surrounding agricultural land, groundwater discharge has been displaced from the quagfen areas. Presently, boezem water is used to supplement water deficits in the quagfens. This water differs greatly from the calcareous oligotrophic to mesotrophic seepage water with respect to nutrient concentrations and the macro-ionic composition (Beltman *et al.*, 1988; Verhoeven *et al.*, 1988). This is a serious threat to these wetlands. For the development of quagfens the inlet of boezem water should be avoided. Therefore, additional measures of water level management and water conservation in surrounding areas are necessary.

The availability of autochtonous water can be advanced by:

1. using the summer water surplus from low-lying seepage areas, such as reclaimed areas and former lakes;
2. storage of the winter water surplus from lakes and wetlands;
3. creating buffering zones.

These applications are briefly explained below.

12.5.3 USING THE WATER SURPLUS FROM LOW-LYING SEEPAGE AREAS

The applicability of artificial wetlands for the purification of supplementary water which is derived from seepage in reclamation areas is illustrated in Figure 12.4. In this case, species-rich quagfens are situated in the Westbroek and Molenpolder swamps. The quagfens, situated in (former) seepage areas, are threatened by drought because of a diminishing quantity of seepage water. Polluted water from the river Vecht

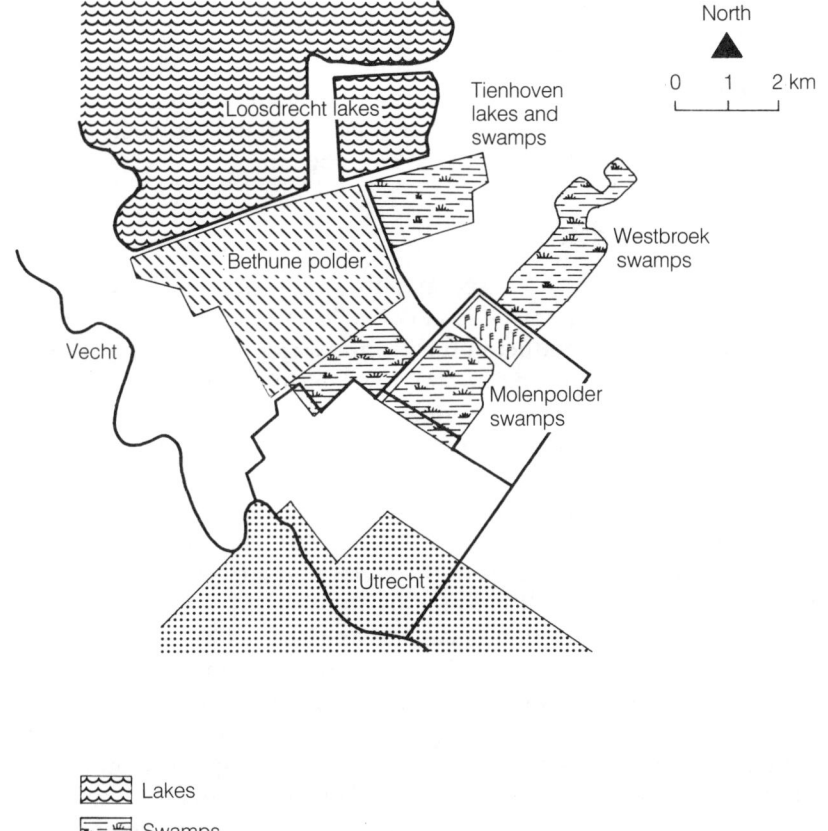

Figure 12.4 Plan view of the artificial wetlands for treatment of seepage water supplied to the Westbroek and Molenpolder Swamps.

has to be supplied. Instead of river water, seepage water from the reclaimed area Bethune polder can be applied as supplementary water to the swamps during summer. However, during transport, this seepage water can become loaded with nutrients from local sources. Before the seepage water enters the swamps, artificial wetlands can be used effectively for nutrient removal (Figure 12.4).

12.5.4 STORAGE OF THE WINTER WATER SURPLUS FROM LAKES AND WETLANDS

In wintertime, artificial wetlands function as water conservation systems. They function as water treatment systems during summer. The mean water storage capacity of artificial wetlands is 5000 m^3/ha, if seasonal water level fluctuation is about 50 cm.

12.5.5 BUFFERING ZONES

Artificial wetlands can be used as buffering elements in two different ways. First, they can be applied between low-lying reclaimed areas and natural wetlands for reducing the groundwater discharges in the natural wetland. A transitional water level is to be maintained for this application. Seepage water from a reclaimed area can be supplied to this artificial wetland to diminish the downward seepage from the natural wetlands to the reclaimed area. During summer, water losses occur in the natural wetlands due to evapotranspiration. To maintain the water levels in the natural wetlands, water suppletion will remain necessary. The water surplus of the reclaimed area can be used for this purpose. This system is illustrated in Figure 12.5.

Second, artificial wetlands can be applied, as a buffer between a boezem and a natural wetland, for the prevention of chemical alterations of the wetland water. Applications can be found in places where natural wetlands are part of an agricultural water management system. In periods of heavy rainfall, the water surplus is quickly pumped out into the boezem to keep fluctuations of the water level at a minimum. In periods of drought, the inlet of boezem water soon becomes neces-

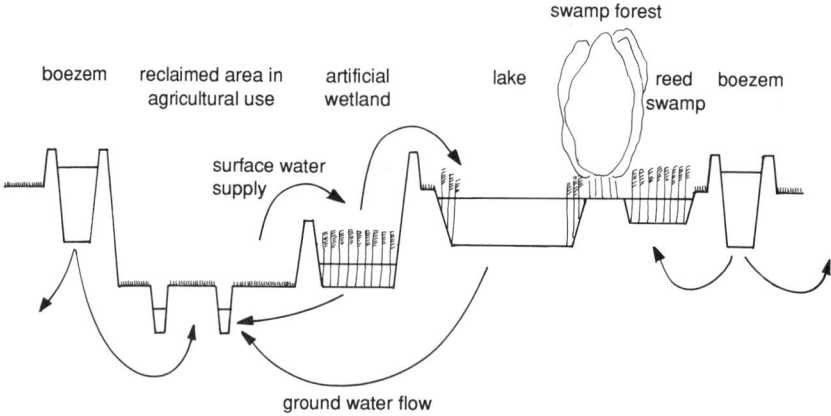

Figure 12.5 Artificial wetlands as hydrological buffering zones between agriculturally used reclaimed areas and natural wetlands.

sary to avoid water shortages in the pastures. In these situations the inlet from and the pumping out into the boezem frequently alternate, about twice a month. Instead, the water surplus in periods of heavy rainfall may be temporarily stored in artificial wetlands. In this way polder water is kept available for supplementary purposes.

12.6 CONCLUSIONS

In The Netherlands, special attention is paid to the applicability of artificial wetlands for treatment of eutrophic surface waters in rural areas. In the Green Heart, the use of artificial wetlands in eutrophication control is a very promising approach additional to reducing the nutrient loading of natural wetlands. Mesotrophic wetlands in which hydrochemical alterations are a natural phenomenon, can be effectively protected from eutrophic boezem and river water by artificial wetlands. To purify the supplementary water for 100 ha of natural wetland, 5 to 10 ha of artificial wetland are required (Table 12.4). Ditch vegetations in pasture areas with *Stratiotes aloides*, the characteristic species of the ditches in the Green Heart, can also be protected in this way, 2–5 ha of artificial wetland being required for this application.

An adequate protection of wetlands in areas of groundwater discharge cannot be achieved by purification of supplementary boezem water. The application of artificial wetlands as buffering zones between reclaimed areas and natural wetlands, or between boezem and natural wetland, can in these situations be combined with the water treatment function. A less effective purification performance as a result of water level fluctuations in the artificial wetland, should be taken for granted.

This chapter is almost entirely theoretical and hypothetical. The feasibility analysis of artificial wetlands in the Green Heart is based on results of field and laboratory experiments in North America and Western Europe. In the Green Heart, large-scale experiments are planned, but not yet realized. In the next decade, the empirical results of these projects will show whether the expectations are justified.

REFERENCES

Andersen, F.O. (1981) Oxygen and nitrate respiration in a reedswamp sediment from eutrophic lake. *Holartic Ecology*, **4**, 66–72.

Andersen, J.M. (1977) Rates of denitrification of undisturbed sediment from six lakes as a function of nitrate concentration, oxygen and temperature. *Archives Hydrobiologia*, **80**, 47–159.

Beltman, B., Duel, H., Van Der Bie, M. *et al.* (1988) Ecohydrology in polders: the Noorderpark. *Landschap*, **5**, 152–69.

Best, E.P.H., De Vries, D. and Reins, A. (1984) The macrophytes in the Loosdrecht Lakes: a story of their decline in the course of eutrophication. *Verhandlungen der Internationalen Vereins für Limnologie*, **22**, 868–75.

Brix, H. and Schierup, H.H. (1989) The use of aquatic macrophytes in water pollution control. *Ambio*, **18**, 100–7.

De Nie, H.W. (1987) The decrease in aquatic vegetation in Europe and its consequences for fish populations. Food and Agriculture Organization of the United Nations. *EIFAC Occasional Paper*, no 19, Rome.

Deuffel, J., Löffler, H. and Wagner, B. (1986) Auswirkungen der Eutrophierung und anderer anthropogener Einflüsse auf die Laichplätze einiger Bodensee-Fischarten. *Österreichische Fischerei - Zeitung*, **39**, 325–36.

Duel, H. (1991) *De mogelijkheden voor toepassing van helofytenfilters in het veenweidegebied Groene Hart*. TNO-report, Delft.

Duel, H and During, R. (1990) Artificial wetlands: systems for treating eutrophic water. *Landschap*, **7**, 269–78.

Duel, H. and Saris, F.J.A. (1986) Purification of surface water by artificial wetlands. *Landschap*, **3**, 295–305.

Duel, H., During, R. and Jongman, R.H.G. (1991) *De mogelijkheden voor toepassing van helofytenfilters in Nederland*. TNO-report, Delft.

Godfrey, P.J., Kaynor, E.R., Pelczarski, S. and Benforado, J. (eds) (1985) *Ecological Considerations in Wetland Treatment of Municipal Wastewaters*. Van Nostrand Reinhold Company, New York.

Goldstein, A.L. (1987) Utilization of wetlands as BMPs for the reduction of nitrogen and phosphorus in agricultural runoff from South Florida water sheds, in *Aquatic Plants for Water Treatment and Resource Recovery* (eds K.R. Reddy and W.H. Smith), Magnolia Publishing, Orlando, FL, p. 300.

Good, B.J. and Patrick, W.H. jr (1987) Root-water-sediment interface processes, in *Aquatic Plants for Water Treatment and Resource Recovery* (eds K.R. Reddy and W.H. Smith), Magnolia Publishing, Orlando, FL, pp. 359–72.

Hammer, D.A. (1989) *Constructed Wetlands for Wastewater Treatment: Municipal Industrial and Agricultural.* Lewis Publishing, Chelsea, MI.

Hoffmann, C.C. (1985) Nitrate reduction in a reedswamp receiving water from an agricultural watershed. *Proceedings 13th Nordic Symposium on Sediments,* Aneboda.

Hoogheemraadschap van Rijnland (1991) *Waterbeheersplan 1992.* Voorontwerp. Leiden.

Hosper, S.H. and Jagtman, E. (1990) Biomanipulation additional to nutrient control for restoration of shallow lakes in the Netherlands, in *Wetlands for the Purification of Water.* The Utrecht Plant Ecology News Report 11, pp. 130–46.

Kadlec, R.H. (1987a) Northern natural wetland wastewater treatment systems, in *Aquatic Plants for Water Treatment and Resource Recovery* (eds K.R. Reddy and W.H. Smith). Magnolia Publishing, Orlando, FL, pp. 83–98.

Kadlec, R.H. (1987b) The hydrodynamics of wetland water treatment systems, in *Aquatic Plants for Water Treatment and Resource Recovery* (eds K.R. Reddy and W.H. Smith) Magnolia Publishing, Orlando, FL, pp. 373–92.

Koerselman, W. (1990) Assessment of nutrient removal efficiency in wetlands, in *Wetlands for the Purification of Water.* The Utrecht Plant Ecology News Report 11, 22–43.

Kwakernaak, C., Klijn, F., Fiselier, J.L. and Duel, H. (1991) Alternatives for water distribution: tools for an integrated regional policy on the distribution of river water in the Netherlands. *Landschap,* 8, 93–108.

Lammens, E.H.R.R. (1986) *Interaction between fishes and the structure of fish communities in Dutch shallow, eutrophic lakes.* Thesis. University for Agriculture, Wageningen.

Melzer, A. (1976) *Macrophytische Wasserpflanzen als Indikatoren des Gewässerzustands oberbayerischer Seen.* Dissertation, Cramer Verlag, Vaduz.

Ministerie van Verkeer en Waterstaat (1989) De derde nota waterhuishouding. Regerigsbeslissing. SDU, The Hague.

Nichols, D.S. (1983) Capacity of natural wetlands to remove nutrients from wastewater. *Journal of Water Pollution Control Federation,* 55, 495–505.

Patrick, W.H. jr (1990) Microbial reactions of nitrogen and phosphorus in wetlands, in *Wetlands for the Purification of Water.* Utrecht Plant Ecology News Report 11, pp. 52–63.

Reddy, K.R. and Smith, W.H. (eds) (1987) *Aquatic Plants for Wastewater Treatment and Resource Recovery.* Magnolia Publishing, Orlando, FL.

Reddy, K.R. and Patrick, W.H. (1984) Nitrogen transformations and loss in flooded soils and sediments. *CRC Critical Reviews and Environmental Control* 13, 273–310.

Reed, S., Bastian, R., Black, S. and Khettry, R. (1984) Wetlands for wastewater treatment in cold climates, in *Proceedings of the Water Reuse Symposium III,* San Diego, CA, pp. 962–72.

Tchobanoglous, G. (1987) Aquatic plant systems for wastewater treatment: engineering considerations, in *Aquatic Plants for Water Treatment and Resource Recovery* (eds K.R. Reddy, and W.H. Smith). Magnolia Publishing, Orlando, FL, pp. 27–48.

Tourbier, J. and Pierson, W.H. jr (eds) (1976) *Biological Control of Water Pollution.* University of Pennsylvania Press, Philadelphia, PA.

Van der Aart, P.J. (ed.) (1985) *Wetlands for the purification of wastewater.* Utrecht Plant Ecology Report 5, University of Utrecht, Utrecht.

Van Liere, L., Gulati, R.D., Lammens, E.H.R.R. and Wortelboer, F.G. (1990) From nutrient to fish and vice versa. *Landschap,* 6, 33–46.

Van Wirdum, G. (1991) *Vegetation and hydrology of floating rich-fens.* Thesis. University of Amsterdam. Datawyse, Maastricht.

Verhoeven, J.T.A., Koerselman, W. and Beltman, B. (1988) The vegetation of fens in relation to their hydrology and nutrient dynamics, in *Vegetation of Inland Waters* (ed. J.J. Symoens,), *Handbook of Veg. Sc.* vol. 15. Junk, The Hague, pp. 249–82.

Verhoeven, J., Kemmers, R.M. and Koerselman, W (1993) Nutrient enrichment of freshwater wetlands, in *Landscape Ecology of a Stressed Environment* (eds C.C. Vos and P. Opdam), Chapman & Hall, London, pp. 33–59.

Vos, C.C. and Zonneveld, J.I.S. (1992) Patterns and processes in a landscape under stress: the study area, in *Landscape Ecology of a Stressed Environment* (eds. C.C. Vos and P. Opdam), Chapman & Hall, London, pp. 1–28.

Whitton, B.A. (ed.) (1984) *Ecology of European Rivers.* Blackwell Scientific Publications, Oxford.

Water relations in urban systems: an ecological approach to planning and design

13

Sybrand P. Tjallingii

13.1 INTRODUCTION

13.1.1 WATER PROBLEMS

Urban and rural areas face growing water problems. Floods, droughts and pollution increase with urbanization and intensified land use. The Dutch, who have struggled against water throughout their history, are now forced to solve the problem of a shortage of good water. In The Netherlands groundwater withdrawal will surpass the yearly recharge in the next decade and national policy is concentrating on water supply and aims to rely more on the river Rhine (Ministerie van Verkeer en Waterstaat, 1983). Yet the river Rhine, which has accumulated huge quantities of toxic substances in the delta sediments, may still be considered one of Europe's main sewers. Thus polluted and eutrophied river water penetrates all parts of the country, threatening not only specific natural vegetation and wildlife but also the functional and aesthetic qualities of rural and urban waters (Barendregt *et al.* 1993, this volume; Vos and Zonneveld, 1993, this volume).

In urban areas large investments are needed to improve water supply: groundwater pumping stations, reservoirs, infiltration projects,

Landscape Ecology of a Stressed Environment
Edited by Claire C. Vos and Paul Opdam
Published in 1993 by Chapman and Hall, London. ISBN 0 412 44820 3

pipelines, etc. Many of these projects disturb the landscape, natural vegetation and wildlife. Yet at the same time that these efforts are being made to meet increasing demands, water is being squandered. Drinking water is used to flush toilets, wash cars and water lawns and precious fresh water is used to flush urban canals. Rain water is removed quickly from huge paved surfaces by improved storm sewer systems, while parks and gardens suffer from drought.

It is time to come up with 'self-reliant, self-responsible' water systems, i.e. systems that are less dependent on neighbouring systems for supply and discharge, that save on water use, and that solve their own problems of water quality. The search for more sustainable methods of water use and water management has to be made at all levels, starting with the smallest units. This chapter explores ways of designing a 'self-reliant, self-responsible' water system for a small new residential area. Alternative water systems for domestic use and their relations to regional level systems are taken into account. The work of Hough (1984) and Spirn (1984), German experiences like Bröker (1987) and Geiler and Hildebrandt (1985), and numerous Dutch studies have contributed to the approach. General guidelines for the design of water systems were formulated in earlier papers (Tjallingii, 1981, 1988a, b).

13.1.2 DESIGN AND ECOLOGY

Related to environmental problems, design is the stage of prevention. Therefore, it is very important to pay attention to the problem-solving proposals and the procedure of the design process.

The usual role of ecologists in planning is to prepare basic maps or to evaluate plans. This is clearly demonstrated by the proceedings of international meetings of landscape ecologists (Tjallingii and De Veer, 1981; Schreiber, 1988). Few ecologists actually take part in the making of planning proposals and if they do, their main concern is usually to design for nature (that is, for natural vegetation and wildlife). The conservation and development of ecological infrastructures is a major planning objective. Van Selm (1988) and Harms and Knaapen (1988), among others, describe the theory and practice of this 'organism approach'. However, the problems of modern landscape management also ask for an ecological 'resource approach', planning for flows of energy, water, nutrients and materials as an integrated part of land-use and physical planning. Rather than design **for** nature this means 'design **with** nature' (McHarg, 1969). McHarg and other landscape architects like Hough (1984) have shown that there are interesting practical prospects of resource-oriented planning, but in recent decades few landscape ecologists have helped elaborate this approach.

Some professional urban planners like Alexander *et al.* (1975) and

Lynch (1981) use the concepts of ecology in their approach to design. To them, 'ecology' primarily stands for human ecology as a branch of social science. However valuable and indispensable their concepts may be, the integration of the ecology of abiotic and biotic nature in the planning process remains to be explored.

The 'theory of basic operations' (Van Leeuwen, 1966, 1981) offers some concepts that seem to be useful for ecologists and planners in this context. In this theory, four basic categories of operations are discerned as conditioning flows in the ecological 'field'. (1) Time, including the natural rhythms and storage. (2) Orientation in space, guiding the direction of natural flows. (3) Position in space, related to the use of the existing landscape for the site and size of the new habitat. (4) Instrumental or technical devices that may enhance the operation of the desired habitat. These concepts are applicable not only to the planning of habitats for plant and animal species, but may be equally useful in planning for specific human functions like those in a residential area. In that case the specific planning objectives should be formulated before dealing with the four basic conditional operations that have to be regulated in the plan. This led to the PROSA design steps in this case study that deals with the planning of a new residential area.

The flow of water was chosen to structure the plan both spatially and functionally for three reasons. (1) Water connects the site with regional recources like climate, groundwater and surface water. (2) The water system reflects the management plan for nutrients and pollutants. (3) Water unites the ecological, the functional and the aesthetic qualities of the plan.

Table 13.1 shows the step-by-step procedure. Each step consists of three parts. First, the existing landscape or system is described. Second, the guiding principles are formulated. In the third part ideas, sketches and calculations for the design are presented. The main reason for adopting this sequence of steps is to make optimal use of the temporal and spatial dimensions of the ecosystem at a given site. Technical devices are subservient to the plan. In this way a number of environmental problems may be prevented and the ecological potential of the site is used to benefit the quality of the new urban neighbourhood as a habitat for humans, plants and animals.

13.1.3 PURPOSE OF THE CASE STUDY

In this 'investigation by design' the design method is used to explore the following questions.

1. Is it possible to design a 'self-reliant, self-responsible' water system at the level of scale of a residential scheme for 400 dwellings?

Table 13.1 PROSA: Steps in designing surface water systems

PROGRAMME
existing system (spatial structure, water balance, problems)
guiding principle (the self-reliant, self-responsible system)
planned system (schemes, water balance, planned land use)

RHYTHM
existing rhythm (daily, seasonal/irregular)
guiding principle (synchronizing, storage)
planned rhythm (storage calculation)

ORIENTATION
existing orientations (directions of flow)
guiding principle (from clean to polluted)
planned orientations (flow and zoning)

SITUATION
existing situation (abiotic, biotic)
guiding principle (using and creating site qualities)
planned situation (first design sketch)

ADJUSTED TECHNOLOGY
existing devices (operation problems)
guiding principles (adapting to time, space, management)
planned technology (operation, steering, normal and emergency)

2. Is the PROSA approach, linking landscape ecological theory to planning practice, a fruitful one? Does it produce structural planning proposals, i.e. feasible spatial ideas that not only refer to specific elements but may structure the whole plan?
3. Are these proposals useful to steer the concepts introduced by other disciplines and resulting from practical use and image considerations?

The chapter is structured according to the PROSA steps. At each step a figure summarizes the most important items explained in the text.

13.2 STEP 1: PROGRAMME (Figure 13.1)

In this first step of the design process a general outline of the main objectives and conditions is described. In this case the design of a water system and a small park are part of the planning of a new residential area for approximately 400 dwellings.

13.2.1 EXISTING CONDITIONS

The planning site, 'Vrijenban', is a 8 ha pasture area on the urban fringe of the city of Delft. The surface waters of Vrijenban are part of the 'Hoge

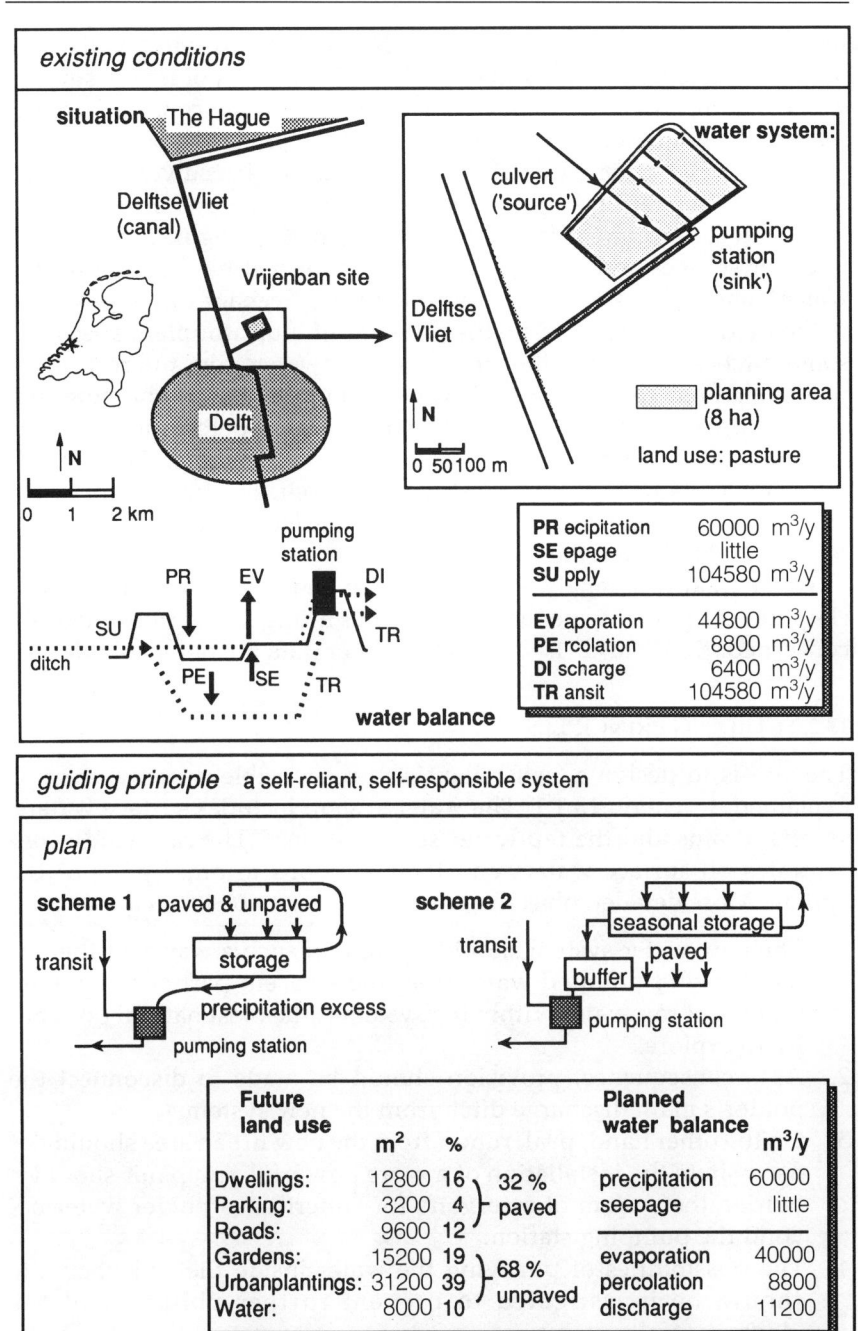

Figure 13.1 Programme.

Broek polder' between Delft and The Hague. This low lying polder, 1–2 m below sea level, is surrounded by dikes. From this polder the surplus water is pumped into the 'Delftse Vliet', the main water connection to the sea. In dry periods an inlet sluice in the northern part of the polder is used for supply. The water enters the study area through a culvert under the motorway separating Vrijenban from the rest of the polder. This point is a 'source' for our system, whereas the pumping engine is a 'sink'.

Other sources and sinks may best be identified by drawing up the water balance (Figure 13.1, *existing conditions*). Seepage can be observed in the field and confirmed by the presence of indicator plant species in some ditches. Supply is the surplus of rainwater from the rest of the polder that has to pass through Vrijenban on its way to the pumping engine (transit function). There is considerable percolation because groundwater is withdrawn for cooling a nearby factory. Discharge is the rainwater surplus from the study area itself. The figures are calculated on the basis of data provided by the local water board. They represent averages of the last 50 years.

Water quality is analysed regularly in the area. The water board classifies the polder water as 'bad' as far as nitrogen and phosphate are concerned. (N-Kjeld>2, P>0.2mg/l) Further data are not available.

13.2.2 GUIDING PRINCIPLES

The aim is to design a 'self-reliant, self-responsible' water system as explained in section 13.1.1. The water system includes surface waters, sewers, drains and the tap water supply system. The case study concentrates on surface waters and their relations to other parts of the system. More detailed objectives and measures to be taken are:

1. The new water system should operate in such a way that there is no inflow of polluted water from the adjacent part of the polder. Storage of rainwater within the system is the most natural possibility to explore.
2. As a consequence, provision should be made to disconnect the polder's main discharge ditch from the new system.
3. On the other hand, peak runoff from the new urban area should not necessitate the installation of a more powerful pump nor should it burden the system of reservoirs for superfluous polder water beyond the pumping station.
4. The possibilities of purifying the water inside the Vrijenban site should be investigated to prevent further pollution of the 'downstream' systems.

A few questions have to be answered before proceeding: What is the quality of rainwater compared with surface water? An analysis of rainwater composition in The Netherlands (Van den Eshof *et al.*, 1986) shows considerable contamination of the precipitation as a result of air pollution. Generally the percentage of sulphate is even higher than in surface waters. However, rainwater contains significantly lower percentages of eutrophying substances, heavy metals, and organic pollutants. This is even more pronounced when rain water is compared with surface waters that are strongly influenced by the river Rhine, as is the case in the study area. Acidification by SO_4-rich rainwater is less important because the soils in this area are alkaline. Preference for rainwater therefore seems to be justified.

But how is rainwater runoff quality affected by the urban system itself? Before rainwater is collected it has already picked up pollutants from roofs and streets. Data on the quality of runoff water from different places over the last 30 years are summarized and discussed by Hengeveld and De Vocht (1982) and Uunk (1985). Results vary considerably from one place to another. In The Netherlands a research programme has been started to collect more precise data from specific paved areas (Van Campen *et al.*, 1986). Various authors, including Roos (1986) have indicted that stormwater discharges may have considerable adverse effects on plant and animal life in the receiving urban canals.

Clearly, runoff from urban areas may be heavily polluted. But the quality is not always inferior to the quality of water discharged by intensively farmed areas like those surrounding the Vrijenban site. 'Point' sources of pollution in urban areas can be readily detected. The best way to abate them is at source. More difficult are the 'non-point', diffuse, sources. In residential areas the most important of these are:

1. Motorized traffic, responsible for oil, heavy metals, organic micropollutants and salt;
2. Corrosion of gutters, fences, drainage pipes, etc., aggravating the contamination with heavy metals;
3. Pesticides applied to paved surfaces, public urban green and gardens;
4. Faeces from dogs, pigeons etc. contributing to the nitrogen and phosphate load of surface water and to bacterial contamination.

A number of technical and design answers to these problems have been proposed:

1. Reducing runoff by minimizing the impervious surface.
2. Using building materials that are less susceptible to corrosion, avoiding pesticides and using phosphate-free washing powders.
3. Concentrating pollution such as dog faeces and waste oil in certain places, in order to facilitate treatment and re-use.
4. Separating discharges from highways, parking lots and market

places from the main surface water system.

5.   Introducing special devices on the overflows of combined sewer systems or on road side gully pots. Such devices include 'first flush' systems that carry the first polluted flush of water to the waste water sewers. Recent experiments testing 'first flush' techniques applied to combined sewer overflows have produced promising results (Balmforth, 1986; Kruitwagen *et al.*, 1986).
6.   Improving waste water treatment facilities and using wetland treatment systems that exploit the properties of plants and soil to assimilate wastes (Van der Aart, 1985; Mönninghof, 1986).

Will runoff water in the planning area be less polluted than inflowing water from the 'source' indicated in Figure 13.1 (*existing conditions*)? The answer to this question largely depends on the extent to which the technical and design solutions mentioned above are implemented. When planning new residential areas the designer should look at two levels of integration: the buildings and their environments and the water system of the area as a whole.

It is assumed that a better system can be created if it is not burdened by pollution coming from outside. The final result will largely depend on the actions taken at the level of houses and private gardens.

In most cases storm water and waste water are removed by one 'combined sewer system'. Overflows of these systems cause serious problems to surface water quality as described by Hengeveld and De Vocht (1982) and CHO-TNO (1986). More recently 'separate-systems' have been introduced in a number of urban areas. And though the quality of the runoff remains a problem, dealing separately with rainwater and wastewater offers better opportunities for the treatment of both.

13.2.3. PLAN

In the Vrijenban site water from the polder upstream should be able to flow to the pump. This water is of such poor quality that it is not connected with the newly designed system and has therefore been omitted from the planned water balance. Inside the areas enough storage should be provided to make it unnecessary to let water flow in during the summer. A storm water retention pond near the pumping station should be part of the plan. The starting point is a scheme for 400 dwellings, 320 of which are single-family houses with gardens. The other 80 are upstairs flats without gardens. The scheme also includes a public park, a small part of which will be designed as a 'nature park' (Londo, 1977).

In Figure 13.1 (*plan*), a first estimate of the future land use is formed.

The relation between paved and unpaved surfaces is critical for the water system. Next, a rough water balance of the planned situation can be calculated. As a consequence of the paved surface, evaporation in urban areas is 40% less than in pasture. This assumption and other empirical reduction factors applied to the potential evaporation data (Penman) are taken from the annual report of the local water board. With the help of these data it will be possible to assess some quantitative effects of choices made during the design process.

In the guiding principles discussed above, a number of strategies for abating water problems are mentioned. Assuming these proposed actions are implemented at the level of buildings and their surroundings, the water system for the whole area may be conceived of as shown in scheme I (Figure 13.1, *plan*). If this is not the case the paved and unpaved surfaces should be disconnected as in scheme II (Figure 13.1, *plan*), in which seasonal storage and runoff retention are separated. Both schemes might be elaborated. Scheme II, however, seems to be rather complicated and in addition there is a more fundamental objection: problems of runoff water pollution are not abated at the source but later in the town. There, treatment becomes more difficult and separation and discharge is the only alternative. Therefore the track indicated by scheme I will be followed.

13.3 STEP 2: RHYTHM (Figure 13.2)

Next, the time-related aspects should be considered.

13.3.1 EXISTING CONDITIONS

The first graph of Figure 13.2 (*existing conditions*) shows the rhythm of precipitation compared with the sum of evaporation and percolation. Between April and August there is a deficit but over the year this is more than compensated by the total rainfall. In the low lying and intensively used polders of The Netherlands a major objective of water management is to maintain fixed water levels in surface waters. Consequently, a yearly rhythm of supply and discharge is reacting to the periods of shortage and excess, as is shown by the second graph. As a result, relatively clean rain water is removed in the winter period and water of poor quality is used for supply in the summer.

Another rhythm, not given in Figure 13.2, is the rainfall frequency curve, used to calculate stormwater retention. In The Netherlands it is not uncommon to take a curve showing a critical rainfall depth of 58 mm after 24 h. This occurs once per ten years and only once per 100 years will this depth reach 85 mm.

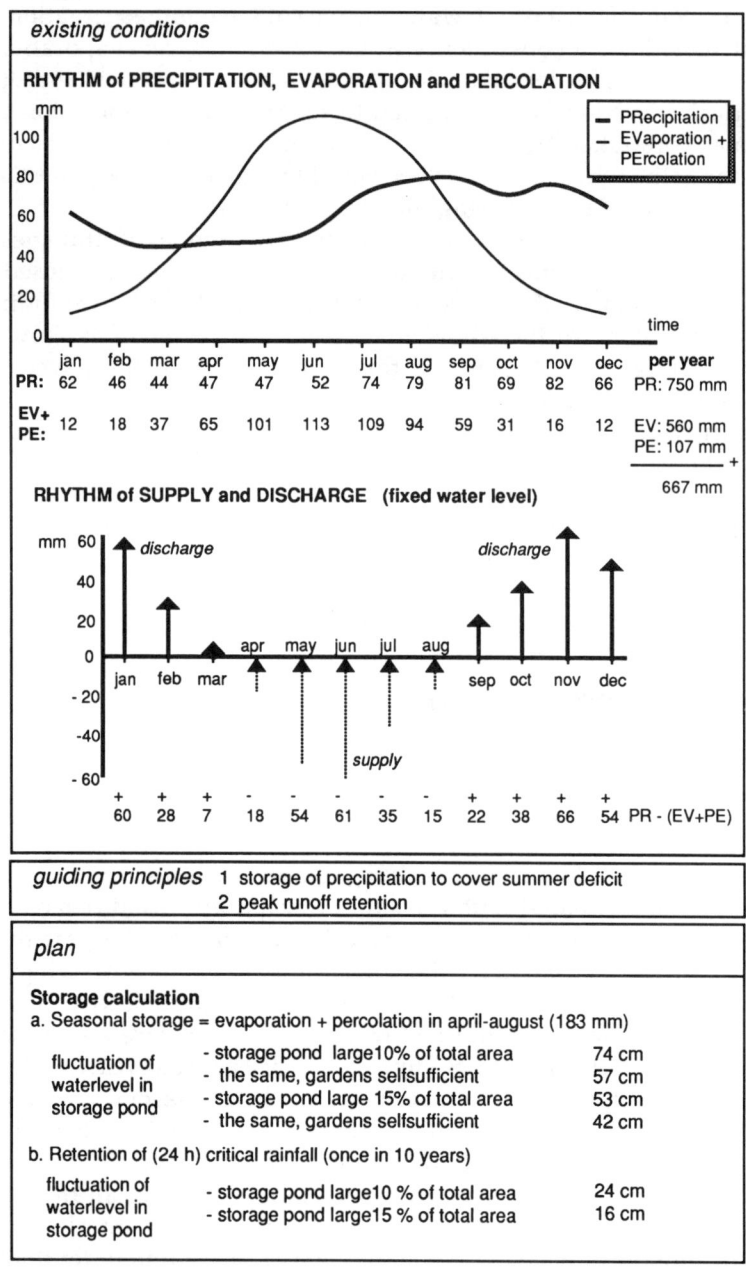

Figure 13.2 Rhythm.

13.3.2 GUIDING PRINCIPLES

The key principle is to retain enough water to make up for the summer deficit from the system's own storage. As a consequence the rhythm of supply and discharge should be replaced by a rhythm of fluctuating water levels in the urban waters. Only the remaining annual surplus should be discharged. A second guiding principle is related to storm runoff. The system should provide for the 24 h retention of heavy rainfall, occurring once in 10 years. It seems reasonable to take the 58 mm limit corresponding to this event.

13.3.3 PLAN

Scheme I, given in Figure 13.1 (plan), does not have to change if the rhythms are studied more closely. But by doing so we are able to calculate how much storage capacity is needed for seasonal retention and for peak-runoff retention. From the graphs in Figure 13.2 (existing conditions) it can be seen that the excess evapotranspiration during the period of April to August is 183 mm. This figure is based on data for pasture. Water from urban plantings and private gardens might evaporate slightly more but in the absence of reliable detailed information, the pasture figure is taken as an indication. The pasture evapotranspiration has been calculated from figures given by the local waterboard. The resulting summer deficit for open water is 245 mm. The storage need can be calculated from these two evaporation figures.

The storage capacity of the soil is estimated, assuming a groundwater fluctuation of 30 cm. If we consider the 32% paved and 68% unpaved ratio mentioned in the Programme stage (Figure 13.1, plan) then a retention pond covering 10% of the total planning area will have to fluctuate by 74 cm in level to provide the storage required (Figure 13.2, plan).

This fluctuation is fairly large if we want the pond site to develop as an interesting and attractive ecosystem. It might cause a silty fringe along the pond. But various other measures can be taken to increase the storage capacity: First the paved area may be reduced. In this case the proposed paved area of 32% is already a modest assumption. But further improvement can be expected if porous pavements are used wherever possible. Another practical proposal is to increase groundwater fluctuations up to 60 cm, which is a normal value found in these soils. Built-up and non-built-up areas can be arranged carefully so that this groundwater movement will not cause nuisance to buildings.

Providing private gardens with cisterns is another way of increasing storage. Rainwater from roofs can be stored and used (Bredow, 1985; Bröker, 1987). In this way private gardens could be made self-sufficient in water. In that case, the storage capacity for the whole area could be met by a pond fluctuating by 57 cm in level (Figure 13.2, *plan*). Enlarging the storage pond to 15% of the total area could reduce the fluctuation to 53 cm or even to 42 cm if combined with gardens self-sufficient in water. Enlarging the pond to this extent, however, would be at the expense of other land use.

The extra storage capacity needed to retain urban runoff caused by a heavy storm can be calculated according to current practice. The results are shown in Figure 13.2 (*plan*).

From these exercises it becomes clear that the storage basin should be located in a place low enough to permit a safely fluctuating water level. Further measures to increase storage capacity may be explored and their effects can be measured. In this case a storage pond of 10% of the total area is considered in subsequent stages of the design.

13.4 STEP 3: ORIENTATION (Figure 13.3)

The third step is to analyse orientation. Flow directions of water, solar radiation and prevailing winds are key factors for the operation of the ecological field in this particular area. Orientation decisions for the plan may be attuned to them.

13.4.1 EXISTING CONDITIONS

Waterflow in the main ditch to the pumping station is in one direction. But as shown on the map in Figure 13.3, *existing conditions*, all the other ditches have water flows in two directions. As a result, relatively clean rainwater is mixed with polluted polder water. Further information about sun (north arrow) and prevailing winds is also given on the map.

13.4.2 GUIDING PRINCIPLE

Stable gradients are a key condition for ecological diversity in surface waters. Upstream pollution sources negatively affect all aquatic communities and the functional and aesthetic qualities in the water course. Therefore the guiding principle adopted is that water should flow from clean to polluted or from nutrient-poor to nutrient-rich. In this way a stable gradient situation can be maintained. This principle was first formulated by Van Leeuwen (1966) and elaborated by Van der Maarel (1975). Both authors investigated the conditions for stable ecological diversity in landscape. The basic idea has been successfully applied to habitat construction in 'nature parks' by Londo (1977). Its implementation in this case requires a zoning of polluting activities. General ideas about zoning in cultural landscapes proposed by ecologists (Van der Maarel, 1981) may now be translated into concrete and detailed design proposals.

13.4.3 PLAN

First the relevant activities are listed according to their possible effects on water quality (Figure 13.3, *plan*). If we combine this sequence from 'clean' to 'polluting' with direction of water flow, a preference can be formulated for the zoning of land use in the urban system.

Scheme I in Figure 13.1 (*plan*) may now be elaborated. In Figure 13.3 (*plan*) two basic models are drawn using the sequence of land use as legend. In both A and B water flows from the storage pond through zones of relatively clean land use to the central zone of the built-up area. A represents a 'waterfront' idea, whereas in B the pond is far away in a park. Clearly both models will only work if certain purification measures are included in the design of the storage pond. This might be a realistic

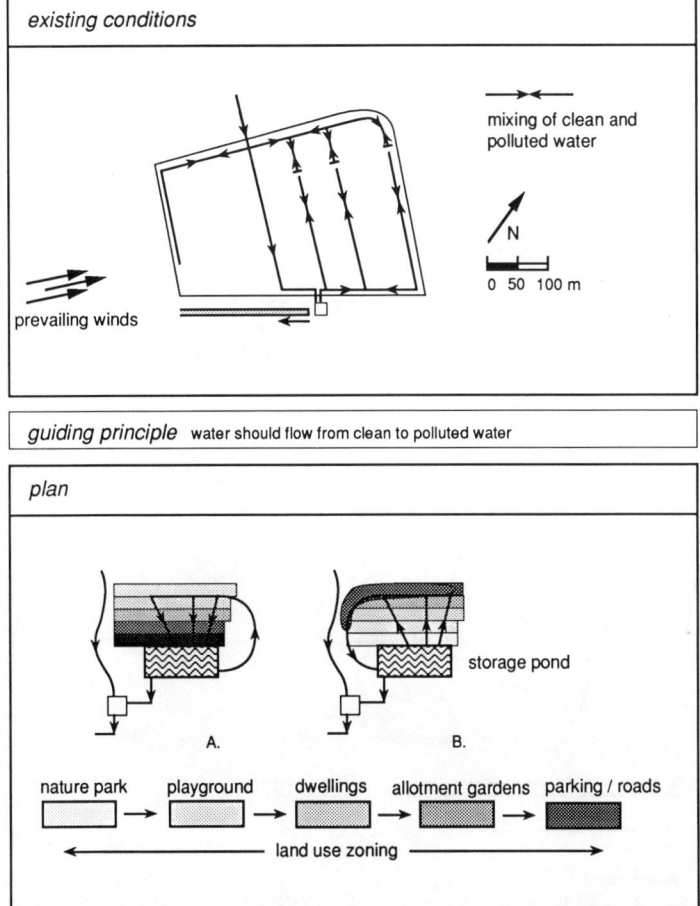

Figure 13.3 Orientation.

possibility in view of the recent revival of wetland purification systems. Further discussion will follow under the adjusted technology step.

13.5 STEP 4: SITUATION (Figure 13.4)

Both the rhythm and the orientation steps deal with a general optimization of the system. Now we have to look at the particular conditions of the site so that we can fit the right system to the right situation.

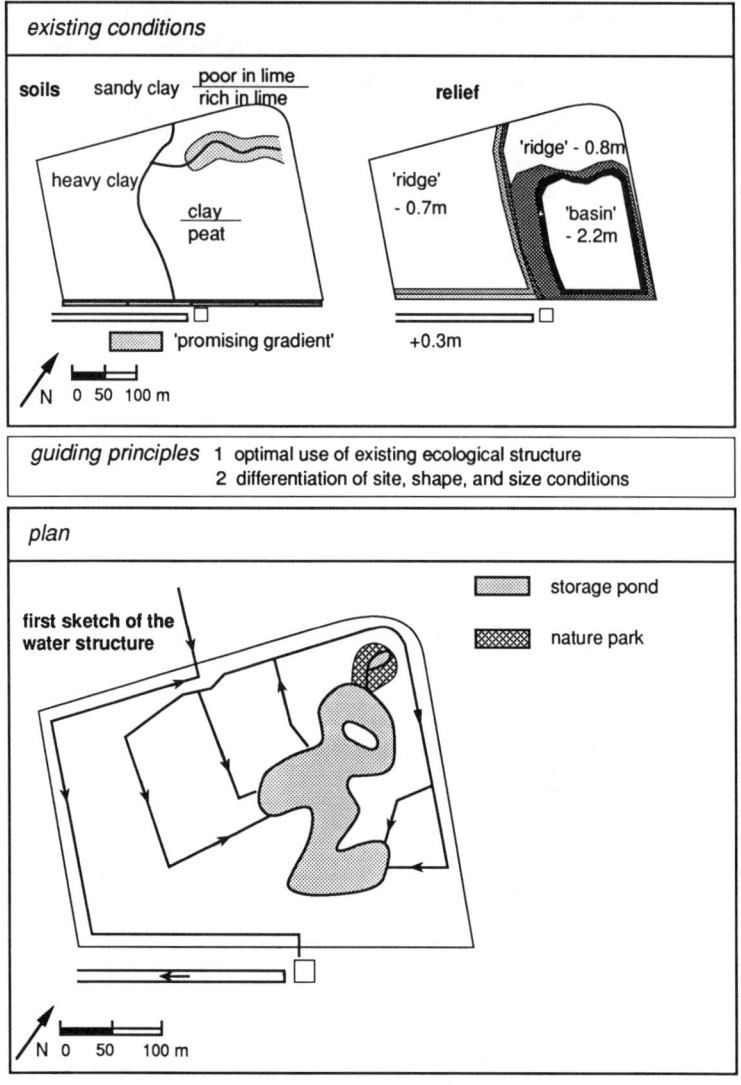

Figure 13.4 Situation.

13.5.1 EXISTING CONDITIONS

Figure 13.4, *existing conditions*, presents schematic versions of soil and relief maps. The first is derived from the 1:50 000 Soil Map of The Netherlands, Sheet 37O. Data on relief are taken from the 1:25 000 Ordnance survey map. In both cases a further field survey was required to complete the information. In the field a remarkable boundary can be seen separating the 'ridge' and the 'basin'. Within 10 metres a higher, drier, sandy-clay environment with a lime-rich subsoil passes into a lower, wetter clay-over-peat environment. This is interpreted as an interesting 'stable gradient' as mentioned above, and called a 'promising gradient' in accordance with Baaijens (1985). In this zone conditions exist for the development of a flourishing and diverse nature park, as demanded by the programme. Similar soil and water conditions at other sites around Delft show interesting plant and animal communities.

Soils also differ in their suitability for building. The clay soils are good to build on but the peat in the 'basin' poses problems of subsidence: to prevent distortion, buildings have to be erected on piles, but the maintenance of roads, parks and gardens is more difficult. To create dry topsoil, water tables can be lowered but this causes the peat to oxidize and subside. More frequently the soil is improved by bringing up a thick layer of sand. However, the weight of the sand again causes subsidence. In many residential areas built on peat in the Western part of The Netherlands subsidence is 10 cm per year or even more. If there is a thick layer of peat, as is the case in the study area, these problems may last for centuries. In similar places every five years 50 cm of sand is brought up, causing, of course, further subsidence. Clearly we cannot expect good quality urban plantings under these conditions.

13.5.2 GUIDING PRINCIPLES

First, optimal use should be made of the existing ecological structure. Here the abiotic structure (soils, relief) comes first because it is less flexible than the biotic. But biotic elements such as hedges, and other connecting 'corridors' should also be considered. Man-made structures like dikes, ditches and buildings are equally important. The second guiding principle is related to the new situation. Here the location, form and size of the elements in the plan should be designed as a spatial configuration that creates optimal conditions for the new system.

13.5.3 PLAN

At this stage, for the first time during the design process, the different parts of the plan are put in their proper place and given suitable proportions (Figure 13.4, *plan*). Given the situation of the study area it seems logical to concentrate building on the clay soils in the western and northern parts. The park and the storage ponds may best be situated in the 'basin'. The 'promising gradient' is best suited for the nature park, designed as a northern extension of the park.

The water system of the adjacent polder can be connected to the pumping station by a main water course in the western fringe. The existing ditch should be extended for this purpose. The design of the other water courses is also based on existing ditches. From the storage pond the water may best be channelled to a central place in the encircling canal dividing the water over the built up area from which it runs back to the storage pond.

The discharges into the pond are situated as far as possible from the nature park area. They are also clustered in order to allow for wetland purification treatment.

The shape of the pond is designed to take into account the prevailing wind as the major ecological factor in the differentiation of life in shallow lakes. With the shape chosen, the fetch varies from place to place. If desirable this shape may be simplified considerably, while retaining the general idea.

At this stage the design seems to correspond to model B of Figure 13.3, *plan*, the storage pond being part of the park and situated some distance from the built-up area. Looking more closely it appears that it will be possible to build close to the water along the western shore (model A). On the northeastern shore the design resembles model B. There, however, no water runs from pond to buildings or vice versa. The nature park at this site can be provided with its own source of clean water, where seepage water from the 'ridge' comes to surface. If a one-way flow is guaranteed, water quality will remain good.

13.6 STEP 5: ADJUSTED TECHNOLOGY (Figure 13.5)

Technology is not a dominating force but a set of simple tools that makes the system work as it should.

13.6.1 EXISTING SITUATION

Technical selectors in the existing situation of the water system are modest but effective for the functioning of the pasture.

13.6.2 GUIDING PRINCIPLES

Technical constructions or devices that enhance the selective operation of rhythms, orientation and situation should be introduced. Further-

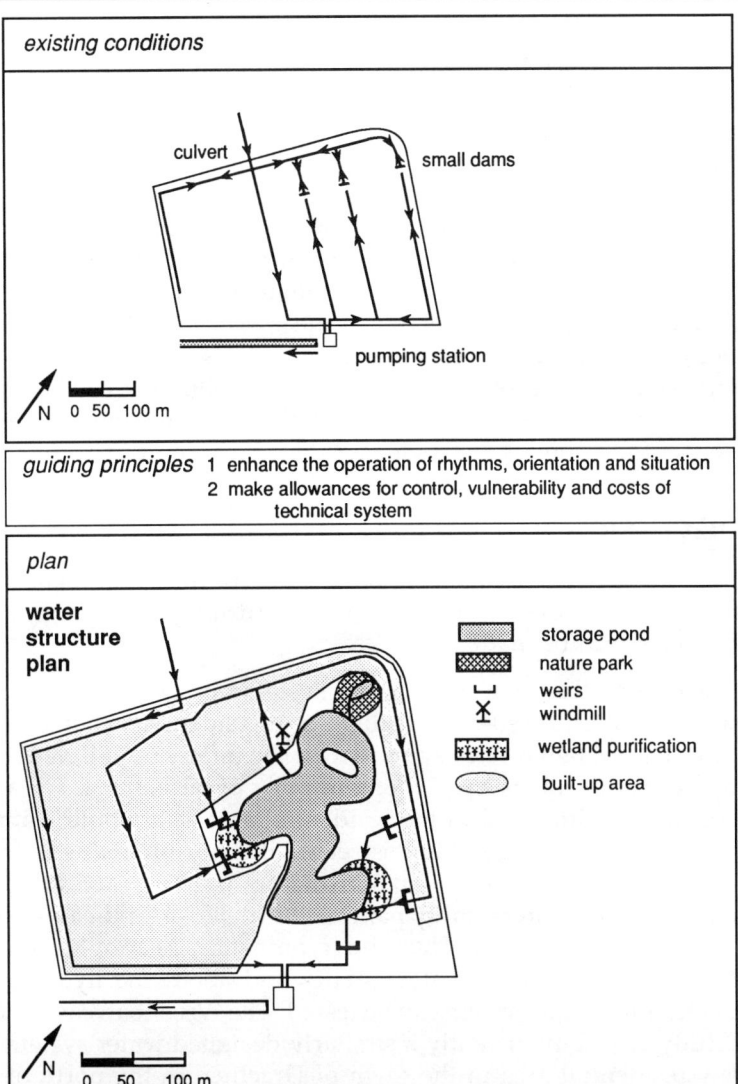

existing conditions

culvert   small dams

pumping station

N   0 50 100 m

guiding principles   1 enhance the operation of rhythms, orientation and situation
2 make allowances for control, vulnerability and costs of
technical system

plan

water
structure
plan

storage pond
nature park
weirs
windmill
wetland purification
built-up area

N   0   50   100 m

Figure 13.5 Adjusted technology.

more, the technology should be simple to operate, easy to maintain and robust.

13.6.3 PLAN

Power is required to bring the water from the storage pond to the higher level of the urban water courses (Figure 5, *plan*). A windmill or an electric pump will supply the necessary power. But the windmill may also operate as a generator for the electric pump. An additional advantage of a windmill is that it is a visible sign to the inhabitants that the water system is functioning.

Wetland purification is planned at the two sites where the urban water courses discharge into the pond. Various wetland systems are available (Van der Aart, 1985; Mönninghof, 1986; Duel, *et al.*, 1993, this volume). The 'Rootzone' system developed by Kickuth (1984) seems to be most effective, removing organic substances as well as phosphate, nitrogen and other contaminants. Phosphate and toxic substances are absorbed by the soil and it may thus be necessary to remove the sludge after some time. If control measures taken at the scale of buildings and parking places are effective, concentrations of pollutants will be low. In that case a more simple wetland purification by a marshy zone along the shores of the pond will provide a satisfactory solution.

13.7 DISCUSSION

In section 13.1 it was stated that the purpose of this 'investigation by design' was to explore three questions. In returning to these questions a few remarks can be made.

1. It seems possible to design a water system that can be described as 'self-reliant' and 'self-responsible' for a residential scheme at this scale. The system is not dependent on the inflow of polluted water from outside and, as a result of a number of measures taken at the level of buildings and in the residential area as a whole, minimal pollution is discharged 'downstream'. But of course this is only a plan. As such it offers a good starting point for a practical experiment in which water quality parameters including indicator organisms can be monitored. Thus the hypothetical assumptions about the relative effects on water quality and about the hydrological functioning of the system can be tested. The 'Vrijenban case' is only a study case, but currently a similarly designed water system in a new residential area in the town of Drachten in the north of The Netherlands is being realized. Here it will be possible to monitor the system and test the hypotheses in a real world experiment.

2. The PROSA approach links theoretical assumptions to practice explicitly, step by step. The result is a structural planning proposal for the whole area that can be discussed both for internal consistency and for feasibility. Systematically scanning the practical implications of general ideas reveals new problems, details and alternative solutions.

3. By formulating guiding principles at each step the design process is open to both public and expert discussion about facts and value

judgements. The resulting water structure plan may guide architects, town planners and civil engineers in the design team. In a case like this the early introduction of a water design proposal seems to be an appropriate way to integrate ecology in urban planning and design.

The resource approach chosen in this article aims to learn sustainable use of resources from natural processes and from the nature of the site. Thus 'normal' actions like the planning of urban water systems, are done in a different way. As a result, better ecological conditions can be created for human health and well-being and for plant and animal life.

ACKNOWLEDGEMENT

The drawings in this chapter were prepared by Jelle Andela of the Urban Design and Environment Group of the Delft University of Technology.

REFERENCES

Alexander, C., Ishikawa, S. and Silverstein, M. (1975) *A Pattern Language. Towns, Buildings, Construction*. Oxford University Press, New York.

Baaijens, G.J. (1985) Over grenzen. *De levende natuur*. **86**, 102–10.

Balmforth, D.J. (1986) Effectiveness of storm sewage overflow structures in handling gross polluting solids, in *Urban Stormwater Quality*. CHO-TNO Proceedings and Information, nr.36, (eds F.H.M. Van der Ven and J.C. Hooghart), TNO, The Hague. pp. 119–33.

Barendregt, A., Wassen, M.J. and De Smidt, J.T. (1993) Hydroecological modelling in a polder landscape, in *Landscape Ecology of a Stressed Environment* (eds C.C. Vos and P. Opdam), Chapman & Hall, London, pp. 79–100.

Bredow, W. (1985) *Regenwasser-Sammelanlage*. Ökobuch Verlag, Freiburg, 6th edn.

Bröker, A. (1987) *Wasserversorgung Alternativ*. Umwelt Aktuell Bd 16, Verlag C.F. Müller, Karlsruhe, 2nd edn.

Commissie Hydrologisch Onderzoek-TNO (1986) *Water in Urban Areas* (ed. J.C. Hooghart), CHO-TNO, Proceedings and Information, nr.33. TNO, The Hague.

Duel, H., During, R. and Kwakernaak, C. (1993) Artificial wetlands: a device for restoring natural wetland values, in *Landscape Ecology of a Stressed Environment* (eds C.C. Vos and P. Opdam), Chapman & Hall, London, pp. 263–80.

Geiler, N. and Hildebrandt, R. (1985) *Wasser konkret, Schritte zur Neuorientierung Hessischer Wasserpolitik*. Die Grünen im Hessischen Landtag, Wiesbaden.

Harms W.B. and Knaapen J.P. (1988) Landscape planning and ecological infrastructure: the Randstad study, in *Connectivity in Landscape Ecology* (ed. K.F. Schreiber), Schöningh, Paderborn, pp. 163–9.

Hengeveld, H., and De Vocht, C. (eds) (1982) *The Role of Water in Urban Ecology*. Elsevier Scientific Publishers, Amsterdam.

Hough, M. (1984) *City Form and Natural Process*. Croom Helm, London and Sydney.

Kickuth, R. (1984) Das Wurzelraum verfahren in der Praxis. *Landschaft + Stadt*, **16**, 145–53.

Kruitwagen, D.G., Zuidervliet, J. and Brandjes, M. (1986) The applicability of improved overflows in the Netherlands, in *Urban Stormwater Quality* (eds. F.H.M. Van der Ven and J.C. Hooghart) CHO-TNO Proceedings and Information, nr. 36 TNO, The Hague. pp. 339–45.

Londo, G. (1977) *Natuurtuinen en -parken*, Thieme, Zutphen.

Lynch, K. (1981) *Good City Form*. MIT Press, Cambridge MA/London.

McHarg, I. (1969) *Design with Nature*, Doubleday/Natural History Press, New York.

Ministerie van Verkeer en Waterstaat (1983) *De waterhuishouding van Nederland*. Staatsuitgeverij, The Hague.

Mönninghof, H. (1986) *Naturnahe Abwasserreinigung*. Ökobuch Verlag, Freiburg.

Roos, C., (1986) Biological effects of stormwater discharges in urban canals, in *Urban Stormwater Quality* (eds. F.H.M. Van der Ven and J.C. Hooghart) CHO-TNO Proceedings and Information, nr. 36 TNO, The Hague. pp. 311-23.

Schreiber, K.F. (ed.) (1988) *Connectivity in Landscape Ecology*. Proceedings 2nd International Seminar IALE. Münstersche Geogr. Arb, 29. Schöningh, Paderborn.

Spirn, A.W. (1984) *The Granite Garden*. Basic Books, New York.

Tjallingii, S.P. (1981) Environmental strategy and research in an urbanized area, in *Perspectives in Landscape Ecology* (eds. S.P. Tjallingii and A.A. De Veer), Pudoc, Wageningen. pp. 175-83.

Tjallingii, S.P. (1988a) Design steps and ecology, in *Connectivity in Landscape Ecology*, (ed. K.F. Schreiber), Schöningh, Paderborn. pp.69-73.

Tjallingii, S.P. (1988b) Strategies in urban water design, in *Urban Water 88* (ed. J.C. Hooghart), Nat. Comm. UNESCO-IHP, FRG/Netherlands, Duisburg/Zoetermeer. pp. 791-8.

Tjallingii, S.P. and De Veer A.A. (ed.) (1981) *Perspectives in Landscape Ecology*. Proc. Int. Congr. N.S.L.E., Pudoc, Wageningen.

Uunk, E.J.B. (1985) Urban stormwater runoff pollution, in *Water in Urban Areas*. (ed. J.C. Hooghart) CHO-TNO, Proceedings and Information, nr.33 TNO, The Hague. pp. 77-98.

Van Campen, A.L.B.M., Oldenkamp, S.A., Scücs, F.G. and Woelders, J.A. (1986) The pollution of stormwater runoff from specific paved areas and roofs, in *Urban Stormwater Quality*, (eds. F.H.M. Van der Ven and J.C. Hooghart) CHO-TNO Proceedings and information, nr.36 TNO, The Hague. pp. 331-3.

Van den Eshof, A.J., Frantzen, A.J., Reijnders, H.F.R., Vrijhof, H. and Wegman, R.C.C. (1986) The chemical composition of rainwater in the Netherlands, in *Water in Urban Areas*, (ed. J.C. Hooghart) CHO-TNO, Proceedings and Information, nr.33 TNO, The Hague. pp. 25-46.

Van der Aart, P.J.M. (ed.) (1985) *Wetlands for the Purification of Wastewater*. Utrecht plant ecology news report, Dept. Plant Ecology, Utrecht University, Utrecht.

Van der Maarel, E, (1975) Man-made natural ecosystems in environmental management and planning, in *Unifying Concepts in Ecology* (eds. W.H. Van Dobben and R.H. Lowe-McConnell), Junk Publ./Pudoc, The Hague/Wageningen. pp. 263-74.

Van der Maarel, E. (1981) Biogeographical and Landscape-ecological planning of nature reserves, in *Perspectives in Landscape Ecology* (eds. S.P. Tjallingii and A.A. De Veer), Pudoc, Wageningen. pp. 227-37.

Van Leeuwen, C.G. (1966) *A Relation Theoretical Approach to Pattern and Process in Vegetation*. Wentia, 15, North Holland Publ. Comp., Amsterdam. pp. 25-46.

Van Leeuwen, C.G. (1981) From ecosystem to ecodevice, in *Perspectives in Landscape Ecology* (eds. S.P. Tjallingii and A.A. De Veer) Pudoc, Wageningen. pp. 29-35.

Van Selm, A.J. (1988) Ecological infrastructure: a conceptual framework for designing habitat networks, in *Connectivity in Landscape Ecology* (ed. K.F. Schreiber), Schöningh, Paderborn. pp. 63-7.

Vos, C.C. and Zonneveld, J.I.S. (1993) Patterns and processes in a landscape under stress: the study area, in *Landscape Ecology of a Stressed Environment* (eds C.C. Vos and P. Opdam), Chapman Hall, London. pp. 1-280.

Index